Microbial Sensing in Fermentation

Microbial Sensing in Fermentation

Edited by

Satinder Kaur Brar
Institut national de la recherche scientifique, Centre - Eau Terre Environnement, Québec, Canada

Ratul Kumar Das
TERI-Deakin Nanobiotechnology Centre, Biotechnology and Management of Bioresources Division, The Energy and Resources Institute, Haryana, India

Saurabh Jyoti Sarma
Department of Biotechnology, Bennett University, Greater Noida, Uttar Pradesh, India

WILEY Blackwell

Registered Office(s)
John Wiley & Sons, Inc., 111 River Street, Hoboken, NJ 07030, USA
John Wiley & Sons Ltd, The Atrium, Southern Gate, Chichester, West Sussex, PO19 8SQ, UK

Editorial Office
The Atrium, Southern Gate, Chichester, West Sussex, PO19 8SQ, UK

For details of our global editorial offices, customer services, and more information about Wiley products visit us at www.wiley.com.

Wiley also publishes its books in a variety of electronic formats and by print-on-demand. Some content that appears in standard print versions of this book may not be available in other formats.

Library of Congress Cataloging-in-Publication Data

Names: Brar, Satinder Kaur, editor. | Das, Ratul Kumar, 1979– editor. |
Sarma, Saurabh Jyoti, editor.
Title: Microbial sensing in fermentation / edited by Prof. Satinder Kaur Brar,
Dr. Ratul Kumar Das, Dr. Saurabh Jyoti Sarma.
Description: First edition. | Hoboken, NJ : John Wiley & Sons, 2018. |
Includes bibliographical references and index. |
Identifiers: LCCN 2018023655 (print) | LCCN 2018036992 (ebook) |
ISBN 9781119247982 (Adobe PDF) | ISBN 9781119247975 (ePub) |
ISBN 9781119247968 (hardcover)
Subjects: LCSH: Fermentation–Analysis. | Molecular microbiology.
Classification: LCC TP511 (ebook) | LCC TP511 .M53 2018 (print) | DDC 664/.024–dc23
LC record available at https://lccn.loc.gov/2018023655

Cover Design: Wiley
Cover Image: © Kateryna Kon/Shutterstock

Set in 10/12pt Warnock by SPi Global, Pondicherry, India

10 9 8 7 6 5 4 3 2 1

Contents

List of Contributors

Aishvarya Agrawal
Institute of Bioinformatics and Biotechnology, Savitribai Phule Pune University, Pune, India

Shadab Ahmed
Institute of Bioinformatics and Biotechnology, Savitribai Phule Pune University, Pune, India

Devangana Bhuyan
TERI Deakin Nanobiotechnology Centre, Biotechnology and Management of Bioresources Division, The Energy and Resources Institute, Haryana, India

Jean François Blais
Institut national de la recherche scientifique, Centre - Eau Terre Environnement, Québec(QC), Canada

Satinder Kaur Brar
Institut national de la recherche scientifique, Centre - Eau Terre Environnement, Québec, Canada

Ratul Kumar Das
TERI-Deakin Nanobiotechnology Centre, Biotechnology and Management of Bioresources Division, The Energy and Resources Institute, Haryana, India

Satya Eswari
Department of Biotechnology, National Institute of Technology, Raipur, Chhattisgarh, India

Ana M. Finco
Bioprocess Engineering and Biotechnology Department, Federal University of Paraná, Curitiba, Brazil

G. Gallastegui
Department of Chemical and Environmental Engineering, Faculty of Engineering Vitoria-Gasteiz, University of the Basque Country (UPV/EHU), Spain

Rosa Galvez-Cloutier
Université Laval, Department of Civil Engineering and Water Engineering, Quebec, Canada

Rachna Goswami
Biosciences Department, Rajiv Gandhi University of Knowledge Technologies, India

Tayssir Guedri
Institut national de la recherche scientifique, Centre - Eau Terre Environnement, Québec(QC), Canada

Krishnamoorthy Hegde
INRS-ETE, Université du Québec, Québec, Canada

Michele Heitz
Université de Sherbrooke, Canada

Susan Grace Karp
Bioprocess Engineering and Biotechnology Department, Federal University of Paraná, Curitiba, Brazil

Guneet Kaur
School of Energy and Environment, City University of Hong Kong, Kowloon Tong, Hong Kong

Azadeh Kermanshahi Pour
Biorefining and Remediation Laboratory, Department of Process Engineering and Applied Science, Dalhousie University, Halifax, Canada

Preetika Rajeev Kuknur
Institut national de la recherche scientifique, Centre - Eau Terre Environnement, Québec, Canada

Pratik Kumar
INRS-ETE, Université du Québec, Québec, Canada

A. Larrañaga
Department of Mining-Metallurgy Engineering and Materials Science & POLYMAT, Faculty of Engineering of Bilbao, University of the Basque Country (UPV/EHU), Spain

Luiz A. J. Letti
Bioprocess Engineering and Biotechnology Department, Federal University of Paraná, Curitiba, Brazil

Sara Magdouli
Institut national de la recherche scientifique, Centre - Eau Terre Environnement, Québec, Canada

Vijay Kumar Mishra
Biosciences Department, Rajiv Gandhi University of Knowledge Technologies, India

Somnath Nandi
Department of Technology, Savitribai Phule Pune University, Pune, India

Shreyas Niphadkar
Institute of Bioinformatics and Biotechnology, Savitribai Phule Pune University, Pune, India

Juliana de Oliveira
Bioprocess Engineering and Biotechnology Department, Federal University of Paraná, Curitiba, Brazil

Carlos S. Osorio-González
Institut national de la recherche scientifique, Centre - Eau Terre Environnement, Québec, Canada

Vinayak Laxman Pachapur
Institut national de la recherche scientifique, Centre - Eau Terre Environnement, Québec, Canada; Université Laval, Department of Civil Engineering and Water Engineering, Quebec, Canada

Maria G. B. Pagnoncelli
Bioprocess Engineering and Biotechnology Department, Federal University of Technology - Paraná (UTFPR), Dois Vizinhos, Brazil

Vishal Pandey
Institute of Bioinformatics and Biotechnology, Savitribai Phule Pune University, Pune, India

Gilberto V. de Melo Pereira
Bioprocess Engineering and Biotechnology Department, Federal University of Paraná, Curitiba, Brazil

Thi Than Ha Pham
Centre National en Électrochimie et en Technologies Environnementales, Shawinigan, Québec, Canada; Université de Sherbrooke, Sherbrooke, Québec, Canada

Ha Thi Thanh Pham
Université de Sherbrooke Canada; Centre National en Électrochimie et en Technologies Environnementales, Shawinigan, Québec, Canada

Radhika Pilli
Microbiology Department, Bharathiar University, Coimbatore, TN, India

Maria Puig-Gamero
Universidad de Castilla La Mancha, Spain

Antonio Avalos Ramirez
Centre National en Électrochimie et en Technologies Environnementales, Shawinigan, Québec, Canada

Keyur Raval
Department of Chemical Engineering, National Institute of Technology, Surathkal, Karnataka, India

Ritu Raval
Department of Biotechnology, Manipal Institute of Technology, Manipal, Karnataka, India

Thana Saffar
Université du Québec en Abitibi-Témiscamingue, Rouyn-Noranda (Québec), Canada

Paula Sánchez
Universidad de Castilla La Mancha, Spain

Luz Sanchez-Silva
Universidad de Castilla La Mancha, Spain

Joseph Sebastian
INRS-ETE, Université du Québec, Québec, Canada

Aishwarya Shankapal
Institute of Bioinformatics and Biotechnology, Savitribai Phule Pune University, Pune, India

Carlos Ricardo Soccol
Bioprocess Engineering and Biotechnology Department, Federal University of Paraná, Curitiba, Brazil

Vanete Thomaz Soccol
Bioprocess Engineering and Biotechnology Department, Federal University of Paraná, Curitiba, Brazil

Rouissi Tarek
Institut national de la recherche scientifique, Centre - Eau Terre Environnement, Québec, Canada

José Luis Valverde
Universidad de Castilla La Mancha, Spain

Mausam Verma
INRS-ETE, Université du Québec, Québec, Canada

1

Biochemical Aspects of Microbial Product Synthesis: a Relook

G. Gallastegui[1], *A. Larrañaga*[2], *Antonio Avalos Ramirez*[3], *and Thi Than Ha Pham*[3,4]

[1] *Department of Chemical and Environmental Engineering, Faculty of Engineering Vitoria-Gasteiz, University of the Basque Country (UPV/EHU), Spain*
[2] *Department of Mining-Metallurgy Engineering and Materials Science & POLYMAT, Faculty of Engineering of Bilbao, University of the Basque Country (UPV/EHU), Spain*
[3] *Centre National en Électrochimie et en Technologies Environnementales, Shawinigan, Québec, Canada*
[4] *Université de Sherbrooke, Sherbrooke, Québec, Canada*

1.1 Introduction

Microbes are living unicellular or multicellular organisms (bacteria, archaea, most protozoa, and some fungi and algae) that must be greatly magnified to be seen. Despite their tiny size, they play an indispensable role for humanity and the health of ecosystems. For instance, until the discovery of an artificial nitrogen fixation process by the German chemists Fritz Haber and Carl Bosch in the first half of the 20th century, some soil microbes on the roots of peas, beans, and a few other plants were the solely responsible for the nitrogen release necessary for plants growth (Hager, 2008). This invention allowed to feed billions more people than the earth could support otherwise.

Besides, humanity has exploited some of the vast microbial diversity like miniature chemical factories for thousands of years in the production of fermented foods and drinks, such as wine, beer, yogurt, cheese and bread. In fact, the use of yeast as the biocatalyst in foodstuffs making is thought to have begun around the Neolithic period (ca. 10 000-4000 BCE), when early humans transitioned from hunter-gatherers to living in permanent farming communities (Rasmussen, 2015). Vinegar, the first bio-based chemical (not intended as a beverage) produced at a commercial scale was known, used and traded internationally before the time of the Roman Empire (Licht, 2014).

The staggering transformation undergone by biotechnology from serendipity and black-box concepts to rational science and increasing understanding of biological systems has led to not only a direct influence of microbes on human lives, but the emergence of new industries that take advantage of these organisms in

Microbial Sensing in Fermentation, First Edition. Edited by Satinder Kaur Brar, Ratul Kumar Das, and Saurabh Jyoti Sarma.

large-scale processes devoted to the manufacture of high value-added compounds, energy production and environmental protection. Nevertheless, scientists and engineers are still discovering the broad array of complex signalling that microorganisms have developed to ensure their survival in a wide range of environmental conditions, and making their utmost effort to direct them towards our own ends (Manzoni et al., 2016). In this chapter, a brief summary regarding the historical production of microbial products, their niche in the current global market and the importance of microbial sensing (and other new disciplines) to convert biological systems in industrially relevant actors is presented.

1.2 History of Industrial Production of Microbial Products

In the 1800s, Louis Pasteur (and later Eduard Buchner) proved that fermentation was the result of microbial activity and, consequently, the different types of fermentations were associated with different types of microorganisms. In more recent times (1928), Alexander Fleming understood that the *Penicillium* mould produces an antibacterial bio-chemical (antibiotics discovery), which was extracted, isolated and named penicillin. Subsequent periods of conflicts (e.g., World Wars I and II) intensified the needs of the population and, at the same time, the creativity and inventiveness of scientists and engineers, who developed large-scale fermentation techniques to make industrial quantities of drugs, such as penicillin, and biofuels, such as biobutanol and glycerol, giving rise to industrial biotechnology. In 1952, Austrian chemists at Biochemie (now Sandoz) developed the first acid-stable form of penicillin (Penicillin V) suitable for oral-administration and achieved an extraordinary success in the treatment of infections during World War II (Williams, 2013).

Biobutanol production is recognized as one of the oldest industrial-scale fermentation processes. It was generated by anaerobic ABE (acetone–butanol–ethanol) fermentation of sugar extract using solventogenic clostridia strains, with a typical butanol:acetone:ethanol mass fraction ratio around 6:3:1. Until the 1920s, acetone was the most sought-after bioproduct of commercial interest. An emerging automotive paint industry and the need of quick-drying lacquers, such as butyl acetate, changed the economic landscape and by 1927 butanol displaced acetone as the target product (Rangaswamy et al., 2012). From 1945 to 1960, about two thirds of the butanol production in North America was based on the conventional ABE fermentation. Nevertheless, butanol yield by anaerobic fermentation remained sub-optimal, and this biobased product was progressively replaced by low cost petrochemical production (Maiti et al., 2016).

When Watson and Crick (with the valuable help from Wilkins and Franklin) worked out the structure of DNA in 1953, they barely imagined that this latter discovery supposed a milestone in the development of modern industrial biotechnology. Thus, in the following decades traditional industrial biotechnology merged with molecular biology to yield more than 40 biopharmaceutical products, such as erythropoietin, human growth hormone and interferons (Demain, 2000). Since then, biotechnology has steadily developed and now plays a key role in several industrial sectors, such as industrial applications, food and beverages, nutritional and pharmaceuticals or plastics and fibers, providing both high value products and commodity products (Heux et al., 2015).

Although, as shown in the previous paragraphs, the use of microorganisms and enzymes for the production of essential items has a long history, the recent linguistic term "white biotechnology" has been assigned to the application of biotechnology for the processing and production of chemicals, materials and energy. It is based on microbial fermentation processes and it works with nature in order to maximize and optimize existing biochemical pathways that can be used in manufacturing. The development of cost effective fermentation processes has allowed industry to target previously abandoned fermentation products and new ones which used to be of small interest for the naphtha-relying chemical industry, such as succinic acid or lactic acid. In the latter case, and although the chemical synthesis of lactic acid from petrochemical feedstock is more familiar to chemists, approximately 90% of its production is accomplished by microbial fermentation (Wang et al., 2015). Nowadays, this platform molecule is used as a building block for the synthesis of chemicals such as acrylic acid and esters (by catalytic dehydration), propylene glycol (by hydrogenolysis) and lactic acid esters (by esterification) (Figure 1.1).

1.2.1 Advances of Biochemical Engineering and Their Effects on Global Market of Microbial Products

Economic viability of bio-derived products, especially in the case of biofuels, has been traditionally limited to a large extent by the selection of cheap carbon-rich raw materials as feedstock, applied production mode, downstream processing and the scarcity of naturally occurring microorganisms that are able to deliver the desired compounds at a high production-rate. Conventional bio-based products ultimately turned out so expensive to compete with petroleum-derived chemicals that they were hardly worth producing.

Despite these drawbacks, advances in biotechnology in recent years have enabled the reengineering of the bioprocesses incorporating several transformation or purification steps into only one, reducing time and operating costs. This has involved the increase of bioprocesses yield, boosting production of biobased materials. Currently, biotechnology advances (microbial, enzymatic and biology engineering) can be considered among the new technological revolutions, having huge impacts in industry, society and economy, as nanotechnology-materials, informatics and artificial intelligence.

Therefore, a resurgence in the production of fermentation chemicals including biofuels, chemical building blocks, such as organic acids, amino acids, alcohols (diols, thiols) and specialty chemicals, such as surfactants, thickeners, enzymes, antibiotics and fine chemicals (pigments, fragrances, etc.) is expected in the years to come. The global fermentation chemicals market was $51.83 \cdot 10^6$ tons in 2013 and is expected to reach $70.76 \cdot 10^6$ tons by 2020, growing at a Compound Annual Growth Rate (CAGR) of 4.5% from 2014 to 2020, with North America emerging as the leading regional market and accounting for 33.8% of total market volume (Grand View Research, 2014).

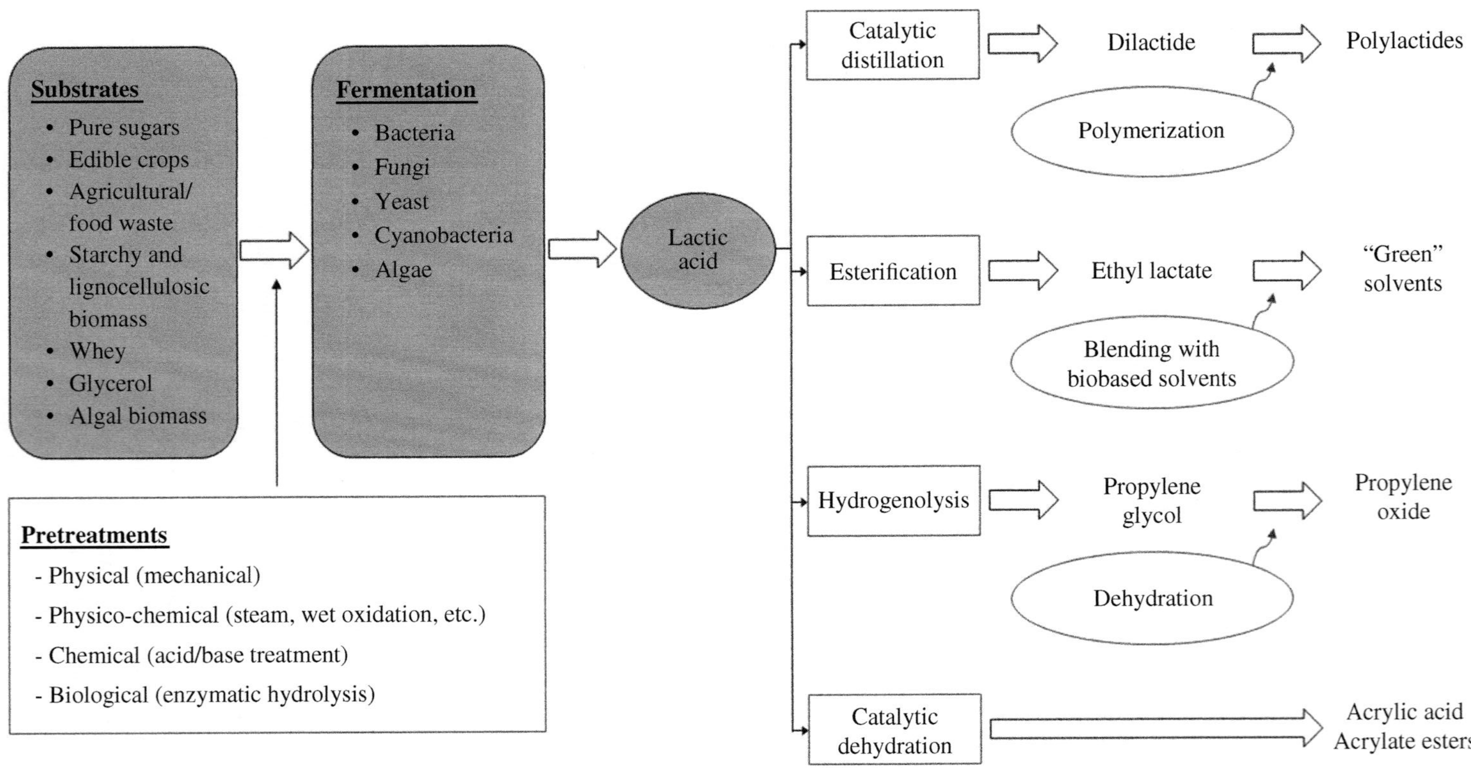

Figure 1.1 Production of lactic acid by microbial fermentation and its derivatives.

Among all the possible products and value streams obtained from biomass in the biorefineries, the chemical market (both commodity and fine chemicals) is expected to grow at a rate almost double to that of biofuels, since chemicals are on average priced 15 times higher than energy (Deloitte, 2014), which will entail that by 2025 at least a 45% share of chemicals will be accounted by biorenewable chemicals in the USA (Bardhan et al., 2015). In Europe, biobased chemicals account at present time for 5.5% of total turnover for chemicals produced in the EU, and they are expected to grow up by over 5% per year, until reaching a total proceeding of sales of about $44 billion in 2020 (Schneider et al., 2016).

Additionally, compared to the production of first-generation biofuels, the production of more bio-based materials will not have a price enhancing effect on food products (van Haveren et al., 2008) since it would be based on the utilisation of the carbohydrate fraction of lignocellulosic biomass (i.e., cellulose and hemicellulose) and inedible oil seed crops or algal oil as feedstock. In the report edited by Deloitte, the authors estimated that replacing all petrochemicals would require just 5% of agricultural biomass production and global arable land, which is about 60 times less than what would be required to replace all fossil energy (Deloitte, 2014). Straathof (2014) reported in his extensive review about the biochemical formation of commodity chemicals from biomass that 21 of the compounds cited are already commercially produced (including carboxylic acids, alcohols and amino acids), and at least 9 others have been tested at pilot scale. Frost & Sullivan (2011) calculated that the global market for fermentation derived fine chemicals was $16 billion in 2009.

However, as with all the main human inventions, modern biotechnology presents contradictions and confronts the ethic principles of our societies. It is at the same time a tool to face the main human challenges (energy needs, environment conservation, human health, food supplying, etc.), but it also represents high risks to the environment and to human health if it is not properly used. Thus, even if the use of genetically modified microorganisms (GMM) has offered advantages over traditional methods of improving chemical selectivity and the supply of desired bioproducts thus reducing production cost (Bullis, 2013), their implementation has been controversial among the general public, especially when these microorganisms contain genes introduced from other species. Taking into account that newly isolated strains of microorganisms and GMMs can be patented, pressing questions arise regarding whether these organisms have any place in our ethical considerations and how they should be treated (Cockell, 2011).

Microbial sensing, microbial nanocontrol, smart fermentations, smart enzymatic systems, and the bioinformatics can be included amongst the main new developments which will revolute the biotechnology itself. The discovery in the 1970s of sophisticated cell-cell communication mechanisms (quorum sensing), became evident that microbial populations are synchronized at a certain cell density, by means of diverse signalling molecules that are synthesized and secreted by the microbes themselves (Bassler and Losick, 2006). Thus, the deep knowledge of the quorum sensing regulation on microbial metabolism and the control of microbial sensing will allow the complete redesign of all bioprocesses in terms of microbial signalization. We will be able to control better the bioprocesses (shortening residence times, controlling contamination, increasing production yields), to change the way to fight against microbial illnesses with new molecules (other than antibiotics) or antagonist bacteria, to improve life quality of livestock, to protect better the environment, etc.

As most of technological advances, several of these improvements obtained by microbial sensing appear so far or impossible to develop with time. But the biotechnological revolution is highly associated to the technological advances of other science branches, such as materials science, photonics, electronics, microscopy, and others. The development of new powerful and high-sensitive analytic equipment is essential to identify microbial signals and to construct mapping interactions (Moon et al., 2010).

In the near future, transformation of biomass into chemicals using enzymes or cells will be implemented with success only if the production process is more attractive than for alternative options (petrochemical route) to produce these chemicals based on their ecologic, social, and economic value. The present book tries to show a brief portrait of the state of the art of "four magic e" bioproducts (large-scale microbial fermentation products considering economic, ethical, environmental, and engineering aspects) and how microbial sensing has a main role in their present and future production.

1.2.2 Importance of Microbial Sensing in Product Formation

However, microbial sensing is so wide that it is necessary to delimit the goal of this work. This book presents a comparison among the different control concepts for the carbon transformation by microorganisms, analysing microbial, biochemical and molecular biology control concepts. The microbial sensing concept is emphasized showing the potentiality to use it for fermentation control and predict the scaling up.

According to the combination signals' origin-cell sensor, the microbial sensing defined as the identification of internal and external signals by microbial sensors can be classed in five main categories (Figure 1.2), as follows:

1) Internal signals. The molecules to be captured by microbial sensors are produced by the same cell in its cytoplasm. These signals are employed by the cell to control the production of functional cell structures (proteins, enzymes, organelles, etc.), as well as to control the cell aging.
2) Signals in a homogenous microbial community. They are produced by cells of the same species in a homogenous microbial community to control the interactions among them, for example the quorum sensing to conglomerate and begin the formation of homogenous biofilm.
3) Signals in a heterogeneous microbial community. They are produced by the cells of the different species present in a heterogeneous microbial community to develop synergistic or antagonistic interactions among them. For example, production of toxic molecules to inhibit the growth of competitive species.
4) Signals produced by the effect of environmental factors. They are caused by the effect of extracellular environmental factors such as light, humidity, ionic strength, pH or temperature.
5) Signals in host bodies. They are produced by both cells and infected bodies. The interactions among them can be both synergistic (e.g., probiotic microorganisms in human or animal gut) or antagonistic (e.g., pathogen infections).

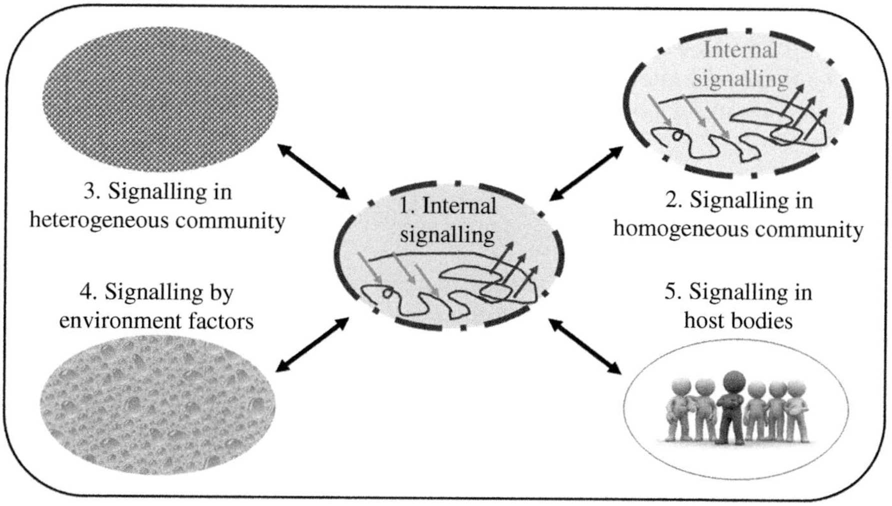

Figure 1.2 Microbial sensing classification according to signal origin. (*See insert for color representation of this figure.*)

The effect of environmental factors is the most known signalling mechanism of microbial sensing since their role has been clearly defined for several bioprocesses, such as alcoholic fermentations. Besides, the major efforts undertaken regarding the understanding of infections by pathogenic agents and host health and homeostasis, has contributed to gain appropriate and reliable information about the signals produced in host bodies (5th mechanism) (Kendall and Sperandio, 2016).

The remaining categories (1-3) became important at the end of last century, and it is precisely these categories which represent the state-of-the-art in this field. However, empiric and scientific data related to the first three cases is scarce, and there are still many gaps and uncertainties in the relevant scientific knowledge about signalling processing. In addition, the experimentation with animal or human models is a very sensitive subject constrained by ethic rules which must be respected, limiting the number and the quality of the scientific research. Therefore, even if all kind of microbial sensing is now studied around the world, there is a lack of updated reviews showing the most important advances done during the last 5-10 years.

1.3 Conclusion

The present book documents and critiques those aspects related to microbial production and performance, including the type of carbon source, cellular and biochemical control over the microbial products, etc. from the perspectives of molecular biology and biochemistry. Together with these aspects, the ways to quantitatively and qualitatively control the microbial products as well as approaches to scale-up and optimize these processes along with specific future market perspectives and policy initiatives are thoroughly reviewed in this book. Accordingly, it will be of particular interest for those researchers working in the field of microbial biotechnology but it will also cover those areas related to molecular biology, biochemistry and materials science, among others.

Acknowledgments

The authors wish to acknowledge the financial support received from the State Agency for Research (AEI) of the Spanish Government and the European Regional Development Fund (FEDER) (Project CTM2016-77212-P). We thank the Basque Government (Department of Education, Language Policy and Culture) and the Fonds de recherche du Québec – Nature et technologies (FRQNT) for the postdoctoral grant to Dr. Larrañaga and the postdoctoral scholarship of short duration to Dr. Gallastegui, respectively. We thank also the Discovery-NSERC program for the funds to finance the postdoctoral internship of Dr. Pham.

References

Bardhan, S.K., Gupta, S., Gorman, M.E., and Haider, M.A. (2015) Biorenewable chemicals: Feedstocks, technologies and the conflict with food production. *Renewable & Sustainable Energy Reviews.* 51: 506–520.

Bassler, B.L. and Losick, R. (2006) Bacterially speaking. *Cell.* 125: 237–246.

Bullis, K. (2013) *Genetically Modified Bacteria Produce 50 Percent More Fuel.* MIT Technology review. Link (Date accessed 31/07/2017): https://www.technologyreview.com/s/519791/genetically-modified-bacteria-produce-50-percent-more-fuel/

Cockell, C. (2011) Microbial rights? *EMBO reports.* 12: 181.

Deloitte. (2014) Opportunities for the fermentation-based chemical industry. An analysis of the market potential and competitiveness of North-West Europe. (Date accessed 31/07/2017): https://www2.deloitte.com/content/dam/Deloitte/nl/Documents/manufacturing/deloitte-nl-manufacturing-opportunities-for-the-fermentation-based-chemical-industry-2014.pdf

Demain, A.L. (2000) Microbial biotechnology. *Trends in biotechnology.* 18: 26–31.

Frost & Sullivan. (2011) *Advances in Fermentation Technologies--An Industry Overview* (Technical Insights). (Date accessed 31/07/2017): http://www.ieabioenergy.com/wp-content/uploads/2013/10/Task-42-Biobased-Chemicals-value-added-products-from-biorefineries.pdf

Gran View Research. (2014) Fermentation Chemicals Market Analysis By Product (Alcohols, Organic Acids, Enzymes), By Application (Industrial, Food & Beverages, Nutritional & Pharmaceuticals, Plastics & Fibers) And Segment Forecasts To 2020. (Date accessed 31/07/2017): http://www.prnewswire.com/news-releases/fermentation-chemicals-market-analysis-by-product-alcohols-organic-acids-enzymes-by-application-industrial-food--beverages-nutritional--pharmaceuticals-plastics--fibers-and-segment-forecasts-to-2020-300232947.html

Hager, T. (2008) *The alchemy of air*. Broadway Books. New York.

Heux, S., Meynial-Salles, I., O'Donohue, M.J., and Dumon, C. (2015) White biotechnology: State of the art strategies for the development of biocatalysts for biorefining. *Biotechnology Advances* 33: 1653–1670.

Kendall, M.M. and Sperandio, V. (2016) What a dinner party! Mechanisms and functions of interkingdom signaling in host-pathogen associations. *mBio* 7: e01748–15.

Licht, S. (2014) Fermentation for biofuels and bio-based chemicals. In: *Fermentation and Biochemical Engineering Handbook, 3e* (ed. C.C. Todaro and H.C. Vogel), Ch. 5, pp. 59–82. Elsevier.

Maiti, S., Gallastegui, G., Sarma, S.J. et al. 2016. A re-look at the biochemical strategies to enhance butanol production. *Biomass and Bioenergy*, 94:187–200.

Manzoni, R., Urrios, A., Velázquez-García, S. et al. (2016) Synthetic biology: insights into biological computation. *Integrative Biology.* 8: 518–532.

Moon, I., Yi, F., and Javidi, B. (2010) Automated three-dimensional microbial sensing and recognition using digital holography and statistical sampling. *Sensors.* 10: 8437–8451.

Rangaswamy, V., Isar, J., and Joshi, H. (2012) *Butanol fermentation using acid pretreated biomass.* Patent US20140178955.

Rasmussen, S.C. (2015) From honey wine to the cultivation of the grape: An early history of fermented beverages. In: *Chemical technology in antiquity.* (ed. Rasmussen, S.C.) Ch.4, pp 89–138. ACS Symposium Series.

Schneider, M.N., Iaconi, A., and Larocca, S. (2016) Oleochemical biorefinery. In: *Chemicals and fuels from bio-based building blocks.* (ed. Cavani, F., Albonetti, S., Basile, F., Gandini, A.) Chapter 19. pp 519–534. Wiley-VCH Verlag GmbH & Co.

Straathof, A.J.J. (2014) Transformation of biomass into commodity chemicals using enzymes or cells. *Chemical Reviews.* 114: 1871–1908.

van Haveren, J., Scott, E., and Sanders, J. (2008) Bulk chemicals from biomass. *Biofuels, Bioproducts & Biorefining.* 2: 41–57.

Wang, Y., Tashiro, Y, and Sonomoto, K. (2015) Fermentative production of lactic acid from renewable materials: recent achievements, prospects, and limits. *Journal of Bioscience and Bioengineering.* 119: 10–18.

Williams, C. (2013) *Medicinal Plants in Australia*, Vol. 4: An Antipodean Apothecary. Rosenberg Publishing. Dural, NSW, Australia.

2

Cellular Events of Microbial Production: Important Findings So Far

Devangana Bhuyan and Ratul Kumar Das

TERI Deakin Nanobiotechnology Centre, Biotechnology and Management of Bioresources Division, The Energy and Resources Institute, Haryana, India

2.1 Introduction

Microbes are the key manufacturers of a product, be it beer, bread, solvents or antibiotics. Microbial metabolism is a process by which a microorganism utilizes the available nutrients and generates energy to survive and proliferate. Being one of the most fundamental cellular characteristics, it involves a complex biochemical processes implemented through the coordination of different metabolic reactions and their interactions with environmental factors. (Liao et al., 2015). Different strains of microbes are used to obtain different products. Moreover, different internal and external factors also govern the rate of production and the quality and quantity of the product. The oldest and most commonly used method for large scale production from microbial sources is fermentation technology. However, our understanding of the cellular events for the cellular adaptation due to exposure to multiple stresses is still limited despite multiple studies in the past decade (Walker et al., 2014). This chapter highlights the basic metabolic processes in bacteria and yeast, and the extracellular and intracellular factors affecting microbial metabolism and their response and adaptation mechanism in the case of different types of environmental stresses. The microbial sensing involved in the cellular mechanisms required for protecting yeast cells from multiple stress factors is essential to understand and unravel multi-stress tolerance in microbial metabolism (Kitichantaropas et al., 2016).

Microbial Sensing in Fermentation, First Edition. Edited by Satinder Kaur Brar, Ratul Kumar Das, and Saurabh Jyoti Sarma.

2.2 Microbial Metabolism and Evolution of Metabolic Pathways

Metabolism can be referred as the sum of all chemical reactions that occur within each cell of a living organism to supply energy for vital physiological processes. This includes processes such as the building and breakdown of complex molecules that occur through a series of metabolic pathways. In the case of sugar metabolism, the anabolic pathway is an energy requiring process that synthesizes sugar from smaller molecules, and the catabolic pathway is an energy yielding pathway that breaks down sugar into smaller moieties. In the case of bacterial metabolism, these respective exergonic (energy-yielding) and endergonic (energy-requiring) reactions are catalyzed within the living bacterial cell by integrated enzyme systems, the end result being the multiplication of the cell. The ability of microbial cells to manifest and replicate in a suitable culture medium and the chemical changes that result during this transformation constitute the scope of bacterial metabolism (Jurtshuk, 1996).

Organisms have evolved and are believed to have adapted to anaerobic metabolism billions of years ago in an unfavourable atmosphere lacking oxygen in order to survive. This has resulted in varying complexities of metabolism in different organism across species. But despite the differences in the process of metabolism at the molecular level, evolutionary biologists have discovered that all life forms in the planet have similar basis for the metabolic processes. But even with the diverging pathways of metabolism in different organisms, the underlying principle for the survival of all organisms is that they have to derive energy from their environment and convert it to ATP in order to carry out essential cellular functions. The primary pathway for harvesting energy in photosynthetic plants and organisms like algae and cyanobacteria is photosynthesis, wherein solar energy is used as the raw material to manufacture the carbohydrates required for the growth of the organism. Oxygen released as a by-product of photosynthesis is utilized by other cells to carry our cellular respiration, during which O_2 aids in the breakdown of carbohydrates. The breakdown of complex carbohydrates also yields O_2 and ATP. But some eukaryotic organisms also use anaerobic metabolism, i.e. they can metabolize in the absence of oxygen (Prescott et al., 2002).

2.3 Microbial Fermentation

Fermentation is basically an example of heterotrophic metabolism that requires an organic substrate and an electron acceptor, which produces organic products by anaerobic dissimilation of glucose or some other carbohydrate. In fermentation, the pyruvate made in glycolysis does not continue through oxidation and the citric acid cycle and electron transport chain discontinues in fermentation. Since the electron transport chain isn't active, the NADH produced during glycolysis does not convert to NAD^+. The purpose of the extra reactions in fermentation, then, is to regenerate the electron carrier NAD^+ from the NADH produced in glycolysis. The extra reactions accomplish this by letting NADH drop its electrons off with an organic molecule (such as pyruvate, the end product of glycolysis). This drop-off allows glycolysis to keep running by ensuring a steady supply of NAD^+. Enzymatic breakdown of glucose through the dehydrogenation reactions produces energy in the form of ATP. The organic substrates are incompletely oxidized by bacteria, yet yield sufficient energy for microbial growth. Glucose is the most common hexose used to study fermentation reactions. Fermentations occur when microorganisms consume susceptible organic substrate as part of their own metabolic processes (Stanbury et al., 1995; Prescott et al., 2002).

Microbes play a central role in the production of a number of primary and secondary metabolites, enzymes, antibiotics, etc. The commercial production of microbial biomass by fermentation based on product can be divided into two types: production of bakers' yeast and production of single cell protein (SCP), which is generally used as a food supplement (Stanbury et al., 1995). Although yeast was produced as food on a large scale in Germany during the First World War (Laskin, 1977), the concept of utilizing microbial biomass as food was not thoroughly investigated until the 1960s. Since the 1960s, a large number of industrial companies have explored the potential of producing SCP from a wide range of carbon sources. Almost without exception, these investigations have been based on the use of continuous culture as the growth technique. An overview of the major classes of fermentative microorganisms is briefly summarized in Table 2.1.

The successful use of microorganisms in fermentation technology can be attributed to a number of factors like high surface to volume ratio, genetic adaptability, the ability to use both carbon and nitrogen as an energy source. The high surface to volume ratio allows the maximum area for molecular diffusions to occur between the microbe and its environment. This efficiency in nutrient assimilation facilitates rapid synthesis of new cells, which in turn shows a high metabolic rate, yielding increased biomass. This is seen in *Saccharomyces cerevisiae* where the production of protein manifolds itself faster than in plants. Another favorable trait of microbes is their metabolic versatility to use different energy sources and to use different types of terminal electron acceptors. An array of diverse food products is produced due to the organism's ability to use different substrates for energy. When microorganisms ferment food constituents, they derive energy in the process and increase in numbers. The production of alcohol by yeast from malt or fruit pulp has been carried out on a large scale since the earliest recorded human civilizations, making it the first "industrial" process for the production of a microbial metabolite. Thus industrial microbiologists use the term "fermentation" to describe any process for the production of product by the mass culture of a micro-organism. Foods as diverse as yogurt, hard sausages, and sauerkraut are all a result of fermentation. Some of the important microbes used in fermentation include bacteria, yeasts, fungi, etc. Fermentative organisms largely convert carbohydrates to alcohols, with acids and CO_2 being the by-products. These by-products do not generally adversely affect the organoleptic properties of the product; rather with sufficient accumulation they inhibit the growth of lipolytic and proteolytic organisms that are essentially food spoilage organisms. This is the underlying principle of food preservation by

Table 2.1 Overview of two classes of fermentative microorganisms.

Microorganism	Industrially important species	Substrate types	Products formed	Factor affecting fermentation	
Fungi	Aspergillus, Penicillium,	Sugars like glucose and amylose	Penicillin, Citric acid, Cheese, Enzymes etc.	Nutrient stress, Ethanol content, Extracellular stress factors	High surface to volume ratio, Genetic adaptability, C and N use efficiency, Metabolic versatility.
Yeast	Saccharomyces, Candida	Sugars like glucose, sucrose, maltose, molasses etc.	Bread, Beer, Wine, Citric acid etc.		
Bacteria	Lactic Acid Bacteria (LAB), Stretoccocus, Acetobacter etc.	Ethanol, sugars	Dairy products, Vinegar, Antibiotics, Enzymes	Auto-acidification, ROS generation	

fermentation–to suppress the growth of spoilage-causing microbes and encourage the synthesis and metabolism of fermentative organisms. Consequently, the fermentative microbes quickly grow in number and inhibit the growth of other types of microbes due to the production of alcohol, in addition to competing for nutrients (Stanbury et al., 1995; Gibson et al., 2007).

The diversity of fermentation products released by the microorganisms is attributed to the rich diversity of microorganisms which have a diverse metabolism that can yield various types of fermentation products. In the case of yeast fermentation, the selection of yeast strains for efficient fermentation performance in the industrial production of wine or beer typically focuses on attributes such as predictable fermentation at the relevant process temperatures, desired fermentation vigour and the extent of sugar attenuation with efficient conversion to ethanol. Alcoholic fermentation occurs by fermenting sugars to ethanol and CO_2 with the help of some bacteria, fungi, yeast and algae. It is the process that yields beer, wine and other spirits. During this process, pyruvate is decarboxylated to acetaldehyde and subsequently reduced to ethanol by the enzyme alcohol dehydrogenase with NADH as the electron donor. This is also the basis of leavening of bread by yeast. Alcohol from yeast-fermented cider, in the presence of oxygen, will be further fermented by bacteria such as *Acetobacter aceti* to produce vinegar. The reduction of pyruvate to lactate is the basis of lactic acid fermentation. Lactose (milk sugar), fermented by *Streptococcus lactis* bacteria, gives lactic acid, which curdles the milk to yield cottage cheese and curd from which other cheeses can be made. Lactic acid fermenters can be separated into two groups. Homolactic fermenters use the glycolytic pathway and directly reduce almost all their pyruvate to lactate with the enzyme lactate dehydrogenase. Heterolactic fermenters form substantial amounts of products other than lactate; many produce lactate, ethanol, and CO_2 by way of the phosphor-ketolase pathway (Nelson et al., 2008).

2.4 The Microbial Cellular Events

The complete sequence of events in a cell starting from its formation, replicating itself to the next generation of daughter cells and finally its death, is referred to as the cell cycle. Microbial growth is described as an orderly increase in all chemical components in the presence of suitable culture medium. There are four types of microbial growth: bacteria grow by binary fission, yeast divide by budding, fungi divide by chain elongation and branching and viruses grow intracellularly in host cells.

Bacterial binary fission is a relatively simple type of cell division: the cell elongates, replicates its chromosome, and separates the newly formed DNA molecules so there is one chromosome in each half of the cell. Finally, a septum (cross wall) is formed at mid-cell, dividing the parent cell into two progeny cells, each having its own chromosome and a complement of other cellular constituents. Cell division causes an exponential increase in the number of cells in a population. Population growth can be analyzed quantitatively as a doubling of the cell number per unit time for bacteria and yeasts or a doubling of biomass per unit time for filamentous organisms such as fungi. Microorganisms are generally grown in batch culture when they need to be grown in a liquid medium. In batch culture, they are incubated in a closed vessel with a single batch of medium, either aerobically or anaerobically. Since the medium is supplied only once during the incubation period, with the increase of product biomass, nutrient concentrations start declining and concentrations of waste materials start increasing. The growth of microorganisms reproducing by binary fission can be plotted as the logarithm of the number of viable cells versus the incubation time. The resulting curve has four distinct phases.

When microorganisms are introduced into a fresh culture medium, usually no immediate increase in cell number occurs. This period is called the lag phase. However, cells in the culture start synthesizing new components. A lag phase can be necessary for a variety of reasons. The cells may be old and depleted of ATP, essential cofactors, and ribosomes; these must be synthesized before growth can begin. Here new enzymes would be needed to use different nutrients. Possibly the microorganisms have been injured and require time to recover. Whatever the causes, eventually the cells begin to replicate their DNA, increase in mass, and finally divide. During the log phase or exponential phase, with suitable culture medium, microorganisms exhibit growth and cell division at the maximum rate. Organisms grow at a constant rate in this phase, with uniform population kinetics, in terms of physiological and biochemical properties. Exponential growth is balanced growth, with all the cellular components being manufactured at constant rates relative to each other. The rate of microbial growth is influenced by the concentration of nutrients present in the medium, increasing with an increase of nutrient concentration, in a hyperbolic manner. At sufficiently high nutrient levels, the transport systems are saturated, and the growth rate does not rise further with increasing nutrient concentration. In a closed system such as a batch culture, population growth eventually ceases after the log phase and the growth curve becomes horizontal. This stationary phase is usually attained by bacteria at a population level of around 10^9 cells per ml. A significant factor determining the size of the final population is the nutrient availability. In the stationary phase, the total number of viable microorganisms remains constant. This may result from a balance between cell division and cell death, or the population may simply cease to divide but remain metabolically active. A number of factors are at play for microbial populations to enter the stationary phase such as:

a) Depletion in nutrient concentration: If an essential nutrient is severely depleted, population growth will decline;
b) O_2 concentration: Aerobic organisms are often limited by O_2 availability. Oxygen is not very soluble and may be depleted so quickly that only the surface of a culture will have an O_2 concentration adequate for growth; and
c) Concentration of waste by-products: Population growth may also cease due to the accumulation of toxic waste by-products. This factor seems to limit the growth of many anaerobic cultures. One good example of this is the production of lactic acid and other organic acids by *Streptococci* during sugar fermentation that their medium becomes acidic and growth is inhibited (Prescott et al., 2002).

Due to the depletion of required nutrients and the accumulation of waste by-products, cell death occurs. In the case of long-term growth experiments, it is seen that there is a gradual decline in the number of viable cells. This decline phase can last up to when the population is diminished. During this time, the bacterial population continually grows so that actively reproducing cells are best able to use the nutrients released by their dying brethren and best able to tolerate the toxins. This dynamic process is marked by successive waves of genetically distinct variants.

The growth curve of a specific microbe also describes the products formed at different phases of growth. For instance, the primary products required for microbial growth and metabolism, like amino acids, proteins, nucleotides, carbohydrates, lipids, etc. are produced during the log phase or the exponential growth phase. During the stationary phase and the decline phase, some microbes synthesize secondary metabolites, products which are not essential to normal metabolism, but have accessory functions (Bu'Lock et al., 1965). It is important to note that secondary metabolites sometimes tend to be intermediates or derived from products of primary metabolism. Additionally, not all microorganisms undergo secondary metabolism, it is more common in filamentous bacteria and fungi but not seen in the *Enterobacteriaceae* group. The commonly produced secondary metabolites include products having antimicrobial activity, growth promoter molecules, and products with pharmacological importance. Therefore, the production of secondary metabolites has formed the basis of a number of fermentation processes. The production of secondary metabolites also varies based on their culture conditions.

Bacterial growth rates during the phase of exponential growth, under standard nutritional conditions (culture medium, temperature, pH, etc.) define the bacterium's generation time. Generation times for bacteria vary from about 12 minutes to 24 hours. The generation time for *E. coli* in the laboratory is 15–20 min. Symbionts such as *Rhizobium* tend to have a longer generation time. Some pathogenic bacteria, e.g. *Mycobacterium tuberculosis* have especially long generation times and this is thought to be an advantage to their virulence.

When growing exponentially by binary fission, the increase in a bacterial population is by geometric progression. The generation time is the time interval required for cells (or population) to divide:

$$G = t / n$$

Where G is generation time, n is number of generations and t is time in min/hours.

Measuring techniques used for cell counting are:

a) The first method is to measure the dry weight of the cell material in a fixed volume of the culture by measuring the dry weight of the cell material in a given volume of the culture. The cells need to be removed from the medium and dried;
b) another method is to use the spectrophotometer to estimate absorbance of cell suspensions. The absorbance at a particular wavelength is proportional to the cell concentration. By plotting a standard curve of absorbance versus cell concentration, the cell concentration of an unknown sample can be calculated;

c) several plating methods can be used to determine the number of viable microbes in a sample. These are referred to as viable counting methods (plate counts) because they count only those cells that are able to reproduce when cultured. Each viable unit will form a colony and each colony that can be counted is called a colony forming unit (cfu) and the number of cfu's is related to the viable count in the sample. Usually the count is more accurate with the help of a colony counter. The samples should ideally be between 30 and 300 colonies for most accurate counting; and
d) in the case of cultures maintained in liquid medium, turbidity measurement is a fast and non-destructive method especially for counting large numbers of bacteria in clear liquid media and broths – but cannot detect cell densities less than 10^7 cells per ml.

Fungi include eukaryotic, spore-bearing organisms, lack chlorophyll and reproduce sexually and asexually. Fungi are primarily terrestrial organisms; however, a few species can be found in fresh water and marine water environment as well. Out of the approx. 90 000 species of fungi that have been studied, many are beneficial to humans. Fungi act as decomposers in the ecosystem, playing a very significant role. Fungi can help degrade complex organic matter to simpler compounds and inorganic molecules. They help in the breakdown of dead organic organisms to their constituent elements like nitrogen, phosphorous, and carbon to make it available for absorption. They then absorb the soluble products. They are chemoorganoheterotrophs and use organic compounds as a source of carbon, electrons, and energy. Glycogen is the primary storage polysaccharide in fungi. Most fungi use carbohydrates (preferably glucose or maltose) and nitrogenous compounds to synthesize their own amino acids and proteins. Fungi are usually aerobic. Some yeasts, however, are facultative anaerobic and can obtain energy by fermentation, such as in the production of ethyl alcohol from glucose.

Yeast is a unicellular fungus having a single nucleus that reproduces either asexually by budding and transverse division or sexually through spore formation. Each bud that separates can grow into a new yeast cell, and some group together to form colonies. Generally yeast cells are larger than bacteria, vary considerably in size, and are commonly spherical to egg shaped. They do not have flagella but do possess most of the other eukaryotic organelles. *Saccharomyces cerevisiae* strains, also commonly referred to as "Baker's yeast" or "Brewer's yeast" are extensively used for making bread, production of wine, beer, sake, as well as bioethanol. A typical yeast cell like *S. cerevisiae* is capable of living a "dual life", as a haploid organism in the case of scarcity of food when the yeast cell is malnourished; and fuse together to make diploid organisms in the case of abundance of food matter for the cell. In this case of food availability, the yeast mother cell undergoes budding giving rise to diploid daughter cells. However, in case of food scarcity, a starved yeast cell undergoes meiosis, wherein four cells develop within the mother cell and they persist until the cell obtains a food source. Additionally, two haploids can fuse and make a diploid. Once formed, it buds even before it attains its final shape. In the case of wine fermentation, *Saccharomyces cerevisiae* consumes the grape sugars and converts them to produce alcohol, CO_2 and secondary products at both aerobic and anaerobic conditions that impart the wine with its characteristic colour, flavour and aroma. At aerobic conditions, respiration is possible; respiration occurs with oxygen being the final electron acceptor. But *S. cerevisiae* continues alcoholic fermentation until low sugar levels are reached. This phenomenon is called the "Crabtree effect" (De Deken, 1966), and the yeasts showing this property are called Crabtree-positive yeasts. In contrast Crabtree-negative yeasts lack fermentative products and under aerobic conditions, biomass and carbon dioxide are the sole products (Dashko et al., 2014). But in industrial grade fermentations, the required yeast strains are propagated initially in specially designed vessels before scaling up to pilot scale. Yeast is typically propagated in wort under aerobic growth conditions by introducing sterile air into the wort. After the propagation of the yeast slurry, it is transferred into a special cylindro-conical fermentation vessel. After the initial lag phase, yeast growth occurs at an exponential rate with rapid depletion of the oxygen and nutrient resources, thus creating an anaerobic environment in the vessel. The nutrients and sugars are utilized rapidly, eventually resulting in carbon and nutrient deficit, making the strain enter a dormant phase. Typically, during this stage, with the depletion of the carbon source, a steady rise in ethanol concentration is observed. On the completion of the fermentation process, the yeast that forms as sediment at the bottom of the fermentation vessel is stored partially under low temperature conditions to be used in subsequent fermentations. Industrial grade fermentation in breweries for the production of good quality wine or beer can be achieved by maintenance of vitality of the yeast strain as well as the optimum growth conditions required. Brewery fermentations expose the yeast strains to a number of extracellular stresses such as oxygen concentration, osmotic potential, pH, nutrient availability, and temperature. As a non-motile, unicellular organism, yeast relies on physiological mechanisms to cope with these environmental changes (Gibson et al., 2007; Brice et al., 2014).

2.5 Cell Signalling in Microorganisms

Cell signalling was first discovered when it was observed that the marine bioluminescent bacterium *Vibrio fischerii* had the ability to glow by producing a molecule termed as autoinducer. This autoinducer which was later identified as N-acylhomoserine lactone (AHL),that regulated various responses like bioluminescence, virulence and biofilm formation, antibiotic production, plasmid transfer etc. In many species of bacteria, AHL freely diffuses across the plasma membrane. In high population density bacterial communities, AHL accumulates outside the cell, so that the AHL can enter the cell through the change in the diffusion gradient. Quorum sensing is the phenomenon where the AHL influx and its action is dependent on the cell density. Because the influx of AHL is cell density-dependent, it enables individual cells to assess population density. This is referred to as quorum sensing. When AHL reaches a threshold level inside the cell, it induces the expression of target genes that regulate a number of functions, depending on the microbe. These functions are most effective only if a large number of microbes are present. Cell-to cell signals occur through chemical signals that are released by the cells in the form of small soluble or volatile ligands. Ligands are small molecules that bind to a specific receptor molecule to elicit a response by the population or initiate a specific signalling pathway. Ligands and receptors exist in several varieties for different metabolic and physiological functions of the cellular process. Receptors are protein moieties in target cells and they have high degree of fidelity to the specific ligand to bind to elicit a response. The communication through such signalling pathways can either be between cells called intercellular signalling, or it can be within a cell, called intracellular signalling.

Apart from being the raw materials for synthesizing cellular components for metabolism and growth, some nutrients also partake in complex signalling pathways. Sensing and signalling are especially important in case of unicellular organisms since they are often subjected to drastic shifts in the nutrient levels in their environment, from abundant supply to periods of nutrient limiting conditions. Higher organisms maintain relative homeostasis by typically altering secretion of certain hormones or enzymes. But there have been studies on higher eukaryotes that show that certain nutrients can initiate signalling pathways independently of hormonal balance (Nobukuni et al., 2005). One of the most well explored model organisms for studies in nutrient mediated signalling pathways is *Saccharomyces cerevisiae,* accentuating that several signalling events are highly conserved from yeast to higher eukaryotes (Manning et al., 2002; Swinnen et al., 2006).

Communication by extracellular signals usually involves six steps:

1) Synthesis of signalling molecules;
2) release of the signalling molecule by the signalling cell;
3) transport of the signal to the target cell;
4) detection of the signal by a specific receptor protein;
5) a change in cellular metabolism, function, or development triggered by the receptor-signal complex; and
6) removal of the signal, which often terminates the cellular response.

In many eukaryotic microorganisms (e.g. yeast, slime molds, and protozoans), secreted molecules coordinate the aggregation of free-living cells for sexual mating or differentiation under certain environmental conditions. Chemicals released by one organism that can alter the behavior or gene expression of other organisms of the same species are called pheromones. More important in plants and animals are extracellular signalling molecules that function within an organism to control metabolic processes within cells, the growth of tissues, the synthesis and secretion of proteins, and the composition of intracellular and extracellular fluids. Inside the cell, ligands bind to their internal receptors, allowing them to directly affect the cell's DNA and protein-producing machinery. Using signal transduction pathways, receptors in the plasma membrane produce a variety of effects on the cell. The results of signalling pathways are extremely varied and depend on the type of cell involved as well as the external and internal conditions.

Microorganisms in their natural environment often undergo drastic shifts in their nutrient requirement, from excess nutrient levels to nutrient-starved and the intermediate changes between the two. In the case of fermentation processes, the genetic diversity of the numerous industrial-grade yeast strains contribute to the subtle differences in the organoleptic characteristics of the products formed. These industrial yeast strains are always viable with the ability to perform fermentation in high concentrations of sugar and sensing and adapting to changes in their environment. Generally, changes in their growth medium occurs under anaerobic culture conditions, such as high sugar or organic acids concentration could give rise to high osmotic pressure. In the case of changes in the pH or nitrogen levels, extended fermentation periods with an increase in the concentration of ethanol could occur (Walker et al., 2014). Glucose is the preferred carbon source of yeast cells, due to its rapid fermentation to ethanol, inhibiting the growth of other nutrient-competing microorganisms. Fermentative yeast cells exhibit active glycolysis and sensitivity to environmental stresses like heat, pH, salinity, high osmolarity, nutrient depletion, etc. They also accumulate small amounts of carbohydrates trehalose and glycogen (Versele et al., 2001). An important signalling pathway involved in growth and starvation and its intermediate phases is the kinase target of rapamycin (TOR). TOR enables the eukaryotic organisms to sense the availability of nutrients to promote growth in a favorable environment. In higher multicellular organism TOR has a significant role in maintaining homeostasis and regulates the growth of the organism (Zoncu et al., 2011). Keeping in context to this chapter, it is important to note that *S. cerevisiae* has two TOR genes, whereas almost all eukaryotic organisms, be it plants, flies or mammals have only one TOR gene, which helps in differentiating two different signalling cascades based on differential requirements of the two TORs (Loewith and Hall, 2011). TOR1 and TOR2 are present in two functionally different multiprotein complexes, with TOR1 regulating ribosomal biogenesis, translation, protein synthesis and transcription of starvation-induced stress genes and TOR2 mediates actin cytoskeleton organization and cell polarization (Swinnen et al., 2006; Zoncu et al., 2010).

2.6 Microbial Performance Under Stress Conditions

In glucose limiting conditions, the yeast cells undergo a diauxic shift during which the cells adapt to the unavailability of glucose and shift their mode of metabolism to respiration. This metabolic switch is followed by a phase of slow growth during which alternate sources of energy like ethanol, acetate and other initial products of fermentation is used as a carbon source. Once the alternate carbon sources are exhausted, the cells enter a stationary phase, during which transcriptional and metabolic readjustments occur in the yeast cells. During this period of starvation the cells adapt and acquire stress resistance for survival by activation of the stress genes and accumulation of trehalose. Other adaptation strategies include repression of genes in protein synthesis, resulting in arrested growth. The major stresses in *S. cerevisiae* cultures and their effects and adaptation are elucidated in Table 2.2. Apart from the nutrient-limiting conditions, environmental stress factors (like heat stress) can also prompt the cells into the stationary mode. With this in reference, nutrient deficit and stress signalling pathways converge to Msn2 and Msn4 (Msn2/4). This pair of partially redundant transcriptional factors regulates expression of stress genes containing stress responsive elements (STREs) in their promoters (Reinders et al., 1998; Swinnen et al., 2006).

Table 2.2 Response and adaptation of *Saccharomyces cerevisiae* to major stresses.

Factors	Response and adaptation mechanisms	References
Carbon stress	Metabolic and transcriptional readjustments. Trehalose accumulation and activation of stress genes. cAMP-PKA pathway.	Kim et al., 2011; Brice et al., 2014
Nitrogen stress	Stuck fermentation occurs due to N stress. Rim15 pathway involved in altering the cell metabolism with the available nitrogen. Trehalose and glycogen accumulation.	Jorgensen et al., 2002
Oxidative stress	Release of antioxidant molecules. Trehalose helps with the degenerative effects of ROS.	Higgins et al., 2003; Gibson et al., 2007;
Ethanol stress	Over-expression of transcription factor MSN2 and Ras-cAMP pathway inhibitor 1 (RPI1) promotes ethanol fermentation and tolerance in *S. cerevisiae.*	Sasano et al., 2012; Kim et al., 2013

One of the main controlling factors governing alcoholic fermentation is the concentration of available nitrogen, a key nutrient which is consumed at the early stages of the fermentation process. With the gradual decline in the levels of nitrogen, the process may slow down or stop, which is termed as stuck fermentation. In the case of wine fermentations, the requirement of nitrogen in different strains of yeasts varies from one to other, which reflects in the capacities of different strains to carry out fermentation in nitrogen limiting conditions (Brice et al., 2014). There are multiple reports studying the starvation response of aerobically starved cells in *Saccharomyces cerevisiae,* which a commercially important model organism. But anaerobic starvation is the more commonly observed phenomenon during the production of bread, beer and wine. It was observed that in single nutrient starvation studies of the yeast strain, the cells were more tolerant to carbon starvation as compared to nitrogen starvation (Nilsson et al., 2001; Jorgensen et al., 2002). The intolerance for nitrogen starvation is due to the degradation of the glucose uptake pathway in nitrogen starved cells with an available fermentable carbon source (Busturia, and Lagunas, 1986). However, it was reported that even in the presence of a non-fermentable carbon source, like ethanol the fermentation capacity of the cells was significantly less during nitrogen starvation compared to glucose starvation (Nilsson et al., 2001). A number of factors are at play for sensing the nitrogen deficit in the medium. An adaptive mechanism of the yeast cell is the accumulation of glycogen and Trehalose in nutrient limiting conditions but can be found in bulk during nitrogen starvation with a carbon source present (Lillie and Pringle, 1980; Thomsson et al., 2005). More recently, Brice et al. (2014) conducting a study with "low nitrogen requirement strains" (LNR) and "high nitrogen requirement strain" (HNR) reported that the former expressed biosynthetic genes whereas the latter showed a specific gene expression pattern with the stress genes being overexpressed. It was observed that the HNR strains were more sensitive to nitrogen starvation resulting in lower fermentative capacity. An important signalling pathway involved in nitrogen signalling is the Rim15 pathway that senses the nitrogen status in the cell and makes the metabolism of the cell adapt to the available nitrogen (Swinnen et al., 2006; Brice et al., 2014). Studies have shown that Rim15 pathway is essential in haploid and diploid cells for entry into the stationary phase. Moreover, it was found that Rim15 was required to integrate signals from various nutrient sensory kinases (such as, PKA, TORC1 and Pho85- Pho80) that transmit information on the status of nutrients.

Oxygen is an essential component in breweries for the initial stage of fermentation and manufacturing yeast slurry. However, an excess of exposure of the yeast to oxygen during fermentation can result in an overpopulation of yeast leading to a sharp increase of ethanol production (Briggs et al., 2004). During aerobic respiration, derivative forms of O_2 known as reactive oxygen species (ROS) is produced endogenously by cells under aerobic conditions. ROS includes hydrogen peroxide, hydroxide radical and superoxide radical, can cause damage to cellular components and contribute to ageing and finally leading to cell death (Beckman & Ames, 1998). Studies report more specific effects like protein inactivation, lipid peroxidation and DNA damage, including mtDNA (Girotti, 1998; Cabiscol et al., 2000; Salmon et al., 2004; Ribeiro et al., 2006; Gibson et al., 2007). This contradiction of oxygen requirement for aerobic respiration and metabolic processes simultaneously having toxic effects to the cell is commonly referred to as 'oxygen paradox' (Davies, 1995). In *Saccharomyces cerevisiae*, the antioxidant defences include non-enzymatic molecules like flavohaemoglobin, erythroascorbic acid, glutathione, ubiquinol, polyamines, trehalose and ergosterols. Enzymatic antioxidant molecules include cytochrome C peroxidase, catalase, glutathione peroxidase, glutathione reductase, superoxide dismutase, thioredoxin, and thioredoxin peroxidase (Higgins et al., 2003; Dawes, 2004; Saffi et al., 2006; Gibson et al., 2007). Antioxidant molecules produced by brewing strains of yeast react rapidly to oxidative stress. In a small-scale fermentation of a semi-defined medium it was seen that Cu/Zn superoxide dismutase, Mn superoxide dismutase and catalase activities of the brewing strains of yeast changed rapidly, indicating that O_2 exposure regulated enzymatic antioxidant activity (Clarkson et al., 1991). In a study by Higgins et al. (2003), it was also seen that expression of antioxidant-coding genes was up to 53 folds higher in the first hour of the pilot-scale fermentation, a response typical of oxidative stress. A subsequent second study showed that in the later stages of fermentation a significant upregulation of the stress responsive genes was seen, indicating the effect of residual ROS within cells responding to other stress factors during fermentation (James et al., 2003). In terms of oxidative stress response, antioxidants are seen to be involved in protection of the cells from environmental or ethanol stress, both of which are generally encountered by the cells during industrial-grade fermentation (Rand and Grant, 2006). Trehalose also plays a role in protecting yeast cells from the degradative effects of ROS. The increased accumulation of trehalose in the presence of a hydroxyl radical producing system and the superoxide-generating drug menadione has been associated with better resistance to oxidative stress damage (Benaroudj et al., 2001; Herdeiro et al., 2006).*Saccharomyces cerevisiae* has two stress response pathways, namely; the heat shock response (HSR), activated by the sub-lethal stress mediated by the heat shock transcription factor (HSF)

(Morimoto et al., 1996; Chatterjee et al., 2000). Additionally the global stress response (GSR) is activated by a number of environmental factors, including pH, heat, osmotic, oxidative stress, as well as and nitrogen starvation (Ruis and Schuller, 1995; Schmitt and McEntee, 1996). It is believed that the global stress response (GSR) is an evolutionary adaptation of the yeast cells allowing them to respond and survive in adverse conditions, in an order that cellular fecundity is retained while specific responses are activated (Martinez-Pastor et al., 1996).

Acknowledgment

This work was supported by the TERI Deakin Nanobiotechnology Centre, Biotechnology and Management of Bioresources Division, The Energy and Resources Institute, New Delhi.

References

Beckman, K.B. and Ames, B.N. (1998) The free radical theory of aging matures. *Physiological Reviews*, 78(2), pp.547–581.

Brice, C., Sanchez, I., Bigey, F. et al. (2014) A genetic approach of wine yeast fermentation capacity in nitrogen-starvation reveals the key role of nitrogen signalling. *BMC Genomics*, 15:495.

Brice, C., Sanchez, I., Tesnière, C., and Blondin, B. (2014) Assessing the mechanisms responsible for differences in nitrogen requirements between Saccharomyces cerevisiae wine yeasts in alcoholic fermentation. *Appl Environ Microbiol* doi:10.1128/AEM.03856-13.

Briggs, D.E., Brookes, P.A., Stevens, R., and Boulton, C.A. (2004) *Brewing: science and practice*. Elsevier.

Bu'Lock, J.D. (1965) Biosynthesis of natural products; an introduction to secondary metabolism.

Busturia, A. and Lagunas, R. (1986) Catabolite inactivation of the glucose transport system in Saccharomyces cerevisiae. *Microbiology*, 132(2), pp.379–385.

Cabiscol, E., Piulats, E., Echave, P. et al. (2000) Oxidative stress promotes specific protein damage in Saccharomyces cerevisiae. *J Biol Chem* 275: 27393–27398.

Chatterjee, M.T., Khalawan, S.A., and Curran, B.P.G. (2000) Cellular lipid composition influences stress activation of the yeast general stress response element (STRE). *Microbiology* 146: 877–844.

Clarkson, S.P., Large, P.J., Boulton, C.A., and Bamforth, C.W. (1991) Synthesis of superoxide dismutase, catalase and other enzymes and oxygen and superoxide toxicity during changes in oxygen concentration in cultures of brewing yeast. *Yeast* 7: 91–103.

Dashko, S., Zhou, N., Compagno, C., and Piškur, J. (2014) Why, when, and how did yeast evolve alcoholic fermentation? *FEMS yeast research*, 14(6), pp.826–832.

Davies, K.J. (1995) Oxidative stress: the paradox of aerobic life. In *Biochemical Society Symposia* (Vol. 61, pp. 1–32). Portland Press, London.

Dawes, I.W. (2004) Yeast stress responses. In: *The Metabolism and Molecular Physiology of Saccharomyces cerevisiae (eds)* J.R. Dickinson and M. Schweizer. CRC Press, Boca Raton, FL (2004), xiv, pp. xiv+ 459.

De Deken, R.H. (1966) The Crabtree effect: a regulatory system in yeast. *J Gen Microbiol* 44: 149–156.

Gibson, B.R., Lawrence, S.J., Leclaire, J.P. et al. (2007) Yeast responses to stresses associated with industrial brewery handling. *FEMS Microbiology Reviews*, 31(5), pp.535–569.

Girotti, A.W. (1998) Lipid hydroperoxide generation, turnover, and effector action in biological systems. *J. Lipid. Res.*, 39: 1529–1542.

Higgins, V.J., Beckhouse, A.G., Oliver, A.D. et al. (2003) Yeast genome-wide expression analysis identifies a strong ergosterol and oxidative stress response during the initial stages of an industrial lager fermentation. *Appl. Environ. Microbiol.*, 69: 4777–4787.

Jorgensen, H., Olsson, L., Ronnow, B., and Palmqvist, E. (2002) Fed-batch cultivation of baker's yeast followed by nitrogen or carbon starvation: effects on fermentative capacity and content of trehalose and glycogen. *Appl. Microbiol. Biotechnol.*, 59:310–317.

Jurtshuk, J.P. (1996) Bacterial metabolism In: *Medical Microbiology* 4th ed. (ed.) S. Baron University of Texas Medical Branch at Galveston.

Kim, J.H., Roy, A., Jouandot II, D., and Cho, K.H. (2013) The glucose signaling network in yeast. *Biochimica et Biophysica Acta (BBA)-General Subjects*, 1830(11), pp.5204–5210.

Kitichantaropas, Y., Boonchird, C., Sugiyama, M. (2016) Cellular mechanisms contributing to multiple stress tolerance in *Saccharomyces cerevisiae* strains with potential use in high-temperature ethanol fermentation. *AMB Express*, 6(1), p.107.

Laskin, A.I. (1977) Single cell protein. In *Annual reports on fermentation processes* (Vol. 1, pp. 151–180). Elsevier.

Liao, C., Seo, S.O., Celik, V. (2015) Integrated, systems metabolic picture of acetone-butanol-ethanol fermentation by Clostridium acetobutylicum. *PNAS* 27: 8505–8510.

Lillie, S. and J. R. Pringle. (1980) Reserve carbohydrate metabolism in *Saccharomyces cerevisiae*: responses to nutrient limitation. *J. Bacteriol.* 143: 1384–1394.

Loewith, R. and Hall, M.N. (2011) Target of rapamycin (TOR) in nutrient signaling and growth control. *Genetics*, 189(4), pp.1177–1201.

Manning, G., Plowman, G.D., Hunter, T., and Sudarsanam, S. (2002) Evolution of protein kinase signaling from yeast to man. *Trends Biochem Sci*, 27:514–520.

Martinez-Pastor, M.T., Marchler, G., Schuller, C.et al. (1996) The Saccharomyces cerevisiae zinc finger proteins Msn2p and Msn4p are required for transcriptional induction through the stress response element (STRE). *EMBO. J.* 15: 2227–2235.

Morimoto, R.I., Kroeger, P.E., and Cotto, J.J. (1996) The transcriptional regulation of heat shock genes: a plethora of heat shock factors and regulatory conditions. In *Stress-inducible cellular responses*, pp. 139–163.

Nelson, D.L., Lehninger, A.L., and Cox, M.M. (2008) *Lehninger principles of biochemistry*. Macmillan.

Nilsson, A., Påhlman, I.-L., Jovall, P.A. et al. (2001) The catabolic capacity is preserved to a higher extent during carbon compared to nitrogen starvation. *Yeast* 18:1371–1381.

Nobukuni, T., Joaquin, M., Roccio, M. et al. (2005) Amino acids mediate mTOR/raptor signaling through activation of class 3 phosphatidylinositol 3OH-kinase. *Proc. Natl. Acad. Sci. U.S.A.*, 102:14238–14243.

Prescott, L.M., Harley, J.P., and Klein, D.A. (2002) Microorganisms in aquatic environments In *Microbiology* 5th edition, McGraw-Hill Companies Inc., New York, pp.632–663.

Rand, J.D. and Grant, C.M. (2006) The thioredoxin system protects ribosomes against stress-induced aggregation. *Molecular biology of the cell*, 17(1), pp.387–401.

Reinders, A., Burckert, N., Boller, T. et al. (1998) Saccharomyces cerevisiae cAMP-dependent protein kinase controls entry into stationary phase through the Rim15p protein kinase. *Genes. Dev.*, 12:2943–2955.

Ribeiro, G.F., Corte-Real, M., and Johansson, B. (2006) Characterization of DNA damage in yeast apoptosis induced by hydrogen peroxide, acetic acid, and hyperosmotic shock. *Mol. Biol. Cell*, 17: 4584–4591.

Ruis, H. and Schuller, C. (1995) Stress signaling in yeast. *BioEssays*, 17: 59–965.

Saffi, J., Sonego, L., Varela, Q.D., and Salvador, M. (2006) Antioxidant activity of L-ascorbic acid in wild-type and superoxide dismutase deficient strains of *Saccharomyces cerevisiae. Redox. Rep.*, 11: 179–184.

Salmon, T.B., Evert, B.A., Song, B., and Doetsch, P.W. (2004) Biological consequences of oxidative stress-induced DNA damage in *Saccharomyces cerevisiae. Nucleic. Acids Res.*, 32: 3712–3723.

Sasano, Y., Haitani, Y., Hashida, K. et al. (2012) Simultaneous accumulation of proline and trehalose in industrial baker's yeast enhances fermentation ability in frozen dough. *Journal of bioscience and bioengineering*, 113(5), pp.592–595.

Schmitt, A.P. and McEntee, K. (1996) Msn2p, a zinc finger DNA-binding protein, is the transcriptional activator of the multi-stress response in *Saccharomyces cerevisiae. Proceedings of the National Academy of Sciences*, 93(12), pp.5777–5782.

Seshasayee, A.S., Bertone, P., Fraser, G.M., and Luscombe, N.M. (2006) Transcriptional regulatory networks in bacteria: From input signals to output responses. *Curr. Opin. Microbiol.*, 9, 511–519.

Stanbury, P.F., Whitaker, A., and Hall, S.J. (1995) *Principles of fermentation technology*. Elsevier.

Swinnen, E., Wanke, V., Roosen, J. et al. (2006) Rim15 and the crossroads of nutrient signalling pathways in *Saccharomyces cerevisiae. Cell Division*, 2006, 1:3 doi:10.1186/1747-1028-1-3

Thomsson, E., Gustafsson, L., and Larsson, C. (2005) Starvation Response of *Saccharomyces cerevisiae* Grown in Anaerobic Nitrogen- or Carbon-Limited Chemostat Cultures. *Applied and Environmental Microbiology*, pp. 3007–3013.

Versele, M., Lemaire, K., and Thevelein, J.M. (2001) Sex and sugar in yeast: two distinct GPCR systems. *EMBO reports*, 2(7), pp.574–579.

Walker, M.E., Nguyen, T.D., Liccioli, T. et al. (2014) Genome-wide identification of the Fermentome; genes required for successful and timely completion of wine-like fermentation by Saccharomyces cerevisiae. *BMC genomics*, 15(1), p.552.

Zoncu, R., Efeyan, A., and Sabatini, D.M. (2011) mTOR: from growth signal integration to cancer, diabetes and ageing. *Nat. Rev. Mol. Cell Biol., 12*.

3

Microbial Metabolism in a Refined Carbon Source: Generalities

Vinayak Laxman Pachapur[1,2], *Preetika Rajeev Kuknur*[1], *Satinder Kaur Brar*[1], *and Rosa Galvez-Cloutier*[2]

[1] *Institut national de la recherche scientifique, Centre - Eau Terre Environnement, Québec, Canada*
[2] *Université Laval, Department of Civil Engineering and Water Engineering, Quebec, Canada*

3.1 Introduction

Microbes require energy and nutrients for living, which they obtain during the breakdown of carbon source during biochemical mechanism usually known as microbial metabolism. In nature, microorganisms coexist for proper interaction and assisting each other to perform effectively amongst the groups (Chang et al., 2008). Microorganisms utilize 30–40% of substrate with the remaining 60–70% used for metabolite formation. Microorganisms perform two major functions: transmitting molecular signals and communicating with each other using metabolites. These signals contribute to the division of labor for easy and rapid degradation of the substrate. Thus, the use of microbes helps in substrate utilization and degradation efficiency at a faster rate (Baghchehsaraee et al., 2010). The metabolic flexibility of these microorganisms allows them to develop a synergistic metabolic process which can result in better substrate utilization and elimination of feedback inhibition. (Nishio & Nakashimada, 2004; Hussy et al., 2005).

Microbial metabolism during fermentation plays a very important role during the fermentation. In addition to the desired products during the fermentation, the metabolites produced are referred to as metabolite by-products. In some cases, if the metabolites interfere during the downstream processing of desired product they will be referred to as metabolite impurities. Lastly, if the presence of metabolite during the fermentation causes the inhibition of substrate conversion pathway they will be referred to as product inhibitors. However, the metabolites are the integral part of the microorganisms and are also the information source of the fermentation. During the substrate conversion step, the microorganisms tend to follow a particular pathway to produce the metabolite in order to survive and generate energy in the form of ATPs. Based on the metabolite behavior analysis, the best microorganisms can be selected for the fermentation.

Microbial Sensing in Fermentation, First Edition. Edited by Satinder Kaur Brar, Ratul Kumar Das, and Saurabh Jyoti Sarma.

Microbial fermentation utilizes the substrates (glucose, glycerol) to convert into energy source such as hydrogen, methane or biogas; or production of end metabolites such as acetate, butanol, ethanol, etc. (Pachapur et al., 2016a). Analysis of these end metabolites could be useful indicators for monitoring the utilization of substrate in any fermentation process and also to determine the product formation (Pachapur et al., 2015b; Pachapur et al., 2015c). The microbial metabolism with end metabolites production, differ with changes in the substrate used, inoculum used, operating parameters and fermentation conditions.

In this chapter we have gathered the information related to microbial metabolism during glycerol fermentation for biofuels production. The metabolic pathway of glycerol fermentation for biofuels production as an example is represented in Figure 3.1. The glycerol as a substrate is highly reductive in nature, tend to get reduced into 1,3-propanediol by reductive pathway. During glycerol fermentation, the strict anaerobic bacteria follow the reductive pathway to produce 1,3-PD as main product with decreased hydrogen production (González-Pajuelo et al., 2004). To balance the microbial growth with utilization of glycerol, microorganisms follows oxidative pathway to produce acetate, butyrate, ethanol and hydrogen (Pachapur et al., 2015b; Pachapur et al., 2015c). In the presence of the carbon source, the microbial metabolisms vary in the presence of pure and mixed-culture, co-substrate, at different input parameters and operating conditions.

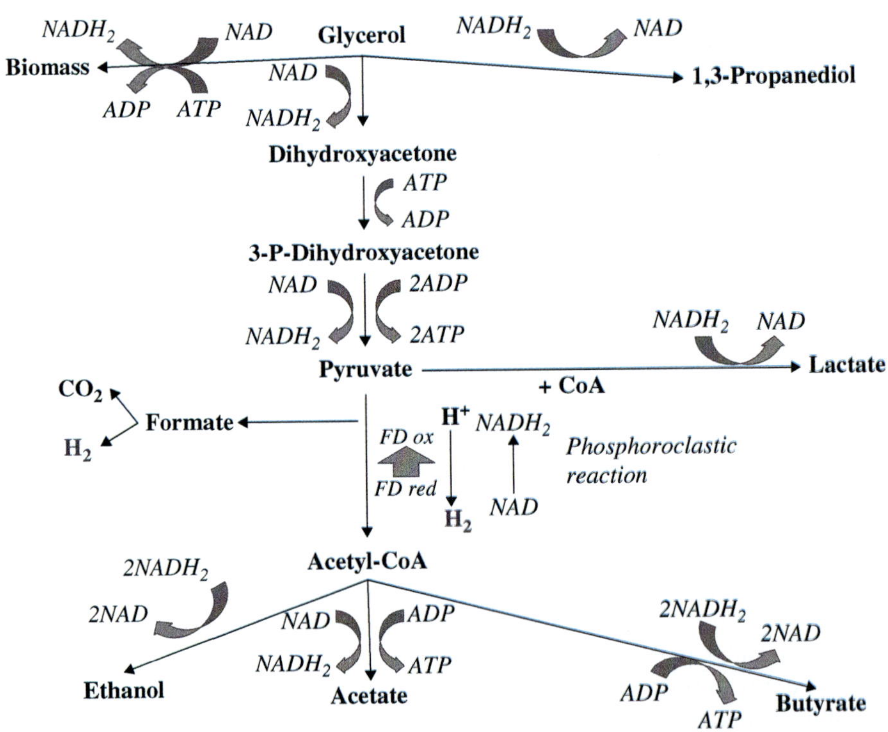

Figure 3.1 The metabolic pathway of glycerol fermentation during biofuels production.

3.2 Microbial Metabolism in Presence of Pure and Crude Substrate

The substrate used during fermentation plays a very important role during biofuel production and also in metabolite distribution. Substrates can be of two kinds, pure and raw/crude substrate. The substrates are selected depending upon availability and cost; pure substrates are costly in nature in comparison to crude substrate, which are abundant and available at low price. Pure substrates have the least amount of impurities in comparison to crude substrates which contain more impurities. The refined carbon sources are high cost carbon source and some are derived from petrochemical routes, which bring additional cost to fermentation together with environment concerns. With advancements in biotechnology along with improvements in the fermentation process, the valorizations of different industrial wastes are replaced with pure carbon source for the production of value-added compounds (Das et al., 2016).

The crude substrates are tested to determine the essential nutrient requirements, low-cost medium for high production and as a supplement for microbial transformation into industrial applications (Himmi et al., 1999). In order to determine the crude substrate as a possible replacement, a test is carried out with pure and crude glycerol on hydrogen production. The presence of impurities in the crude glycerol plays a beneficial role and at times can also play inhibitor roles. In the presence of crude glycerol, 6 h lag phase was observed indicating the adaption time needed for the inoculum in comparison to 2 h for pure glycerol. The metabolite distribution in the presence of pure and crude glycerol determined the fermentation pathway during H2 production. In the presence of crude glycerol with an increase in H2 production the acetate concentration increased. However, in the presence of pure glycerol with a decrease in H2 production the concertation of ethanol increased. During the cell growth phase to meet the energy requirement, cells produce more acetate with an increase in H^+ concentration. In the late exponential stage, cells re-utilize and convert acetate to butyrate and ethanol with decreased H^+ concentration (Mangayil et al., 2012).

The source and the content of the crude substrate also play a very important role during fermentation and metabolite distribution. Growth inhibition of *Clostridium* was tested over pure (87% glycerol) and crude glycerol (65% glycerol) derived from the transesterification process for biodiesel production using rapeseed oil. In this case there was no significant difference in the fermentation patterns on pure and crude glycerol. However, in the presence of 65% of crude glycerol *Clostridium* displayed a higher growth inhibition; this may be due to the presence of impurities in crude glycerol such as sodium salts and heavy metals at concentration to interfere with cell division (González-Pajuelo et al., 2004).

In case of pure glycerol, the H2 increased from (0.34 to 0.63 L H2) with increase in glycerol concentration from 5 to 10 g/L. Moreover, a further increase of glycerol resulted in stabilization and with an increase above 30 g/L resulted in decreased production. However, the production of 1,3-propanediol (1,3-PD) was vice-versa in comparison to increase in glycerol concentration. All the above observations can be related to limiting the biomass growth due to high concentration of the substrate (Seifert et al., 2009). Substrate limiting study was carried

Table 3.1 For the glycerol fermentation in the process of hydrogen production.

Substrate	Composition of substrate	Microorganisms	Temp/pH	Biofuels yield (mol/mol glycerol)	Ref.
Commercial	Pure/Crude glycerol	Anaerobic digested sludge	37 °C/6	0.41/0.71	(Seifert et al., 2009)
Commercial	Pure	*Enterobacter aerogenes*	37 °C/6.8	0.89	(Markov et al., 2011)
Vegetable source (waste vegetable oil)	Crude glycerol 41%	*Enterobacter aerogenes HU-101*	37 °C/6.8	1.12	(Ito et al., 2005)
Multi-feedstock (soybean oil, beef tallow, pork lard)	Crude glycerol 84%	Engineered	37 °C/6.3	1.02	(Yazdani & Gonzalez, 2008)

out using glycerol for production of bulk chemicals such as ethanol, acetate, butyrate and formate. In case of 60% of limitation, ethanol and formate produced at 1:1 ratio. However, at increasing concentration of substrate, the metabolic pathway shifted towards 1, 3-PD and acetate production at 2:1 ration. At limiting substrate concentration, the amount of acetyl-CoA (as represented in Figure 1) derived ethanol concentration was the same as formate yield. However, in the case of an increase in substrate concentration the biomass flux through acetyl-CoA was decreased as 1,3-PD production was not derived from acetyl-CoA (Temudo et al., 2008). The higher reduction state of glycerol favors the excess of reducing equivalents into formation of 1,3-PD at higher concentration (Pachapur et al., 2015c).

Growth inhibition studies were carried out using commercial and raw substrate during the production of 1,3-PD by *Clostridium*. Growth was inhibited by 59% at 100 g/L of commercial substrate in comparison to 65% for raw substrate. The inhibition in the case of raw substrate may be due the presence of substances such as sodium salts and heavy metals, inhibiting cell growth (González-Pajuelo et al., 2004).

H2 production in the presence of pure glycerol was around 0.41 mol/mol glycerol, was half the amount of H2 using crude glycerol 0.71 mol/mol using *Enterobacter* as seen in Table 3.1. The discrepancy between the resulted presented in the Table 3.1, can be explained only due to the difference in the biofuels yield due to source and the content of the crude substrate. Metabolic pathway is not only affected by the source and the content of the crude substrate, it is also affected by the presence of microorganisms.

3.3 Microbial Metabolism in Presence of Pure and Mixed Cultures

Mono culture is defined by the use of a pure single defined microbial strain under sterile conditions. Incubation of two or more defined pure microbial strains under sterile conditions defines a co-culture system. In addition, use of different microbial strains under un-sterile conditions defines mixed-culture (Laxman Pachapur et al., 2015). Across co- and mixed-culture systems, microorganisms communicate with each other, form a complex system, divide the tasks, exchange metabolites to degrade complex substances (Balachandar et al., 2013) (Laxman Pachapur et al., 2015). Mixed microbial consortia capable of metabolic flexibility can survive complex substrates and harsh environment conditions (Agler et al., 2011; Pachapur et al., 2016a).

During dark fermentation (in absence of light) the substrate conversion rate reduces with a change in pH, due to accumulation of by-products, such as acetic acid, butyric acid and lactic acid (Yokoi et al., 1998). The inhibition of organic acids needs to be minimized to have a complete conversion of substrate into the desired product. After dark fermentation, the spent media containing the organic acids was centrifuged and supplied along with media components to carry out photo-fermentation (in presence of light). The photo-fermenting microorganisms tend to utilize the spent organic acids as substrate, so combination of dark

Table 3.2 Increased production across the pure and mixed-culture system under different operating conditions.

Microorganisms	Substrate	Operating conditions	Process	Increased production	Ref.
C. butyricum IFO13949 and *Rhodobacter* sp. M-19.	Starch	50 mL at 36 °C initial pH 6.5 with 5000 lux	Use of photosynthetic bacteria for increased H_2 yield	6.6 mol H_2/mol glucose	(Yokoi et al., 1998)
Rhodobiummarinum A-501 and *Lactobacillus amylovorus*	Algal biomass	65 mL at 30 °C with initial pH 6.5	Utilizing algal starch to produce lactic acid and convert lactic acid to H_2	1.55 mmol/h/L of culture	(Kawaguchi et al., 2001)
Citrobacter freundii 01, *Enterobacter aerogenes* E10 and *Rhodopseudomonas palustris* P2	Sugar cane distillery effluent	1 L to 100,00 L at 37 °C, initial pH 7.0 at 7000 lux with 120 rpm	Improving the H2 production using co-cultures at pilot-scale	21.38 kg of H_2 or 2.76 mol H_2/mol glucose	(Vatsala et al., 2008)
C. butyricum and *Rhodobacter sphaeroides*	Glucose	1 L at 30 °C initial pH 6.25 at 5,000 lux.	Increased COD removal from dark fermented liquid by performing photo fermentation	15.9 mL-H_2/L/h	(Lee & Lee, 2012)
Enterobacter aerogenes and *Rhodobacter sphaeroides*	Oil cake	2 L at 30 °C with initial pH 5.1 at 160 W	Co-culture under dark and photo fermentation	7.95 L H_2/L-media	(Arumugam et al., 2014)

and photo-fermenting organisms help for complete conversion of substrate into desired product. In a two-stage sequential system, dark fermentative bacteria will act on solid organic wastes to produce low-molecular weight organic acids (acetic acid, butyric), which are easily degraded by photosynthetic bacteria for hydrogen production (Hema & Agrawal, 2012). A complained dark and photo-fermentation was carried out on algal biomass by using *Lactobacillus amylovorus* for conversion of algal starch to lactic acid. Later, photosynthetic *Rhodium marinum* bacterium used the lactic acid for the production of hydrogen, which resulted in 1.5 fold increase (Kawaguchi et al., 2001). Organic fatty acids consumption prevented pH drop and the combination helped in utilizing the by-products without adjusting the pH (Fang et al., 2006; Lee & Lee, 2012). The utilization rate of photo-fermentative bacteria must be higher than the rate of acidification rate by *dark fermenting bacteria* for a mutual cooperation to get maximum yield (Liu et al., 2010). The co-culture system of utilizing the organic acids is at faster rate in comparison to mono-culture systems and the yield of two-stage was higher in comparison to combined dark- and photo-fermentation can be seen in the Table 3.2.

3.4 Microbial Metabolism in the Presence of Co-Substrate

It is necessary to evaluate the type of substrate during the fermentation process to produce bulk chemicals. Microorganisms during fermentation prefer sugar substrates, such as glucose and starch. However, pure sugar substrates are costly in nature and organic residues containing large quantities of sugars are the alternative sugar substrates. The addition of co-substrate in the form of organic wastes is carried out to supplement the medium during fermentation (Pachapur et al., 2017; Pachapur et al., 2015c). Use of co-substrates has more advantages in enhancing substrate conversion, to provide an optimum medium composition, maintain the fermentation conditions, decrease the cost of medium and increase the production rate of fermentation (Pachapur et al., 2015c). Once the main substrate is fixed, a co-substrate can be selected depending on the fermentation requirement.

The compatibility of microorganisms (*Clostridia* and *E. coli*) on two different substrates before fermentation for the production of biofuels was determined by simple growth characteristics. The experiment was set with two-step cultivation of bacteria in glycerol or glucose as an initial carbon source, followed by inoculum transfer into exchanged carbon source. Growth of *Clostridia* was strongly suppressed in the presence of glycerol and *E. coli* seemed to utilize glycerol and glucose during the two-step cultivation. In the presence of glycerol, there was a growth lag phase, densities of bacteria decreased suggesting that glycerol metabolism needed adaptive features for the growth of *Clostridia* (Jame et al., 2011). In the presence of pure glucose as a substrate the metabolite distribution pattern produced butyrate (31%), acetate (18%), ethanol (15%) with no production of 1,3-PD. However, in the case of raw glycerol as a substrate the metabolite distribution differed with ethanol (46%), 1, 3-PD (28%), acetate (6%)

Table 3.3 The important details of the co-substrates studies during fermentation.

Material	Microorganisms	Concentration	Purpose	Production increase	Ref.
Acetate	*Clostridium* spp.	3.2 g/L	Additional media component	9%	(Heyndrickx et al., 1991)
Polypeptone	*C. butyricum* and *E. aerogenes*	10 g/L	Additional media component	3%	(Yokoi et al., 2001)
Corn steep liquor	*C. butyricum* and *E. aerogenes*	10 g/L	Media replacement	0.74%	(Yokoi et al., 2002)
Apple pomace hydrolysate	*C. butyricum* and *E. aerogenes*	5 g/L	Co-substrate	2.83 fold	(Pachapur et al., 2015c)
Eggshells (EGS)	*C. butyricum* and *E. aerogenes*	0.25% (*w/v*)	Neutralizing and immobilizing agent	1.5 fold	(Pachapur et al., 2017)

and butyrate (0%) (Temudo et al., 2008). The combination of two different substrates with transition from single to two-substrates was studied to determine the capacity of adaption of reactors. The transition of glucose to glycerol and glycerol to glucose resulted in the same range of production of ethanol, butyrate, and difference in production for acetate and 1, 3-PD. Initial with glucose, with feed of glycerol resulted in increase in NADH to be reoxidized, which resulted with 1, 3-PD production with a decrease in glycolysis flux. In a similar way, a feed of glucose to an initial feed with glycerol resulted in the decrease of NADH to be regenerated with 1, 3-PD and glycolysis increased. The transition of substrates determined the adaption of fermentation process to produce bulk chemicals (Temudo et al., 2008).

However, in the case of H2 production along with starch, polypeptone was used as co-substrate (Yokoi et al., 2001). However, polypeptone is too expensive and an alternative substitute such as corn steep liquor (CSL) resulted with higher hydrogen yield (2.7 mol H_2/mol glucose) in comparison to polypeptone (Yokoi et al., 2002). Use of co-substrate increased the substrate conversion rate along with exchange of metabolite towards finished product (Masset et al., 2012). A proper mixed co-substrate in concentrations of acetate, butyrate and propionate, resulted in increased H2 production. Use of co-substrates helped to decrease the initial concentration of the main substrate and re-consumed the produced formate to transform it into H_2 production substrate (Wang et al., 2014).

In similar ways the co-substrate of glycerol and apple pomace hydrolysate (APH) resulted in H2 production. Use of APH, favored the H2 production pathway for *Clostridium* sps and use of glycerol favored the *Enterobacter* sps together resulted in increased hydrogen production (Pachapur et al., 2015c). The monoculture fermentation test with each of the substrates along with the metabolite analysis, helped to determine the bacteria adaptation towards the substrate for increased H2 production (Pachapur et al., 2015c). In addition to the use of co-substrate as additional nutrient source, use of Eggshells (EGS) along with glycerol also provided an immobilizing agent for bacterial growth and acted as a neutralizing agent during fermentation (Pachapur et al., 2017). Use of co-substrate before fermentation requires initial pretreatment steps to convert a complex substrate into fermentable sugar for the growth of microorganisms. In the case of APH, an initial acid-hydrolysis step was carried out to convert complex carbohydrate into simple monomers. The selection of pretreatment steps depends on the type of the co-substrate to be used for the fermentation process.

The important details of the co-substrates studies for increased production are presented in Table 3.3.

3.5 Microbial Metabolism in the Presence of Input Parameters

During the microbial metabolism physico-chemicals parameters are very important. The physical operating conditions such as temperature, agitation, air/oxygen/nitrogen sparging, etc and chemical operating conditions such as pH, substrate, and co-substrate concentration, media supplement, etc need to be maintained for a better outcome. The optimized parameters build an enriched microbial community and greatly affect production efficiency (Mangayil et al., 2012). Each of the input parameters are monitored and analyzed along with the product of interest or the substrate utilizations or with the by-product concentration.

In the case of hydrogen production using crude glycerol as substrate, cultivation temperature varied between 25–46 °C keeping all other input parameters constant. H2 production increased from 0.3 to 1.0 mol/mol of glycerol consumed along with substrate utilization from 2.9 to 35.7% during the temperature rise from 24 to 45 °C. However, the acetate and ethanol concentration remained unchanged during the rise of temperature from 35 to 40 °C. Further increase in temperature decreased the H2 yield along with acetate concentration, indicating the H2 production was optimum during the mesophilic conditions (Mangayil et al., 2012).

The crucial input parameter is the initial substrate concentration (crude glycerol) varied across (0.5, 1, 2.5, 3.5, 5 g/L) to test the H2 production and metabolite distribution along with substrate utilization. The H2 yield increased with increasing substrate concentration up to 1 g/L, however further increase decreased the H2 yield. The decrease in H2 production, made the way for metabolite production increase with butyrate (2.2 to 4.5 mM) accompanied by ethanol (2.7 to 4.7 mM) production (Mangayil et al., 2012). Increase in substrate concentration shifted the H2 production pathway towards ethanol and butyrate production.

Genetic engineering is gaining importance in metabolic engineering using different genetics tools with inserting/overexpression of genes for solvent production, knocking out genes for unwanted by-products and integrational plasmid technology to disrupt unwanted metabolic pathways. Overexpression of molecular chaperone in *Clostridium* resulted in 40% increase of solvent production in comparison to control strain respectively. The expression of hydrogenase was manipulated in the section of metabolism in *Clostridia* resulted with 1.6-fold acetone production with 76% decrease in butanol production (Zheng et al., 2015). Product inhibition during the fermentation process is considered as potent inhibitor to the growth of microorganism. Overexpression of SpoOA, groESL genes in *Clostridium* resulted to improve tolerance with prolonged metabolism in high butanol concentration. The overexpression of genes improves butanol tolerance by 33%, assist protein folding with uninterrupted growth under butanol stress. In the case of economic viability the final product must be at a higher concentration in comparison to the by-products. Using Targetron technology the production of the unwanted by-product (acetone) was disrupted, resulting in 80% increase in butanol production with a decrease in acetone from 3 to 0.2 g/L respectively (Zheng et al., 2015). These studies suggest the advanced technology before fermentation carried out in metabolic engineering for improving the heterologous metabolic pathways with an increased production of value-added products.

3.6 Microbial Metabolism in the Presence of Varying Fermentation Conditions

Once the kind of pure or mixed-cultures, substrates, co-substrates and input parameters are fixed, fermentation can be further varied for increased production. In order to increase accessibility of substrate to the microorganisms, a pretreatment step is carried out on the substrate. Metabolic dexterity of the microbes can result in the initiation of unwanted metabolic pathways, which can reduce the rate of hydrogen production (Baghchehsaraee et al., 2010), Reports also conclude that bacterial successions may take place in a batch system, irrespective of the complex or simple medium in enhancing the biofuel production. (Huang et al., 2010; Koskinen et al., 2007).

In the case of crude glycerol to remove soap and other impurities, a pretreatment step of the addition of distilled water and pH adjustment using NaCl is carried out. A amount of NaCl used directly affects the microorganisms growth and results in decreased H2 production (Sarma et al., 2014). However, in the absence of NaCl with use of surfactant pre-treatment resulted in increased H2 production (Pachapur et al., 2016b). In the case of anaerobic fermentation during H2 production, the medium is sparged with Nitrogen gas to create an anaerobic condition. However, sparging of gas requires an additional gas station and increase to the production cost. A co-culture system of facultative microorganism was used to create an anaerobic condition along with strict anaerobe to produce H2. Using a co-culture system, the Nitrogen sparging step was eliminated with a process-based improvement resulting in a metabolic shift from reductive (Nitrogen sparging) to oxidative (No sparging) with increased hydrogen production (Pachapur et al., 2015a).

At a large-scale production, the microorganisms are retained at the end of the fermentation using immobilization support, to be recycled during the next fresh batch. At large scale, use of immobilization technique as accounted for increased production and also resulted in higher production cost. Costly immobilization support materials are replaced with Eggshells (EGS), providing better support for bacterial growth and enzymes. Use of EGS as immobilization support increased the glycerol uptake rate (98%) in comparison to absence of EGS (85%) during glycerol fermentation for hydrogen production (Pachapur et al., 2017). Similarly, use of EGS as biofilms during the growth of *Rhizopus oryzae 1526*, resulted in increased productivity with 3.95 times during FA production (Das et al., 2015). Use of EGS replaced the refined supplements and delivered dual role as immobilization support along with source of $CaCO_3$ during the fermentation process. In order to increase the production of hydrogen, different types of fermentor are carried out. In the case of batch fermentation, the H2 production (36 mmol/L) was marginal along with the production of metabolites. Repeated-batch fermentation was carried out, which resulted in increased H2 production (42 mmol/L). However, a second cycle of repeated-batch fermentation resulted with increased metabolite concentration, which acted as metabolite inhibition to the microorganisms during the fermentation (Pachapur et al., 2017). In order to

Table 3.4 Different fermentation conditions carried out to increase final output.

Fermentation conditions	Microorganism used	Final output	Ref.
Pre-treatment step with NaCl addition	*Enterobacter aerogenes*	93.03% decreased production	(Sarma et al., 2014)
Pre-treatment step with Surfactant addition	*Enterobacter aerogenes* and *Clostridium butyricum*	32.1 ±0.03 mmol/L increased production	(Pachapur et al., 2016b)
No Pre-treatment step of Nitrogen sparging	*Enterobacter aerogenes* and *Clostridium butyricum*	1.5 mmol-H2/mol of glycerol increased production	(Pachapur et al., 2015a)
Eggshells as immobilizing agent	*Rhizopus oryzae 1526*	Productivity increase by 3.95 times	(Das et al., 2015)
Eggshells as immobilizing agent	*Enterobacter aerogenes* and *Clostridium butyricum*	glycerol uptake rate (98%)	(Pachapur et al., 2017)
Aqueous two-phase system	*Enterobacter aerogenes*	1.48 to 11.65 mmol/L-medium increased production	(Sarma et al., 2015)

decrease the metabolite concentration, an aqueous two-phase system was developed to extract different organic acids into the organic phase from the aqueous phase containing microorganisms. With the transfer of organic acids from the aqueous phase, the metabolite inhibition on the microorganisms was decreased, which resulted in increased hydrogen production (Sarma et al., 2015).

Depending upon the type of the refined substrates and the microorganisms used, either submerged (SmF) or solid-state fermentation (SSF) can be carried out. Costly carbon source was replaced with pulp and paper solid waste (PPSW) and production of fumaric acid was carried out through submerged and solid-state fermentation. In the case of SmF, dried and ground PPSW along with water and hydrolysate of PPSW was as substrate to produce 23 g/L of FA. With SSF, the dried and ground PPSW was used directly with moisture adjustment to produce 41.45 g/L of FA. The smallest particle size of PPSW in SmF played the supportive role for microorganism growth and resulted in increased productivity in comparison to SSF. The productivity profile of SmF in the presence of PPSW (0.48 g/Lh) is comparable to productivity in the presence of corn starch (0.50 g/Lh) (Das et al., 2016) and see Table 3.4 above.

3.7 Pros and Cons of Refined Substrate for Metabolic Metabolisms

Refined substrates are the automatic choice during the fermentation process for the production of value-added compounds. Use of refined substrate during fermentation has varied advantages and disadvantages. In this section, the pros and cons of refined substrates are explained in comparison to raw substrates.

The refined substrate such as pure glucose, glycerol and cellulose are near to 100% purity and have high market cost. The refined substrates are considered as control set of experiments in comparison to the use of raw substrates experiments. With minimum or no characterization and pre-treatment steps are necessary to determine the substrate composition/availability during the fermentation process. Use of refined substrates release minimum by-products and have minimum interference/inhibition during the reaction of substrate to products. The purification process of recovery of the final product will go through minimum downstream steps using the refined substrates. The refined substrates are mostly used and are first choice, when it comes to producing finishing products used for medical applications. In the case of pharmaceutical compounds, the refined substrates have advantage over the raw substrates in delivering 100% pure compound. At the end of fermentation with minimum by-products in the effluent, the waste management will be of lesser environment concern. These are some of the advantages of using the defined substrates over the raw substrates.

However, with rapid population growth and industrialization, worldwide demand for value-added products is increasingly continuously. Conventional defined substrates are unable to meet the global demand for value-added products due to the high cost and low abundance. Therefore, raw substrates such as crude glycerol, apple pomace and agro-industry wastes are attracting industries for an alternative, low-cost and high abundance feedstocks for value-added productions. Industries are exploring new technologies for valorizing the waste into a raw substrate source. Using raw substrate, the material input cost will be minimized and at the end the process will be economical in comparison to using defined substrates. Using raw substrate, treatment, dumping and management of waste will be reduced and valorization of waste will be a green approach. However, not all finished products can be produced using raw substrates, so there is always a necessary requirement of refined substrates.

3.8 Conclusions

Refined substrates play a very important role in the metabolic pathway of microorganisms during the fermentation for the production of value-added products. The behavior of the microorganisms differs in the presence of refined and raw substrates. The metabolic pathways vary using the refined substrates with lesser by-products and deliver more by-products using raw substrates. The microorganisms can easily access the defined substrates with directly available carbon sources. However, in the case of raw substrate initial pre-treatment, steps are necessary to break down complex substrates into simple carbon sources. In most cases, when using raw substrates for the simple fermentation test using the refined substrate is carried out as control set to compare the results of final end product. However, raw substrates are only considered over defined substrates only to reduce the cost of feedstocks and as valorization of industrial wastes. In the end, not all finished products can be produced by raw substrate and those produced by defined substrate will fall short of market demand.

Acknowledgment

The authors are also thankful to "Post-Doctoral merit scholarship program for foreign students (FQRNT)" for financial assistance to Vinayak Laxman Pachapur.

References

Agler, M.T., Wrenn, B.A., Zinder, S.H., and Angenent, L.T. (2011) Waste to bioproduct conversion with undefined mixed cultures: the carboxylate platform. *Trends in biotechnology*, 29(2), 70–78.

Arumugam, A., Sandhya, M., and Ponnusami, V. (2014) Biohydrogen and polyhydroxyalkanoate co-production by Enterobacter aerogenes and Rhodobacter sphaeroides from Calophyllum inophyllum oil cake. *Bioresource technology*, 164, 170–176.

Baghchehsaraee, B., Nakhla, G., Karamanev, D., and Margaritis, A. (2010) Fermentative hydrogen production by diverse microflora. *international journal of hydrogen energy*, 35(10), 5021–5027.

Balachandar, G., Khanna, N., and Das, D. (2013) Biohydrogen Production from Organic Wastes by Dark Fermentation. 103–144.

Chang, J.-J., Chou, C.-H., Ho, C.-Y. et al. 2008. Syntrophic co-culture of aerobic Bacillus and anaerobic Clostridium for bio-fuels and bio-hydrogen production. *International Journal of Hydrogen Energy*, 33(19), 5137–5146.

Das, R.K., Brar, S.K.,and Verma, M. (2016) Potential use of pulp and paper solid waste for the bio-production of fumaric acid through submerged and solid state fermentation. *Journal of Cleaner Production*, 112, 4435–4444.

Das, R.K., Brar, S.K., and Verma, M. (2015) Valorization of Egg shell biowaste and brewery wastewater for the enhanced production of fumaric acid. *Waste and Biomass Valorization*, 6(4), 535–546.

Fang, H.H., Zhu, H., and Zhang, T. (2006) Phototrophic hydrogen production from glucose by pure and co-cultures of *Clostridium butyricum* and *Rhodobacter sphaeroides. International Journal of Hydrogen Energy*, 31(15), 2223–2230.

González-Pajuelo, M., Andrade, J., and Vasconcelos, I. (2004) Production of 1, 3-propanediol by Clostridium butyricum VPI 3266 using a synthetic medium and raw glycerol. *Journal of Industrial Microbiology and Biotechnology*, 31(9), 442–446.

Hema, R. and Agrawal, P. (2012) Production of clean fuel from waste biomass using combined dark and photofermentation. *IOSR J. Comput. Eng.*, 1, 39–47.

Heyndrickx, M., De Vos, P., Vancanneyt, M., and De Ley, J. (1991) The fermentation of glycerol by Clostridium butyricum LMG 1212 t2 and 1213 t1 andC. pasteurianum LMG 3285. *Applied microbiology and biotechnology*, 34(5), 637–642.

Himmi, E.H., Bories, A., and Barbirato, F. (1999) Nutrient requirements for glycerol conversion to 1, 3-propanediol by Clostridium butyricum. *Bioresource Technology*, 67(2), 123–128.

Huang, Y., Kuan, W., Lo, S., and Lin, C. (2010) Hydrogen-rich fuel gas from rice straw via microwave-induced pyrolysis. *Bioresource technology*, 101(6), 1968–1973.

Hussy, I., Hawkes, F., Dinsdale, R., and Hawkes, D. (2005) Continuous fermentative hydrogen production from sucrose and sugarbeet. *International Journal of Hydrogen Energy*, 30(5), 471–483.

Ito, T., Nakashimada, Y., Senba, K. et al. (2005) Hydrogen and ethanol production from glycerol-containing wastes discharged after biodiesel manufacturing process. *Journal of bioscience and bioengineering*, 100(3), 260–265.

Jame, R., Vilímová, V., Lakatos, B., and Verecka, L. (2011) The hydrogen production by anaerobic bacteria grown on glucose and glycerol. *Acta Chimica Slovaca*, 4(2), 145–157.

Kawaguchi, H., Hashimoto, K., Hirata, K., and Miyamoto, K. (2001) H^2 production from algal biomass by a mixed culture of *Rhodobium marinum* A-501 and *Lactobacillus amylovorus*. *Journal of bioscience and bioengineering*, 91(3), 277–282.

Koskinen, P.E., Lay, C.-H., Beck, S.R. et al. (2007) Bioprospecting thermophilic microorganisms from Icelandic Hot Springs for hydrogen and ethanol production†. *Energy & Fuels*, 22(1), 134–140.

Laxman Pachapur, V., Jyoti Sarma, S., Kaur Brar, S. et al. (2015) Co-culture strategies for increased biohydrogen production. *International Journal of Energy Research*, 39(11), 1479–1504.

Lee, J.Y. and Lee, E.J. (2012) Effects of pH and carbon sources on biohydrogen production by co-culture of Clostridium butyricum and Rhodobacter sphaeroides. *Journal of microbiology and biotechnology*, 22(3), 400–406.

Liu, B.-F., Ren, N.-Q., Tang, J. et al. (2010) Bio-hydrogen production by mixed culture of photo-and dark-fermentation bacteria. *International Journal of Hydrogen Energy*, 35(7), 2858–2862.

Mangayil, R., Karp, M., and Santala, V. (2012) Bioconversion of crude glycerol from biodiesel production to hydrogen. *International journal of hydrogen energy*, 37(17), 12198–12204.

Markov, S.A., Averitt, J., and Waldron, B. (2011) Bioreactor for glycerol conversion into H 2 by bacterium Enterobacter aerogenes. *International Journal of Hydrogen Energy*, 36(1), 262–266.

Masset, J., Calusinska, M., Hamilton, C. et al. (2012) Fermentative hydrogen production from glucose and starch using pure strains and artificial co-cultures of Clostridium spp. *Biotechnology for biofuels*, 5(1), 35.

Nishio, N. and Nakashimada, Y. (2004) High rate production of hydrogen/methane from various substrates and wastes. In: *Recent progress of biochemical and biomedical engineering in Japan I*, Springer, pp. 63–87.

Pachapur, V.L., Das, R.K., Brar, S.K.et al. (2017) Valorization of crude glycerol and eggshell biowaste as media components for hydrogen production: A scale-up study using co-culture system. *Bioresource technology*, 225, 386–394.

Pachapur, V.L., Kutty, P., Brar, S.K., and Ramirez, A.A. (2016a) Enrichment of Secondary Wastewater Sludge for Production of Hydrogen from Crude Glycerol and Comparative Evaluation of Mono-, Co-and Mixed-Culture Systems. *International Journal of Molecular Sciences*, 17(1), 92.

Pachapur, V.L., Sarma, S.J., Brar, S.K. et al. (2015a) Evidence of metabolic shift on hydrogen, ethanol and 1,3-propanediol production from crude glycerol by nitrogen sparging under micro-aerobic conditions using co-culture of

Enterobacter aerogenes and Clostridium butyricum *International Journal of Energy Research.*

Pachapur, V.L., Sarma, S.J., Brar, S.K. et al. (2015b) Evidence of metabolic shift on hydrogen, ethanol and 1, 3-propanediol production from crude glycerol by nitrogen sparging under micro-aerobic conditions using co-culture of Enterobacter aerogenes and Clostridium butyricum. *International Journal of Hydrogen Energy.*

Pachapur, V.L., Sarma, S.J., Brar, S.K. et al. (2015c) Biohydrogen production by co-fermentation of crude glycerol and apple pomace hydrolysate using co-culture of Enterobacter aerogenes and Clostridium butyricum. *Bioresource technology*, 193, 297–306.

Pachapur, V.L., Sarma, S.J., Brar, S.K. et al. (2016b) Surfactant mediated enhanced glycerol uptake and hydrogen production from biodiesel waste using co-culture of Enterobacter aerogenes and Clostridium butyricum *Renewable Energy*, 193, 297–306.

Sarma, S.J., Brar, S.K., Le Bihan, Y., and Buelna, G. (2015) A novel anaerobic two-phase system for biohydrogen production and in situ extraction of organic acid byproducts. *Bioprocess and biosystems engineering*, 1–6.

Sarma, S.J., Brar, S.K., Le Bihan, Y. et al. (2014) Mitigation of the inhibitory effect of soap by magnesium salt treatment of crude glycerol–A novel approach for enhanced biohydrogen production from the biodiesel industry waste. *Bioresource Technology*, 151, 49–53.

Seifert, K., Waligorska, M., Wojtowski, M., and Laniecki, M. (2009) Hydrogen generation from glycerol in batch fermentation process. *International Journal of Hydrogen Energy*, 34(9), 3671–3678.

Temudo, M.F., Poldermans, R., Kleerebezem, R., and van Loosdrecht, M.C. (2008) Glycerol fermentation by (open) mixed cultures: a chemostat study. *Biotechnol Bioeng*, 100(6), 1088–98.

Vatsala, T., Raj, S.M., and Manimaran, A. (2008) A pilot-scale study of biohydrogen production from distillery effluent using defined bacterial co-culture. *international journal of hydrogen energy*, 33(20), 5404–5415.

Wang, R., Cui, C., Jin, Y. et al. (2014) Photo-fermentative hydrogen production from mixed substrate by mixed bacteria. *International Journal of Hydrogen Energy.*

Yazdani, S.S. and Gonzalez, R. (2008) Engineering Escherichia coli for the efficient conversion of glycerol to ethanol and co-products. *Metabolic engineering*, 10(6), 340–351.

Yokoi, H., Maki, R., Hirose, J., and Hayashi, S. (2002) Microbial production of hydrogen from starch-manufacturing wastes. *Biomass and Bioenergy*, 22(5), 389–395.

Yokoi, H., Mori, S., Hirose, J. et al. (1998) H2 production from starch by a mixed culture of Clostridium butyricum and Rhodobacter sp. M [h] 19. *Biotechnology letters*, 20(9), 895–899.

Yokoi, H., Saitsu, A., Uchida, H. et al. (2001) Microbial hydrogen production from sweet potato starch residue. *Journal of bioscience and bioengineering*, 91(1), 58–63.

Zheng, J., Tashiro, Y., Wang, Q., and Sonomoto, K. (2015) Recent advances to improve fermentative butanol production: genetic engineering and fermentation technology. *Journal of bioscience and bioengineering*, 119(1), 1–9.

4

Non-refined Carbon Sources and Microbial Performance

Guneet Kaur

School of Energy and Environment, City University of Hong Kong, Kowloon Tong, Hong Kong

4.1 Introduction

Recent global sustainable development goals with increasing focus on environmental protection and material and energy efficiency have driven the establishment of bioprocesses based on renewable feedstock. Consequently, the trends of utilizing non-refined carbon sources such as agro-industrial wastes including wheat and rice bran, bagasse, apple pomace, crude glycerol, brewery industry waste etc. have been speeded up for various fermentation-based products. With such a concept in mind, it is particularly important to critically understand and analyze the cellular and biochemical complexities of microbial metabolism to enable the design of bioprocesses with high yields and productivities. Knowledge on the principles which guide key decisions made by cells via intricate and tightly regulated metabolic circuits for multiple substrate uptake, assimilation and transformation to products facilitates efficient utilization of non-refined carbon sources. This chapter provides an account of microbial behavior in terms of growth, accumulation and conversions of non-refined carbon sources. It also provides some key examples of rewiring of carbon metabolism in microorganisms including *Escherichia coli* and *Saccharomyces cerevisiae* to illustrate probable product outcomes.

4.2 Non-refined Carbon Sources: a Brief Account

The increasing global concerns of climate change and depletion of fossil fuels have directed immense research efforts towards the development of sustainable processes for the production of fuels, chemicals and materials. In this context, the development of biorefineries has emerged as an attractive alternative to petroleum-based processes. Biorefinery can be defined as the sustainable

Microbial Sensing in Fermentation, First Edition. Edited by Satinder Kaur Brar, Ratul Kumar Das, and Saurabh Jyoti Sarma.

processing of biomass into marketable products (Jojima et al., 2011). Microorganisms are the main characters in industrial bioprocesses wherein they are responsible for catalyzing important reactions giving rise to products. Their versatile nature makes them the ideal players for conversions occurring in biorefineries. For decades, microorganisms have been harnessed by humans to produce products such as fermented foods, beverages, chemicals including solvents and pharmaceuticals such as penicillin using simple substrates such as sugars. However, the present global situation commissions the development of bioprocesses through valorisation of non-food biomass to fulfil the critical requirement both in terms of sustainability and environmental footprint (Saling, 2005). Thus, in more recent applications of microorganisms to build bioprocesses, the utilization of renewable feedstock or biomass has been extensively explored. In this regard, lignocellulosic biomass, derived from agricultural residues or non-food crops emerges as an attractive and renewable feedstock for production of biofuels and biochemicals.

Lignocellulose is the most abundant biomass on earth with an estimated annual production of 150–170 × 10^9 tons. Out of this, only 2 × 10^9 are currently used for energy, paper and pulp, and timber applications, while the remainder is left unused (Saini et al., 2015). This makes it a readily and inexpensively available non-refined carbon source for other uses such as bio-based production of various industrially important products. Materials such as agricultural wastes, forestry residues, grasses and woody materials are all lignocellulosics. Major examples of agro-based lignocellulosic biomass include sugar cane bagasse, sweet sorghum, hardwood, soft wood, corn cobs, corn stover, rice straw, grasses, newspaper, wheat straw, wheat bran and rice bran. Lignocellulosics are more complex and recalcitrant than other conventionally used fermentation substrates, e.g. pure sugars and contain a mixture of carbon sources including cellulose, hemicellulose and lignin (Da Costa et al., 2009; Kim et al., 2010). These can be further broken down into C6 sugars such as glucose, mannose, fructose, C5 sugars such as xylose and arabinose; and cellobiose (dimer of glucose) and be available for microbial nutrition. These C6 and C5 sugars obtainable from lignocellulose can be readily utilized as feedstock for fermentation by several microorganisms including bacteria and yeast to produce biofuels (ethanol, butanol, hydrogen) and value-added chemicals (succinic acid, lactic acid). Lignin is the most abundant by-product and has unique properties to produce chemicals for applications in synthesizing renewable polymers, specialty chemicals, materials and high-quality fuels.

In addition to agricultural and forestry-based lignocellulosics, various industrial wastes such as fruit pomace wastes, sugar processing industry wastes, brewery wastes, dairy industry wastes and starchy substrates are among other abundantly available organic matter (El-Asar 2006; Dhillon et al., 2012). Being rich in carbon and other vital nutrients, these wastes are useful non-refined carbon sources which could be utilized for microbial growth and metabolite production. Furthermore, other non-sugar substrates have also emerged as suitable feedstocks for microbial fermentation. A prominent example is glycerol which is the main by-product of the burgeoning biodiesel industry and accounting for 10% (w/w) of biodiesel produced (da Silva et al., 2009). With such a huge excess of crude glycerol, its market value has reduced greatly and has turned it into a "waste stream" instead of a by-product. Glycerol is a valuable feedstock for production of high-valued chemicals such as 1,3-propanediol, 2,3-butanediol, organic acids, polyol as well as biofuels such as ethanol and n-butanol. In fact, the use of glycerol for microbial utilization is more advantageous because of the highly reduced nature of carbon atoms in glycerol, which generates more reducing equivalents than sugars (glucose or xylose). The development of bioprocesses and/or biorefineries based on glycerol also emerges as an economically viable production chain which valorizes waste glycerol to valuable products and reduces industrial waste disposal.

The use of non-refined carbon sources derived as agricultural and/or industrial wastes is a highly relevant and desirable method for sustainable production of metabolites, however, there are several factors pertaining to the complexities involved in microbial behavior i.e. growth and metabolism; and assimilation of these carbon sources. A critical and thorough understanding of these interactions is important to control, optimize and engineer metabolic networks and develop efficient bioprocesses yielding high product concentrations and productivities. The intricacies of cellular metabolism and its regulation towards sensing and assimilation of non-refined carbon sources is discussed in the following sections.

4.3 Microbial Assimilation of Non-Refined Carbon Sources

Sugars derived from pretreatment of agro-industrial lignocellulosics including rice bran, wheat bran, bagasse, apple pomace are useful feedstocks for industrial microorganisms including bacteria and yeast. Carbon source is by far the most important consideration for cell in the culture environment since it is a source of energy generation and biosynthesis for them. Usually non-refined carbon sources contain a mixture of substrates (e.g. sugars) which vary in composition depending on the feedstock and pretreatment method used. Thus, microbes are usually presented with a scenario in which critical decisions with respect to substrate uptake and transformation are guided by information from metabolic assimilatory and regulatory pathways. Consequently, these decisions lead to different product outcomes in fermentation processes.

While glucose is the simplest carbon source, most microorganisms including bacteria can utilize various other carbon sources available from non-refined feedstocks. However, in most bacteria, a hierarchical use of carbon sources (mostly sugars) is usually observed wherein a preference for glucose as the primary substrate is recognized (Aidelberg et al., 2014). The reason for this phenomenon is glucose repression or carbon catabolite repression (CCR) which leads to preferential uptake of glucose among available carbon sources (Chubukov et al., 2014). CCR is observed in most heterotrophic bacteria including *Escherichia coli*, *Zymomonas mobilis*, *Bacillus subtilis*, *Clostridium acetobutylicum* etc. However, in some other bacteria, e.g. *Corynebacterium glutamicum*, a co-metabolism of glucose and other carbon sources is made under highly regulated mechanisms.

It is indeed interesting to note that glucose is not the preferred carbon source in all bacteria. Bacteria such as *Clostridium thermocellum* prefers cellodextrins to glucose and cellobiose, while *Streptococcus thermophilus* and *Bifidobacterium longum* assimilate lactose more readily than glucose (Zhang and Lynd, 2005; Parche et al., 2006; Frunzke et al., 2008). Considering the environmental diversity, it is reasonable to understand that different bacteria have evolved to utilize nutrients other than glucose as their favorites.

Bacteria use several mechanisms for sugar uptake and assimilation. While primary transport of sugars is mainly driven by ATP, electrochemical gradients of translocated molecules dictate the secondary transport with the aid of various transporter molecules. For e.g. glucose accumulation is driven by electrochemical proton (or sodium ion) gradient. Bacteria have a unique mechanism for group translocation of sugars which is absent in archaea and eukaryotic microorganisms and is involved in the phosphotransferase system (PTS)(Görke and Stülke 2008). Once transported inside the periplasm, glucose is internalized into cytoplasm by PTS. PTS is composed of soluble and non-sugar specific components called Enzyme I (EI) and phosphohistidine carrier protein (HPr) which transfer phosphoryl group from phosphoenolpyruvate (PEP) to the sugar-specific enzyme IIA and IIB. The enzyme EIIC is another component of PTS, which recognizes and transports sugar molecules, where it is phosphorylated by EIIB. EI and HPr are common to all PTS carbohydrates while EII is carbohydrate specific and thus bacteria contain many different EIIs. For example, around 21 different enzyme II complexes are known to exist in *E. coli* which are involved in the transport of 20 different carbohydrates. The EII^{Glu} and EII^{Man} complexes are involved in the transport of glucose, mannose, fructose, N-acetylglucosamines and glucosamine with similar efficiency. Once recruited by PTS, glucose is converted to glucose 6-phosphate and continues to get metabolized in various metabolic pathways including glycolysis. Three different pathways exist for utilization of fructose in bacteria. In the primary pathway, fructose is transported via a membrane spanning protein (FruA) and concomitantly phosphorylated by Fructose PTS system allowing it to enter the cell as fructose 1-phosphate. It is then converted into fructose 6-phosphate (F6P) by D-fructose-1-phosphate kinase. In the second pathway, fructose is recognized by 3,4,5-D-arabino-hexose configuration by a membrane spanning protein and converted to F6P by a specific sucrose-induced D-fructokinase. In yet another pathway, fructose enters the cell by diffusion mediated by a glucose transporter ptsG where it is phosphorylated by ATP to become F6P. In addition to these sugars, the mechanism of xylose and arabinose transport is important with regards to their presence as dominant sugars in lignocellulosic feedstocks. Xylose is converted to D-xylulose by xylose isomerase and then subsequently phosphorylated by xylulokinase to form D-xylose 5-phosphate (Groff et al., 2012). It is important to consider that under anaerobic conditions, xylitol and xylitol 5-phosphate are produced upon induction of xylulose reductase. These compounds are inhibitory for microbial cell growth and hence their controlled regulation in bacteria is critical to allow efficient utilization and conversion of lignocellulose-derived sugars to useful bioproducts. Arabinose follows a similar pathway as xylose and is converted to xylose 5-phosphate after a series of conversions via ribulose and ribulose 5-phosphate. Xylose 5-phosphate is then commissioned via Pentose phosphate pathway (PPP) to conversions in Embden-Meyerhof-Parnas pathway.

The assimilation of carbon sources in yeast is very similar to bacteria in terms of sugar preferences, their uptake and metabolism. Most native yeasts such as *Saccharomyces cerevisiae* and *Yarrowia lipolytica* can consume only C6 sugars (e.g. glucose), while both C5 (e.g. xylose) and C6 sugars can be consumed by some other yeasts such as *Pichia stipites* (Ryu et al., 2016). In yeasts, specific-sugar transporters are used for effective sugar assimilation. These transporters are controlled at both transcriptional and enzymatic levels. Glucose is a preferred carbon and energy source, and the pathway for glucose metabolism is glycolysis. Fructose, galactose and mannose also enter through this pathway. Fructose is readily phosphorylated and enters the pathway directly, whereas galactose and mannose are first converted to glucose-6-phosphate and fructose-6-phosphate respectively before entering glycolysis. For growth on other sugars such as xylose, yeasts need xylose-specific transporters to transport xylose into cytosol and then perform a series of conversions to finally form precursor metabolites of core metabolism required for cell growth (Khankal et al., 2009). After import into cytosol, xylose is converted into xylitol mediated by xylose reductase, followed by conversion of xylitol to xylulose catalyzed by xylulose dehydrogenase and penultimate conversion of xylulose to xylulose-5-phosphate mediated by xylulose kinase. Xylulose-5-phosphate then enters the central metabolic pathway, the PPP through conversion into glyceraldehyde-3-phosphate.

Bacteria are also efficient in consuming other non-sugar carbon sources e.g. glycerol. Bacterial assimilation of glycerol is usually via facilitated diffusion aided by an integral membrane protein, the glycerol facilitator GlpF. After entering cytosol, glycerol is converted to glycerol-3-phosphate by glycerol kinase and then further metabolized. Glycerol is used as a carbon and energy source by myriad microorganisms including *Klebsiella pneumoniae, Citrobacter freundii, Clostridium pasteurianum, Clostridium butyricum, Enterobacter agglomerans* and *Lactobacillus reuteri* (Kaur et al., 2012). It is metabolized both oxidatively and reductively. In the oxidative pathway, glycerol is funneled into glycolysis through conversions from dihydroxyacetone mediated by glycerol dehydrogenase to dihydroxyacetone phosphate catalyzed by dihydroxyacetone kinase. In the reductive pathway, glycerol is converted into a high-valued industrial chemical, 1,3-propanediol through sequential dehydration and reduction steps mediated by glycerol dehydratase and 1,3-propanediol dehydrogenase respectively. The final product 1,3-propanediol, a valuable precursor for fibre and textile industry, is highly specific for glycerol fermentation and cannot be obtained from any other anaerobic conversion.

In *S. cerevisiae* and other yeasts, glycerol uptake occurs either through passive diffusion, a channel protein or active uptake and glycerol is degraded via either dihydroxyacetone or glycerol-3-phosphate. Glycerol-3-phosphate can follow several routes (Cui et al., 2017). It can be either used as a precursor for lipid biosynthesis or converted to glyceraldehye-3-phosphate via dihydroxyacetone phosphate to enter glycolysis or be used as substrate for synthesis of other metabolites.

4.4 Microbial Sensing to Non-Refined Carbon Sources

The emergence and wide availability of non-refined carbon sources as agricultural and/or industrial wastes opens a plethora of opportunities to design processes for utilization of newer substrates and produce new metabolites in a sustainable way. This is further complemented with the advancement of molecular biology approaches together with bioinformatics and mathematical modelling methods which has enabled gaining crucial insights into the complex inner works of cellular metabolism and quantitative analysis. These tools allow specific genetic alterations to be made in the cellular metabolic and regulatory network and direct the fluxes towards the desired product (Bailey 1991). Traditional fermentation processes can now be dramatically improved through provision of production of novel products using newly created pathways in both native and non-native hosts. Since no native strain can possibly entail all desired traits, creation of a platform microorganism which can metabolize a wide range of carbon sources and optimally produce several metabolites is viewed as an interesting and lucrative option for a biorefinery-based economy. Figure 4.1 shows the various probable product outcomes through co-utilization of multiple substrates in an industrial host. However, despite significant recent advances made in the field of synthetic biology and recombinant DNA technology, rewiring cellular metabolism and regulatory circuits is a rather formidable task. The reason is the inherent and overwhelming complexity of cells themselves (Dekel and Alon 2005). An in-depth understanding of microbial metabolism and the associated regulatory circuits is important to understand how key decisions with respect to substrate uptake, assimilation and transformation to product are made by cells. Such information empowers us to control and manipulate microorganisms and design productive industrial fermentation processes based on unconventional and non-refined substrates.

4.4.1 Microbial Metabolism and Regulatory Circuits

Metabolism is central to microbial life and drives all cellular activities through the provision of building blocks and energy. The fundamental metabolic processes and tasks are highly similar among divergent species despite the diverse lifestyles adopted by them. This could be easily illustrated by the fact that scavenging of nutrients and their integration into central metabolism, monomer synthesis and macromolecule polymerization for conducting essential activities including biomass synthesis and proliferation is a conserved feature of all living cells. In addition to and beyond operating cellular activities, metabolism is also instrumental in integrating environmental conditions into intracellular signals through a large and densely connected network of metabolites, enzymatic reactions and regulatory interactions. There exists a tight regulation of central carbon metabolism to ensure that production of cellular components is in balance with energy production and consumption (Nielsen and Keasling 2016). Such a regulation allows cells to maintain metabolic homeostasis under varying environmental and nutrient conditions.

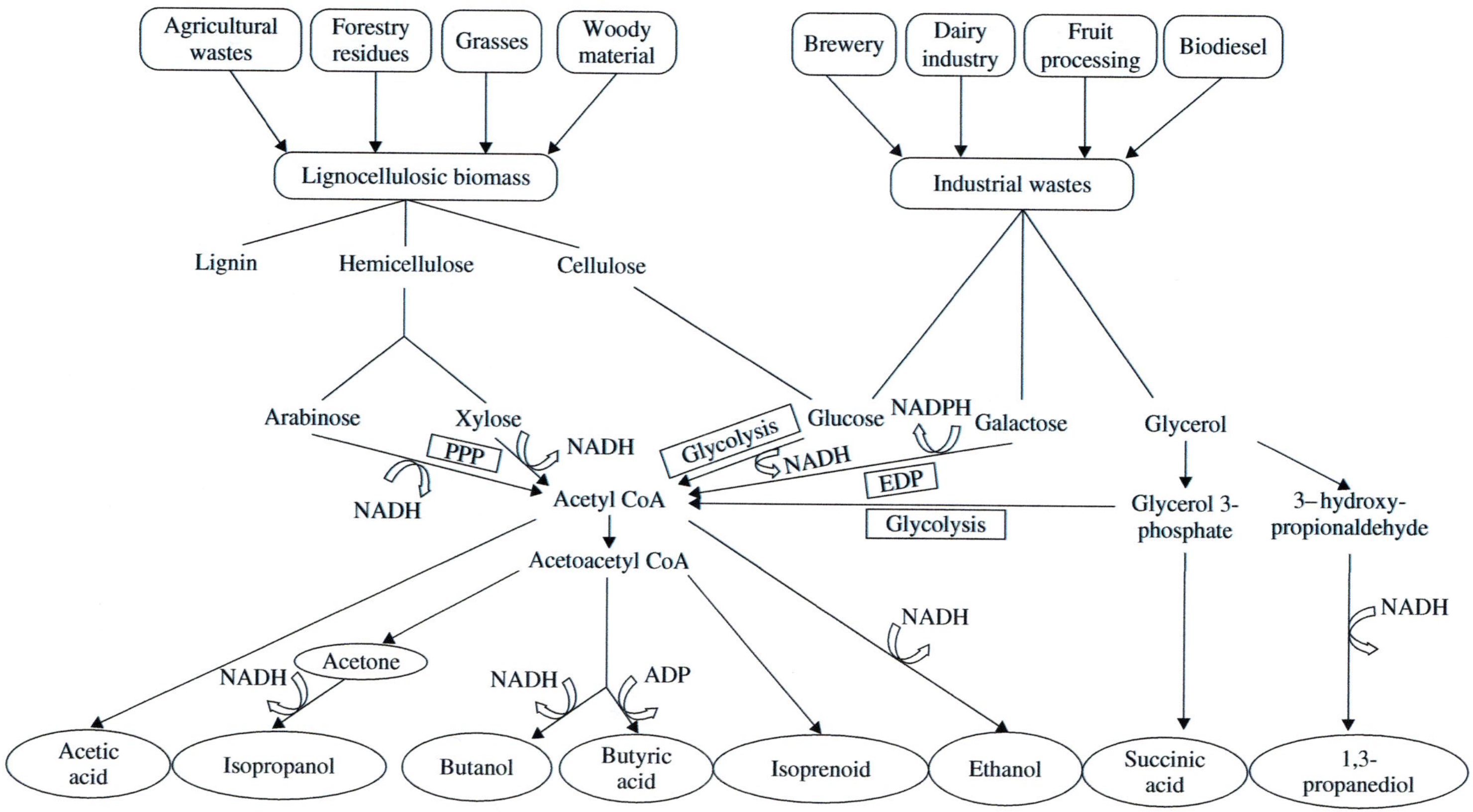

Figure 4.1 Schematic of utilization pathways for various non-refined carbon sources. Operation of multiple pathways simultaneously for several substrates would allow the cell to accumulate precursors such as NAD(P)H and Acetyl CoA for channelling to different products. PPP, Pentose Phosphate Pathway; EDP, Entner-Doudoroff Pathway. Rounded rectangles indicate feedstock while ellipses indicate fermentation products.

As detailed in Section 4.2, most microorganisms can thrive on multiple carbon and energy sources. It is understandable that constitutive expression of all transporters would unnecessarily consume valuable cellular resources and therefore cells have a selective mechanism of expressing transport systems on the basis of extracellular and intracellular signals. In fact, cells have separate kinds of systems to detect different types of signals. The extracellular signals are mainly detected by two-component systems, consisting of a membrane-bound sensor which activates a cytoplasmic regulator. For instance, the *E. coli* genome encodes for 30 such two-component systems which sense nutrients including phosphate, nitrate and phosphorylated sugars. However, for the majority of the carbon sources, intracellular sensors, typically a transcription factor is used which both recognizes the signal and generates a regulatory output (Martinez-Antonio et al., 2006). In this system, transporters and enzymes are only expressed at basal levels. Upon nutrient availability and consequent interaction of transcription factor with these intermediates, an upregulation of transporter and enzyme expression is caused. Uptake of lactose, glucosamine, trehalose, fucose and maltose in *E. coli* are examples of such sensors and indicate that such one-component sensor mechanisms are indeed the dominant nutrient-sensing mechanisms in prokaryotes.

Microorganisms are faced with enormously diverse and constantly fluctuating environments. Such a situation poses a crucial challenge to microorganisms for making appropriate choices to best use their limited resources and obtain crucial fitness benefit. The response of bacteria when presented with multiple carbon sources is usually either co-metabolism or preferential uptake of one of the carbon sources before others. This has been studied for a long time as the classical example of preferential carbon utilization in *E. coli* by the work of Monod on glucose-lactose diauxic shift (Monod 1942). According to Monod's model, glucose is the first carbon source utilized by bacteria when they are exposed to a mixture of carbon sources; and once glucose is exhausted, bacteria switch to other carbon sources, e.g. lactose. Such preferential carbon source uptake has later been demonstrated for various other microorganisms and this prevents the uptake of secondary carbon sources (Stulke and Hillen 1999). Therefore, understanding the regulatory mechanisms which guide cell decision making and allow cells to be optimal under all conditions is of fundamental interest to industrial biotechnology. Carbon Catabolite Repression (CCR) is one such mechanism in microorganisms which helps in their survival and dominance under ever changing nutrient conditions. CCR is a basic mechanism which ensures that the most energy efficient cognate substrate is the most preferred carbon source and thus enables the microorganisms to grow rapidly on their preferred substrate. CCR acts on 5–10% of all genes in many bacterial species and is usually achieved through inhibition of gene expression of enzymes for catabolism of non-preferred carbon sources (Liu et al., 2005).

While CCR provides a way for microorganisms to adapt to varying nutrient conditions, it is a major obstacle in allowing efficient utilization of lignocellulosic biomass, which is a highly recalcitrant substrate composed of a mixture of carbon sources including cellulose, hemicellulose and lignin. However, CCR severely inhibits their uptake, thus negatively affecting the yield and productivity of fermentation processes using renewable feedstock. Additionally, the high adaptability of microorganisms to new environments courtesy CCR presents a rather disadvantageous situation where rationally developed strains would be swept away while microbes resort to optimal physiological state for survival rather than economic viability (Portnoy et al., 2011). It is also reported that CCR continues to operate even upon introduction of heterologous pathway for alternative sugar utilization. This leads to the understanding that rational construction of production strains to promote efficient secondary carbon source uptake requires the system-level understanding of host cell physiology and optimization of bioprocesses through regulating the biochemical circuits.

In addition to the issue of multiple substrate utility, there are other important aspects to consider with respect to utilization of non-refined carbon sources which influence the substrate uptake and product formation. One of these is the presence of inhibitors in non-refined feedstock. For example, during pre-treatment of lignocellulosic biomass by thermochemical, liquefication or pyrolysis, furfural and 5-(hydroxymethyl) furfural (HMF) are two major aldehydes left behind in hydrolysate as inhibitors. For utilization of such complex and toxic substrates and accumulation of desired product in high titres, it is desirable that the microorganism is tolerant to inhibitory substances or can utilize them as sole carbon and energy source. Inhibitor tolerance is one of the most complex phenotype to engineer since many difficult to predict genes are often involved (Guarnieri et al., 2017).

4.4.2 CCR Regulation of Carbon Uptake and Metabolism

CCR is a natural mechanism for prioritization among multiple substrates which controls the uptake and ensuing metabolism of specific carbon source. The molecular mechanism of CCR has been extensively investigated and well known for the class of sugars which are transported through PTS called PTS sugars, e.g. glucose and mannose. As explained in Section 4.2, when these sugars are taken up by PTS, one of the PTS components, EIIA, is dephosphorylated and directly inhibits the transporters for other alternative or non-preferred carbon sources (Figure 4.2a). This mechanism is known as inducer exclusion and at least in some cases, is sufficient to achieve carbon source prioritization (Deutscher et al., 2006). However, additional regulatory systems are found in *E. coli* which can downregulate the expression of genes involved in catabolism of other substrates. This system is called global transcriptional regulation and centres around transcription factor Crp (catabolic repressor protein), which upregulates the carbon metabolism genes. Crp is activated by intracellular messenger cyclic AMP (cAMP) (Kaplan et al., 2008). cAMP is synthesized by enzyme adenylate cyclase, which in turn is activated by phosphorylated EIIA. Upon dephosphorylation of EIIA in presence of glucose, this mechanism is inactive and thus prevents Crp-mediated activation of alternative carbon uptake genes.

A preference amongst sugars other than glucose is also observed in bacteria. For example, in most bacteria, consumption of xylose is inhibited in the presence of arabinose. This is understood by the mechanism by which sugar-responsive transcription factors for xylose (XylR) and arabinose (AraC) interact with each

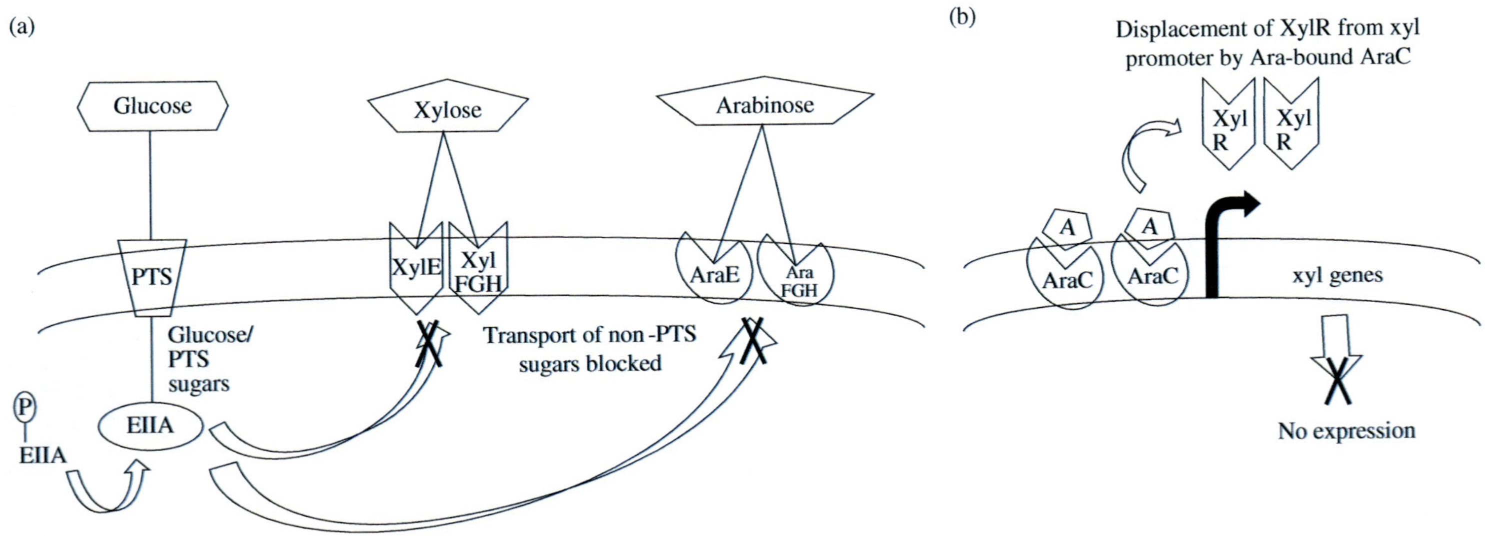

Figure 4.2 Mechanism of substrate prioritization in microbial metabolism. (a) Dephosphorylation of PTS component EIIA upon entry of glucose and/or other PTS sugars causes inhibition of transport of non-PTS sugars via CCR. (b) For xylose/arabinose mixtures, binding of arabinose to its regulator AraC displaces xylose regulator XylR from promoter for xylose utilization genes (*xyl*) thereby leading to repression in gene expression and impaired xylose utilization. PTS, Phosphotransferase system; XylE, xylose-proton symporter; XylFGH, xylose ABC transporter; AraE, arabinose-proton symporter; AraFGH, arabinose ABC transporter; A, arabinose.

other to control transcription of xylose (*xylAB*) and arabinose (*araBAD*) utilization operons. In the event of the binding of arabinose to AraC, the XylR is displaced from *xyl* promoters and the transcription of xylose utilization genes is repressed (Figure 4.2b).

In addition to carbon substrates, there is additional evidence to suggest that Crp is sensitive to other intracellular intermediates which play important roles in both catabolic and anabolic activities. One such finding is that besides the presence of glucose, EIIA phosphorylation is also controlled by ratio of central metabolites phosphoenolpyruvate (PEP) and pyruvate (Hogema et al., 1998). This points out that Crp is not only induced by carbon limitation but also nitrogen and sulphur limitation and that it is the balance between carbon catabolism and anabolism capacity which is perceived by Crp. Indeed, the carbon and nitrogen metabolism is linked involving shared regulation of these key elements. Previous theoretical work has suggested that the key branch point of these two utilization pathways is α-ketoglutarate (αKG) which serves as effective regulators of carbon catabolic flux (as TCA intermediate) and direct carbon precursors for transamination reaction in amino acid biosynthesis (You et al., 2013). Recently, it has been shown that the concentration of αKG acts as an intracellular signal of balance between catabolism and anabolic capacity which controls the activity of Crp. Thus, under nitrogen limiting conditions, the flux of αKG into TCA cycle is reduced and channeled into amino acid biosynthesis, which in turn reduces glucose uptake and cAMP synthesis via adenylate cyclase. This provides a circuit to co-ordinate carbon and nitrogen assimilation in bacteria.

Saccharomyces cerevisiae also exhibits catabolite repression in which glucose represses other sugar uptake, similar to that in bacteria. However, yeasts rely at least partially on external sensing of glucose (Santangelo 2006). This system is advantageous in environments where a small number of external signals are indicators of internal changes and external sensing enables cells to adapt more quickly.

4.5 Guiding Product Outcomes via Rewiring of Cellular Regulatory Circuit

The utilization of non-refined carbon sources such as lignocellulosic wastes, industrial wastes, etc. requires the presence of desirable traits in microorganisms involved in bioconversions. These include high yield and productivity, strong tolerance to substrate (including impurities) and product, and efficient utilization of the entire spectrum of sugars present in such heterogenous feedstocks. It is not surprising that there is no microorganism isolated to date which encompasses all these traits. Therefore, development of novel biocatalysts by genetic engineering approaches has received a great deal of attention in the recent years. With regards to lignocellulosic sugars, two rational approaches can be followed. First, introduction of pentose utilization pathways in a host which can produce one or more high-value products efficiently from glucose. The second alternative approach is to engineer a microorganism which can utilize both pentoses and hexoses to produce a specific product with high yield and productivity

(Gustavsson and Lee 2016). In recent years, the creation of recombinant cell factories through introduction of heterologous pathways has improved biotechnological prospects for production of high-valued products, however realization of full potential of such approaches remains limited due to the challenges posed by CCR.

CCR is an essential regulatory system which allows efficient substrate uptake and survival of microbes under competing conditions. Although a positive selection force in microorganisms, CCR is a major barrier with respect to the development of industrial microbial bioprocesses since it averts the simultaneous utilization of different carbon sources, such as those present in lignocellulosic biomass. Therefore, in order to obtain efficient fed-batch processing of mixed sugars, the hierarchy of sugar consumption needs to be removed and strategies ought to be devised for obtaining high productivities following complete sugar utilization. For instance, a high-level CCR would result in diauxic growth, while a low-level CCR might allow simultaneous utilization of available sugars but usually exhibits a higher uptake rate of one sugar than the other. Such simultaneous assimilation of multiple substrates through independent pathways would be advantageous for biomass-based conversion processes by allowing surplus supply of intermediates needed for bioproduct formation. Another beneficial scenario features a situation in which one substrate be diverted to biomass growth while other is fluxed into metabolite production pathway thus increasing the product yield significantly. Additionally, CCR depressed strains would avoid the problem of unconsumed carbon source which is usually associated with CCR-active strains and poses an additional challenge in downstream processing. Thus, various recombinant and native strains devoid of diauxie and CCR have been developed and utilized for efficiently co-metabolizing sugar mixtures from renewable feedstock. *E. coli* and *S. cerevisiae* are discussed as examples among bacteria and yeast respectively of such engineered strains towards construction of microbial factories.

4.5.1 Cellular Engineering in *E. Coli* for Bioprocessing of Non-Refined Carbon Sources

E. coli is the workhorse of industrial biotechnology. It is one of the most promising candidates for biofuel production from lignocellulosics. The wide availability of genetic tools for easy manipulation of *E. coli* have led to numerous successful attempts to also engineer it extensively for production of a wide range of solvents such as free fatty acids and short chain alcohols. The natural capability of *E. coli* to utilize a multitude of carbon sources implies that CCR system in *E. coli* would be highly complicated. For example, xylose is inhibited when glucose concentration in medium exceeds 40% of total sugars (Lawford and Rousseau 1994). Additional inhibition of xylose consumption is exerted in the presence of arabinose by blocking the binding site of xylose-specific transcription factor. Therefore, strategies have been devised to allow more effective co-utilization of all sugars from complex non-refined waste streams. The main pathway for glucose assimilation in bacteria is PTS wherein dephosphorylation of enzyme EIIA is responsible for causing inhibition of non-PTS sugar uptake in the presence of glucose. This mechanism of inducer exclusion operates to opt-out the so-called non-preferred sugar utilization. Thus, depletion of PTS has been viewed as an effective strategy to enable multiple substrate consumption. Interestingly, it has been reported that glucose transport is still somewhat possible to a varying extent for PTS deleted strains. This is possible through the activation of galactose permease which can transport glucose non-specifically. The strain could therefore co-metabolize glucose and arabinose, while xylose consumption was partially inhibited by glucose. Overall, an increase in sugar utilization by 16% was observed for this strain (Hernández-Montalvo 2001). In yet another attempt of PTS deletion, an improved accumulation of short-chain length polyhydroxyalkanoates (PHAs) by 36% was obtained by co-consumption of glucose and xylose by mutant strain (Li et al., 2007). Similarly, an improvement in lactic acid production was observed upon deletion of PTS domain IIC glucose transporter gene in *E. coli*. The deletion allowed 75% utilization of xylose as opposed to only up to 20% by native strain in the presence of glucose (Dien et al., 2002). A rather different strategy was devised by Nair and Zhao (2010) who achieved reduced glucose impairment and efficient xylose consumption through use of a Crp mutant which remained active irrespective of cAMP level inside the cell. In yet another attempt to address the problem of glucose impairment was introduction of glf-encoding glucose facilitator from *Z. mobilis* into the genome of *E. coli*. This together with strengthening of PPP resulted in high utilization rate of glucose and xylose to produce 97% of theoretical yield of ethanol (29 g/L) and a volumetric productivity of 1.8 g/L/h (Chiang et al., 2013).

Ultimate outcome of CCR exclusion is to achieve improved total sugar uptake by favoring pentose consumption in addition to glucose consumption. An interesting approach is co-culture where one strain specifically consumes one type of sugar while the other strain consumes the second sugar from substrate mixture. By combining a glucose selective *E. coli* strain and a xylose selective strain, diauxic growth could be eliminated while improved sugar utilization rate led to high lactate production (Eiteman et al., 2009). Inhibition of xylose consumption in the presence of arabinose is another impediment to the utilization of xylose which is the dominant sugar in lignocellulosic wastes. In an attempt to improve xylose utilization in presence of arabinose, deletion of *arc* gene which encodes arabinose transcriptional regulator (AraC) enabled co-utilization of glucose and xylose for improved xylitol production (Kim et al., 2015). Yet another strategy to avoid arabinose-xylose diauxie involved the optimization of expression level of XylR in *E. coli* to allow higher transcriptional activity for xylose assimilation. The strategy was indeed successful in improving ethanol production by 36% as compared to the native strain (Groff et al., 2012).

4.5.2 Rewiring *S. cerevisiae* for Accumulation and Conversion of Non-refined Carbon Sources

S. cerevisiae is another important host microorganism for industrial scale production of fuels and chemicals. It has been used since long for the production of alcoholic beverages due to its high and efficient ethanol production capability. However, despite its use for solvent production, *S. cerevisiae* lacks the ability to consume a wide range of substrates, particularly pentose sugars. Therefore, main strategy to employ *S. cerevisiae* for utilization of non-refined complex carbon sources has relied on introduction of heterologous xylose, arabinose and/or cellobiose pathways. It is however, worthwhile to mention that even the heterologous pathways suffer from CCR. The predominant reason for CCR manifestation in heterologous pathways could be explained by non-availability of dedicated xylose (pentose)-specific transporters in *S. cerevisiae. S. cerevisiae* has an array of low and high affinity hexose transport systems which could transport a wide range of substrates including glucose and xylose. Some transporters such as Hxt4p and Hxt7p exhibit a high specificity for xylose, however these are expressed only at low concentrations of glucose. Thus, it can be understood that xylose uptake would happen only when the concentration of glucose diminishes during the course of fermentation.

Nevertheless, efforts have been made to obtain efficient pentose and hexose utilization in *S. cerevisiae.* In one attempt, genes for key enzymes for xylose utilization including xylose reductase, xylitol dehydrogenase and xylulokinase were introduced in *S. cerevisiae* for ethanol production. The expression of hexose transporters was also adjusted to achieve maximal consumption of both glucose and xylose in ethanol fermentation. The results of this study were indeed interesting and revealed that approaches to engineer *S. cerevisiae* should consider the composition of waste stream mainly because of the different affinities of sugar transporters in it. For example, overexpression of low-affinity HXT1 permease was favorable for high consumption of sugars from xylose/glucose blends and achieving high ethanol production rates. On the contrary, improved expression of another transporter, HXT7 permease seemed more suitable to valorize xylose enriched sugar streams (Gonçalves et al., 2014).

Considering the fact that all known xylose transporters are competitively inhibited by glucose, creation of mutant D-xylose transporter which is insensitive to the presence of glucose is another approach to enable efficient xylose consumption (Farwick et al., 2014). One such mutant was reported to be Gal2-N376F which could transport xylose with high affinity without any glucose transport. Such modifications are again best suited for xylose rich streams only. Another carbon source in waste streams, particularly lignocellulosics is cellobiose. With an aim to avoid glucose inhibition and provide sufficient reducing power to xylose reductase, cellobiose (instead of glucose) and xylose utilization pathways were introduced in *S. cerevisiae.* The engineered strain was able to co-ferment cellobiose and xylose to produce ~32% more xylitol as compared to sequential consumption of glucose and xylose by native strain (Oh et al., 2013). Furthermore, an increase in xylitol titres was also observed with high initial concentrations of cellobiose and xylose.

4.6 Conclusions

The intricate and tightly regulated microbial metabolism circuits provide a natural mechanism for microbes to adapt to ever-changing nutrition conditions and are indeed significant for microbial survival and evolution. Development of efficient bioprocesses based on renewable substrates particularly non-refined carbon sources is however impeded by such complicated metabolic networks. The uptake of a specific carbon source, its transport to cytosol, transformations to specific product via a series of enzyme catalyzed reactions, associated inhibitory responses and communication with other nutrient metabolic mediators need a detailed and careful understanding for designing high product yielding bio-based systems. Meanwhile, another pertinent consideration is the composition of non-refined carbon sources in renewable feedstocks which makes it easier to ascertain the optimal cellular engineering targets for obtaining the desired product in high concentrations and productivities.

References

Aidelberg, G., Towbin, B.D., Rothschild, D. et al. (2014) Hierarchy of non-glucose sugars in *Escherichia coli. BMC Systems Biology*, 8, 133–145.

Bailey, J.E. (1991) Toward a science of metabolic engineering. *Science*, 252, 1668–1675.

Chiang, C.J., Lee, H.M., Guo, H.J. et al. (2013) Systematic approach to engineer *Escherichia coli* pathways for co-utilization of a glucose-xylose mixture. *Journal of Agricultural and Food Chemistry*, 61, 7583–7590.

Chubukov, V., Gerosa, L., Kochanowski, K., and Sauer, U. (2014) Coordination of microbial metabolism. *Nature Reviews Microbiology*, 12, 327–340.

Cui, Z., Gao, C., Li, J. et al. (2017) Engineering of unconventional yeast *Yarrowia lipolytica* for efficient succinic acid production from glycerol at low pH. *Metabolic Engineering*, 42, 126–133.

Da Costa Sousa, L., Chundawat, S.P., Balan, V., and Dale, B.E. (2009). 'Cradle-to-grave' assessment of existing lignocellulose pretreatment technologies. *Current Opinion in Biotechnology*, 20, 339–347.

da Silva, G.P., Mack, M., and Contiero, J. (2009) Glycerol: A promising and abundant carbon source for industrial microbiology. *Biotechnology Advances*, 27, 30–39.

Dekel, E. and Alon, U. 2005. Optimality and evolutionary tuning of the expression level of a protein. *Nature*, 436, 588–592.

Deutscher, J., Francke, C., and Postma, P.W. (2006) How phosphotransferase system-related protein phosphorylation regulates carbohydrate metabolism in bacteria. *Microbiology and Molecular Biology Reviews*, 70, 939–1031.

Dhillon, G.S., Brar, S.K., and Verma, M. (2012). Biotechnological potential of industrial wastes for economical citric acid bioproduction by *Aspergillus niger* through submerged fermentation. *International Journal of Food Science and Technology*, 47, 542–548.

Dien, B. S., Nichols, N. N., and Bothast, R. J. (2002) Fermentation of sugar mixtures using *Escherichia coli* catabolite repression mutants engineered for production of lactic acid. *Journal of Industrial Microbiology and Biotechnology*, 29, 221–227.

Eiteman, M. A., Lee, S. A., Altman, R., and Altman, E. (2009) A substrate-selective co-fermentation strategy with *Escherichia coli* produces lactate by simultaneously consuming xylose and glucose. *Biotechnology and Bioengineering*, 102, 822–827.

El-Asar, S.A. (2006). Submerged fermentation of cheese whey and Molasses for CA production by *Aspergillus niger. International Journal of Agriculture and Biology,* 8, 863–867.

Farwick, A., Bruder, S., Schadeweg, V. (2014) Engineering of yeast hexose transporters to transport D-xylose without inhibition by D-glucose. *Proceedings of the National Academy of Sciences U.S.A.,* 111, 5159–5164.

Frunzke, J., Engels, V., Hasenbein, S. et al. (2008). Co-ordinated regulation of gluconate catabolism and glucose uptake in *Corynebacterium glutamicum* by two functionally equivalent transcriptional regulators, GntR1 and GntR2. *Molecular Microbiology,* 67, 305–322.

Gonçalves, D.L., Matsushika, A., de Sales, B.B. et al. (2014) Xylose and xylose/glucose co-fermentation by recombinant *Saccharomyces cerevisiae* strains expressing individual hexose transporters. *Enzyme and Microbial Technology,* 63, 13–20.

Görke B. and, Stülke J. (2008). Carbon catabolite repression in bacteria: many ways to make the most out of nutrients. *Nature Reviews Microbiology,* 6, 613–624.

Groff, D., Benke, P.I., Batth, T.S. et al. (2012) Supplementation of Intracellular XylR Leads to Coutilization of Hemicellulose Sugars. *Applied and Environmental Microbiology,* 78, 2221–2229.

Guarnieri, M.T., Franden, M.A., Johnson, C.W., and Beckham, G.T. 2017. Conversion and assimilation of furfural and 5-(hydroxymethyl)furfural by *Pseudomonas putida* KT2440. *Metabolic Engineering Communications,* 4, 22–28.

Gustavsson, M. and Lee, S.Y. (2016) Prospects of microbial cell factories developed through systems metabolic engineering. *Microbial Biotechnology,* 9, 610–617.

Hernández-Montalvo, V., Valle, F., Bolivar, F., and Gosset, G. (2001) Characterization of sugar mixtures utilization by an *Escherichia coli* mutant devoid of the phosphotransferase system. *Applied Microbiology and Biotechnology,* 57, 186–191.

Hogema, B.M., Arents, J.C., Bader, R. et al. (1998) Inducer exclusion in *Escherichia coli* by non-PTS substrates: the role of the PEP to pyruvate ratio in determining the phosphorylation state of enzyme IIAGlc. *Molecular Microbiology,* 30, 487–498.

Jojima, T., Inui, M., and Yukawa, H. (2011) Metabolic engineering of bacteria for utilization of mixed sugar substrates for improved production of chemicals and fuel ethanol. *Biofuels,* 2, 303–313.

Kaplan, S., Bren, A., Zaslaver, A. et al. (2008) Diverse two-dimensional input functions control bacterial sugar genes. *Molecular Cell,* 29, 786–792.

Kaur, G., Srivastava, A.K., and Chand, S. (2012) Advances in Biotechnological production of 1,3-propanediol. *Biochemical Engineering Journal,* 64, 106–118.

Khankal, R., Chin, J.W., Ghosh, D., and Cirino, P.C. (2009) Transcriptional effects of CRP* expression in *Escherichia coli. Journal of Biological Engineering,* 3, 13–27.

Kim, J.H., Block, D., and Millis, D. (2010). Simultaneous consumption of pentose and hexose sugars: an optimal microbial phenotype for efficient fermentation of lignocellulosic biomass. *Applied Microbiology and Biotechnology,* 88, 1077–1085.

Kim, S.M., Choi, B.Y., Ryu, Y.S. et al. (2015) Simultaneous utilization of glucose and xylose via novel mechanisms in engineered *Escherichia coli. Metabolic Engineering,* 30, 141–148.

Lawford, H. and Rousseau, J. (1994) Relative rates of sugar utilization by an ethanologenic recombinant *Escherichia coli* using mixtures of glucose, mannose, and xylose. *Applied Biochemistry and Biotechnology*, 45–46, 367–381.

Li, R., Chen, Q., Wang, P.G., and Qi, Q.S. (2007) A novel-designed *Escherichia coli* for the production of various polyhydroxyalkanoates from inexpensive substrate mixture. *Applied Microbiology and Biotechnology*, 75, 1103–1109.

Liu, M., Durfee, T., Cabrera, J. E. et al. (2005) Global transcriptional programs reveal a carbon source foraging strategy by *Escherichia coli*. *Journal of Biological Chemistry*, 280, 15921–15927.

Martinez-Antonio, A., Janga, S.C., Salgado, H., and Collado-Vides, J. (2006) Internal-sensing machinery directs the activity of the regulatory network in *Escherichia coli*. *Trends in Microbiology*, 14, 22–27.

Monod, J. (1942) Recherches sur la croissance des cultures bactériennes. Hermann & Cie, Paris, France.

Nair, N. U. and Zhao, H. (2010) Selective reduction of xylose to xylitol from a mixture of hemicellulosic sugars. *Metabolic Engineering*, 12, 462–468.

Nielson, J. and Keasling, J.D. (2016) Engineering Cellular Metabolism. *Cell*, 164, 1185–1197.

Oh, E.J., Ha, S.J., Kim, S.R. et al. (2013) Enhanced xylitol production through simultaneous co-utilization of cellobiose and xylose by engineered *Saccharomyces cerevisiae*. *Metabolic Engineering*, 15, 226–234.

Parche, S., Beleut, M., Rezzonico, E. et al. (2006) Lactose-over-glucose preference in *Bifidobacterium longum* NCC2705: glcP, encoding a glucose transporter, is subject to lactose repression. *Journal of Bacteriology*, 188, 1260–1265.

Portnoy, T., Margeot, A., Linke, R. et al. (2011) The CRE1 carbon catabolite repressor of the fungus *Trichoderma reesei*: a master regulator of carbon assimilation. *BMC Genomics*, 12, 269–281.

Ryu, S., Hipp, J., and Trinh, C.T. (2016) Activating and Elucidating Metabolism of Complex Sugars in *Yarrowia lipolytica*. *Applied and Environmental Microbiology*, 82, 1334–1345.

Saini, J.K., Saini, R., and Tewari L. (2015) Lignocellulosic agricultural wastes as biomass feedstocks for second-generation bioethanol production: Concepts and recent developments,3. *Biotech*, 5, 337–353.

Saling, P. (2005) Eco-Efficiency Analysis of biotechnological processes. *Applied Microbiology and Biotechnology*, 68, 1–18.

Santangelo, G.M. (2006) Glucose signalling in *Saccharomyces cerevisiae*. *Microbiology and Molecular Biology Reviews*, 70, 253–282.

Stulke, J. and Hillen, W. (1999) Carbon catabolite repression in bacteria. *Current Opinion in Microbiology*, 2, 195–201.

You, C., Okano, H., Hui, S. et al. (2013) Coordination of bacterial proteome with metabolism by cyclic AMP signalling. *Nature*, 500, 301–306.

Zhang, Y.H. and Lynd, L.R. (2005) Cellulose utilization by *Clostridium thermocellum*: bioenergetics and hydrolysis product assimilation. *Proceedings of the National Academy of Sciences U.S.A.*, 102, 7321–7325.

5

Cellular versus Biochemical Control over Microbial Products

Carlos S. Osorio-González[1*], *Krishnamoorthy Hegde*[2*], *and Satinder Kaur Brar*[3]

[1] *Institut national de la recherche scientifique, Centre - Eau Terre Environnement, Québec, Canada*
[2] *INRS-ETE, Université du Québec, Québec, Canada*
[3] *Institut national de la recherche scientifique, Centre - Eau Terre Environnement, Québec, Canada*

5.1 Introduction

The depletion of fossil resources currently driving research focus into viable, renewable and environmental friendly alternatives to replace conventional resources (raw materials) as well as processes for obtaining a wide range of Petroleum based chemical alternatives (Jiménez-Quero et al., 2016). The best method of synthesis of renewable chemicals is through the fermentation process, which requires less demanding conditions in terms of energy and uses renewable resources as raw material (Pérez-Bibbins et al., 2013). In this sense, the cellular and biochemical control in the biosynthesis of chemicals from genetically modified microorganisms has emerged as a competitive alternative to traditional processes based on chemical synthesis (Cho et al., 2015). The development of metabolic and genetic engineering promotes the progress in obtaining modified microbial strains for the improvement of the processes and an increase in the yields of the final products through cellular and biochemical control. This is carried out by engineering the biochemical and metabolic pathways in the desired microorganisms often by replacing/assembling several enzymes from different and more efficient organisms, which encode for the overexpression of genes of interest for improvement in the production and yield of the desired products (Moon et al., 2009).

Among different Microorganisms, such as bacteria, yeast and fungi, *Escherichia coli*, a gram-negative bacterium, has been widely used worldwide in production of various platform chemicals (Liu et al., 2016c). The *E. coli* strains has been proven to be an excellent biocatalysis platform for metabolic engineering

Equal contribution authors

Microbial Sensing in Fermentation, First Edition. Edited by Satinder Kaur Brar, Ratul Kumar Das, and Saurabh Jyoti Sarma.

(Wendisch et al., 2016). Some of the particular characteristics and advantages of this strain over other microorgnisms are:

- Sensitive to many antibiotics that can be used as engineering labels to track genetically modified strains;
- clear genomic, proteomic and metabolic information that benefits metabolic engineering efforts;
- extensive knowledge about the metabolism and physiology of the organism; and
- faster growth and higher production yield in comparision to other microorganisms.

Due to these and other advantages, artificial metabolic pathways have been constructed in *E. coli* by recruiting and integrating exogenous genes and enzymes to produce compounds that wild microorganisms can not metabolize (Bozell and Petersen, 2009).

Some of these platform chemicals obtained by genetically modified strains of *E. coli* are described in this review. Further, the chapter summarize the application and cellular and biochemical control of *E. coli* strains for efficient production of an array of platform chemicals.

5.2 3-Hydroxy-propionic Acid

3-hydroxypropionic acid is a chemical compound ranked among the top 12 shelf chemicals selected by the United States Department of Energy (DOE, Kumar et al., 2013). It is a non-chiral, three-carbon organic compound, and a structural isomer of lactic acid (2-hydroxypropanoic acid), also called hydracrylic acid or ethylene-lactic acid (Valdehuesa et al., 2013). It is a high molecular weight compound with thermal and mechanical properties that can be used to improve the properties of stretched films (Kwak et al., 2013). Currently this organic acid is of great importance for several industrial applications, such as chemical industry, food applications and paints. It is used as a precursor for the synthesis of some other commodities, such as 1,3-propanediol or 3-hydroxypropionaldehyde (Matsakas et al., 2014). The raw materials used for the biotechnological production of this compound are diverse (wheat, sugar cane, maize, lignocellulosic compounds). However, glycerol and glucose are the main substrates used (O'Hara et al., 2013; Koutinas; Tingirikari et al., 2016) and its natural production has been reported with different strains, including *Lactobacillus, Geotrichum, Trichoderma, Phomopsisphaseoli* and *Melanconium botulinum,* among others (Tingirikari et al., 2016). However the yields obtained from these biocatalysts are too low (0.18 g/L) to exploit them for commercial production (Luo et al., 2011).

Due to the above constraints, in the last decade research on the development of genetically modified microorganisms to obtain higher yields of this product has intensified, using *E. coli* (David et al., 2017). However, *E. coli* cannot naturally metabolize glycerol to 3-hydroxy-propionic acid as it lacks the enzyme glycerol dehydratase (dhaB) and the negligible expression of aldehyde dehydrogenase (aldH) as both are key enzymes to obtaining this metabolite (Jo et al., 2008). Currently the research on production of 3-hydroxy-propionic acid from glycerol using different strains of recombinant *E. coli* has been intensified (Rathnasingh et al., 2012; Jung et al., 2014; Lim et al., 2016; Niu et al., 2016), due to the fact that this raw material is of very low cost and highly availability in the market (Niu et al., 2016). The main focus in constructing new routes in this strain is the insertion of genes expressing enzymes that stabilize the activity of the enzyme glycerol dehydratase (DHaB) to convert glycerol to 3-hydroxy-propionic acid (Rathnasingh et al. 2009).

Chu et al. (2015), modified *E. coli* DH5α strain, with the inclusion of the enzyme aldehyde dehydrogenase (GabD4) from *Cupriavidus necátor*, yielding up to 71.9 g/L 3-hydroxy-propionic acid from crude glycerol in a fed-batch system (Figure 5.1). This mutation, together with the expression of the dhaB gene, the silencing of competing pathways, the addition of cofactors (IPTG and vitamin B_{12}) and optimization of the fermentation conditions (pH 7.0 and 1 vvm air flow) were the main regulating factors of the biosynthesis of 3-hydroxypropionic acid in *E. coli*.

However, one of the major factors limiting the conversion of glycerol to 3-hydroxy-propionic acid, that needs research attention is the accumulation of 3-hydroxypropionaldehyde (3-HP) during production, which is toxic to host organisms and limits the conversion of the substrate (David et al., 2017).

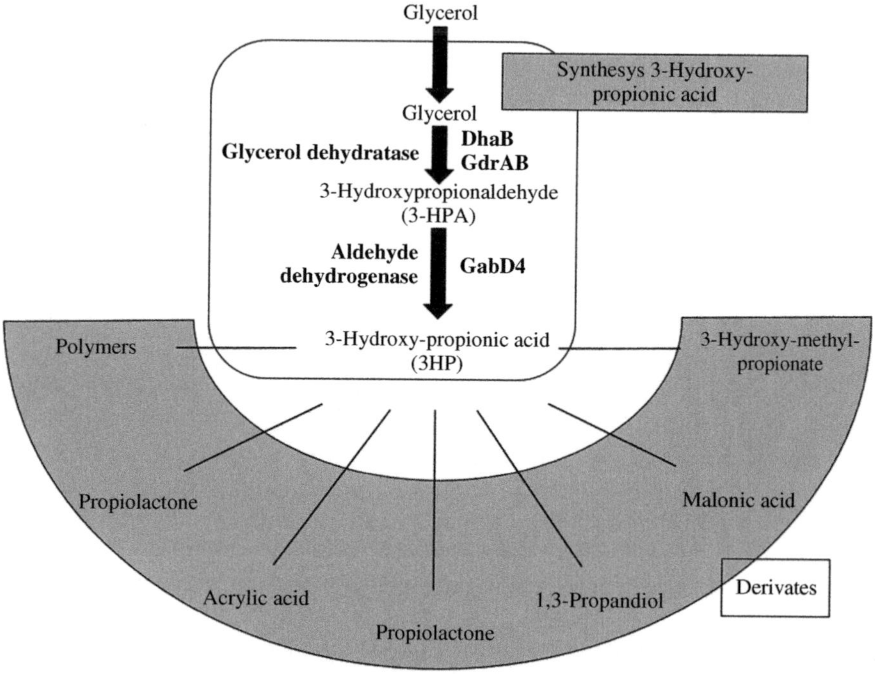

Figure 5.1 Genetically engineered metabolic pathway for the production of 3-hydroxy-propionic acid in *E. coli*. (*See insert for color representation of this figure.*)

The functionality of 3-HP allows conversion to other chemicals of industrial interest, such as malonic acid, 1,3-propandiol, acrylic acid, propiolactone and 3-hydroxymethyl-propionate (Mohan and Sivaprakasam, 2016). However, it is noteworthy that the cost of production of 3-HP in chemical synthesis route is about US$ 1.50/kg. In contrast, with the development of bioprocess technology, the cost of 3-HP synthesized is reported to be approximately US$1.10/kg from glucose and US$0.84/kg from sucrose (Tingirikary et al., 2016).

5.3 Fumaric Acid

Fumaric acid is a four-carbon dicarboxylic acid, currently produced by petrochemical synthesis by the catalytic isomerization of maleic acid (Felthouse et al., 2001; Van Nostrand's, 2006). The main uses of this metabolite are in the food industry, resins, and green chemistry (Mansur et al., 2015; Sunh et al., 2015; Khan et al., 2016, Maghsoodlou et al., 2016). It has now been of great importance for its use as a compound in the synthesis of derivatives with high value in the field of biomedicine (Ahuja et al., 2016; Das et al., 2016b). The raw materials for the production of fumaric acid by fermentation are mainly carbohydrates (glucose and xylose), agrifood residues and lignocellulosic residues (Das and Brar, 2014; Li et al., 2016; Pan et al., 2016, Shah et al. Al., 2016).

Naturally, *Rhizopus* species (*arrhizus, oryzae, formosa, nigricans*) are recognized as the best microorganisms for the production of fumaric acid (Das et al., 2016a; Demain and Martens, 2016; Liu et al., 2016), the production of this metabolite has been reported with different species of fungi (*Penicillium griseofulvum, Aspergillus glaucus and Caldariomyces fumago,* among others; Yang et al., 2011; Zhang et al., 2013). However the *R. arrhizus* and *R. oryzae* strains are the most widely studied. The biosynthesis of fumaric acid in *R. oryzae* is mainly by the reductive pathway of tricarboxylic acids (rTCA), including carboxylation of pyruvate with CO_2 fixation, oxaloacetate hydrogenation and malate dehydration (Ding et al., 2011; Das et al., 2016c).

In a study by Cho et al. (2013) with a metabolically modified *E. coli* strain for the production of fumaric acid, in which the iclR gene (redirecting the carbon flux through the glyoxylate derivative), as well as the fumA, fumB and fumC genes, obtained a yield of 1.45 g/L fumaric acid from 15 g/L glucose. Further, modification for the overexpression of the native plasmid encoding phosphoenolpyruvate carboxylase (PPC), from the tac promoter, showed 4.09 g/L of fumaric acid. In addition, they suppressed arcA, ptsG, and the aspA gene (blocks the conversion of fumaric acid to L-aspartic acid), with the production yield of 28.2 g/L of fumaric acid. This study demonstrated the possibility of efficient production of fumaric acid with genetically modified strains of *E. coli* (Figure 5.2).

The production of fumaric acid until 2008 was 90,000 tons at a cost of US $ 1.63/kg (Engel et al., 2008). Thus, the fumaric acid produced by fermentation from alternative raw materials could become an excellent production option in the near future.

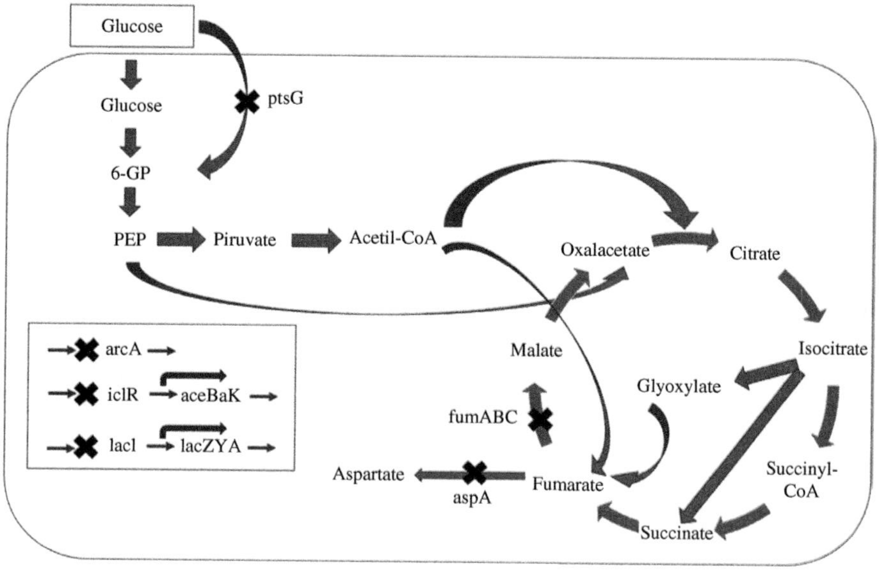

Figure 5.2 Genetically engineered metabolic pathway for the production of fumaric acid in *E. coli*. *Source:* modified from Son et al., 2013.

5.4 Itaconic Acid

The production of itaconic acid by *Aspergillus itaconicus* was discovered by Kinoshita in 1932, today it is commercially produced by the filamentous fungus *Aspergillus terreus* (Klement and Büchs, 2013; Okamoto et al., 2015). This compound can be obtained from different raw materials such as jatropha curcas seed cake (Ahmed El-Imam et al., 2013), potato starch (Tsai et al., 2001), rice husk (Pedroso et al., 2016), glycerol (Vassilev et al., 2012), among others.

In general, the process of production of itaconic acid from fungi is relatively long. Because of this, various genetically modified microorganisms have been developed for the faster and more efficient production of itaconic acid (*A. niger, Saccharomyces cerevisiae, Corynebacterium glutamicum, Yarrowia lipolytica and E. coli* (van deer Straat et al., 2013). However, some of these recombinant microorganisms do not produce the same yields as other commercial strains, such as *A. terreus* (Saha, 2016). The current limitation in itaconic acid production on an industrial scale is production costs (Klement and Büchs, 2013). This limitation can be overcome by improving three processes:

a) Obtaining cheaper raw materials;
b) improving the strains used to produce them (commercial and genetically modified); and
c) optimizing the production processes (Ramos et al., 2016).

Itaconic acid biosynthesis has not yet been fully established and understood, however, what is generally accepted is that biosynthesis occurs in two compartments of the microorganism: the cytosol and the mitochondria (Steiger et al., 2013; Huang et al., 2014).

The biosynthetic pathway begins with glycolysis in the cytosol, after which the pyruvate is transported to the mitochondria and converted to acetyl-CoA, or carboxylated into oxaloacetate in the cytosol, the oxaloacetate is transformed into malate, which is transported in the mitochondria by malate/citrate anti-port carriers in mitochondria are converted to acetyl-CoA and oxaloacetate in citric acid by the enzyme citrate synthase (CS), citric acid is converted to cis-itaconate by the enzyme aconitase (ACO) and returned To the cytosol with the aid of mitochondrial tricarboxylic transporters (MTT), where it is decarboxylated again in itaconate by the enzyme cis-aconitate decarboxylase (CAD), so that finally itaconic acid is exported out of the cell through the acid carrier di-carboxylic acid (Figure 5.3) (Ahmed El-Imam and Du, 2014).

An example of the use of genetically modified strains for the production of this metabolite is *E. coli*, *S. cerevisiae* and *A. niger*, which in their native form do not produce itaconate because it does not have the enzyme cis-aconitate decarboxylase (CadA), which has the function of catalyzing cis-aconitate biosynthesis in itaconate in *A. terreus* (Yahiro et al., 1995). Despite this, the enzyme has been successfully expressed in these strains (Li et al., 1998), but also in the presence of a high level of fermentation (Jeon et al., 1991). In addition, these hosts have many advantages as production tools, which include well-established molecular characteristics and operating conditions are relatively easy at the time of fermentation (Jeon et al., 2016).

Currently, the production of itaconic acid has received great attention because of its promising use in the production of synthetic polymers, detergents, coatings and in the manufacture of absorbents and health care products (Chin et al., 2015; Choi et al., 2015; Hegde et al., 2016). The main producers of this compound worldwide are China, USA, France and Japan; The annual production in the world is 88,000 tons with a sale price of approximately US $2/kg, so the market for this metabolite will increase in the coming decades (Bathia et al., 2106; Harder et al., 2016).

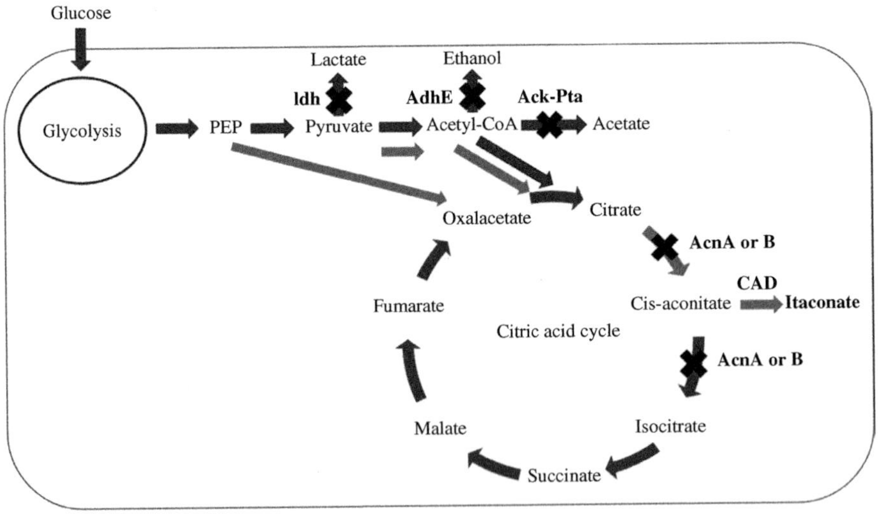

Figure 5.3 Genetically engineered metabolic pathway for the production of itaconic acid in *E. coli*. *Source*: modified from Harder et al., 2016. (*See insert for color representation of this figure.*)

5.5 Glucaric Acid

D-glucaric acid, is a natural and non-toxic compound, which occurs naturally in some vegetables, fruits and mammals (including humans) (Yuan et al., 2017). Conventional production of glucaric acid from glucose is associated with low yields and the production of toxic byproducts (Shiue and Prather, 2014). Because of this, different studies have focused on the biosynthesis of glucaric acid to remove these by-products. The above has been investigated using strains of genetically modified microorganisms (Reizman et al., 2015). Glycolic acid is used in the field of medicine (Walloz et al., 1997; Zoltaszec et al., 2008), in biodegradable detergents (Liu et al., 2016b) and in the production of polymers (Sheldon, 2013).

For the production of this acid in recombinant strains of *E. coli* requires the expression of three heterologous proteins and an additional endogenous enzyme: myoinositol-1-phosphate synthase (Ino1) from *S. cerevisiae*, an endogenous *E. coli* phosphatase, Myoinositol oxygenase (MIOX) from *M. musculus* and uronate dehydrogenase (Udh) from *Pseudomonas syringae* (Figure 5.4)(Lee et al., 2016; Gupta et al., 2016). This is a simple and economical route for the production of D-glucaric acid compared to the long path of more than ten stages of conversion from glucose in mammals (Moon et al., 2009).

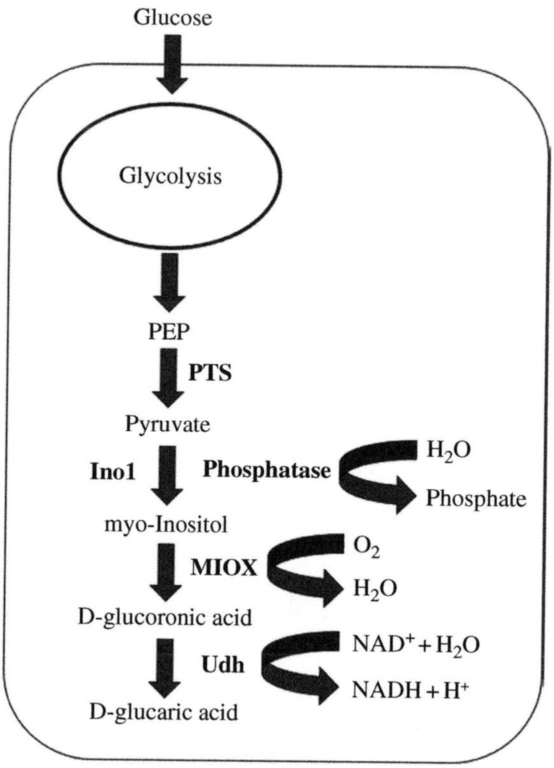

Figure 5.4 Genetically engineered metabolic pathway for the production of glucaric acid in *E. coli*. *Source:* modified from Moon et al., 2009.

The market for glucaric acid, however, is not developed due to limited availability and high price. However, Rivertop Renewables© has developed selective aerobic oxidation in the presence of nitric acid and is pursuing the commercialization of glucaric acid as a chemical platform (Sheldon, 2013).

5.6 Butanol

Commercial production of butanol from fermentation was first performed with the *Clostridium acetobutylicum* strain using glucose and starch as raw materials (Jones and Woods 1986). However, this bio-based production was superseded by the chemical process for synthesis of butanol from petroleum oil (Green, 2011). In spite of this, the bioproduction of butanol has recovered interest due to the need to obtain fuels and chemicals from renewable resources (Branduardi and Porro, 2016). On the other hand, research has intensified in the improvement of the fermentation processes (raw materials, solvent tolerance, product selectivity among others, Chen and Liao, 2016). In relation to the raw materials for obtaining butanol from fermentation the most common are carbohydrates (glucose, xylose, galactose) (Chen et al., 2013; Bhandiwad et al., 2014) and lignocellulosic residues (corn stover, Barley straw, sugarcane bagasse, cassava bagasse) (Qureshi et al., 2014; Pang et al., 2016; Zhang et al., 2016). However, this approach has been hampered by the debate between food and fuel, since traditional fermentation is based on sugar or starchy materials such as substrates, which are important components in human nutrition (Dürre, 2016). Due to this, research has now intensified in the production of this compound from lignocellulose and CO_2 (Fernández-Naveira et al., 2016).

Butanol is a promising next-generation biofuel since it has similar properties to gasoline and can be used in engines designed for this type of fuel without the need for modification (Schadeweg and Boles, 2016). The natural biosynthesis of butanol in *Clostridium* species is produced by fermentation with acetone-butanol-ethanol (ABE) through a series of stages of reduction of the intermediates bound to CoA, while the native producer, *Clostridium* remains the main species for the production of butanol on an industrial scale (Saini et al., 2016). Their production from the fermentation has been demonstrated using native bacterial cell factories or genetic engineering microorganisms, such as *E. coli* (Figure 5.5) (Bond-Watts et al., 2011; Green, 2011).

Wen and Shen, (2016) designed a self-regulated production system in an *E. coli* strain through endogenous fermentation regulatory elements (FRE) to boost the biosynthetic genes of the natural production of 1-butanol (ldhA, frdABCD, adhE y ackA). Fermentation was developed under anaerobic conditions for all cases, in this sense, the enzyme, formate dehydrogenase (Fdh) overexpressed under the FREadhE gene showed a higher yield of 1-butanol (10 g/L in 24 h and a yield of 0.25 g/g glucose). This indicates that subjecting cells to oxygen limiting conditions prior to significant accumulation of biomass is crucial for the induction of enzymatic synthesis and fermentation efficiency.

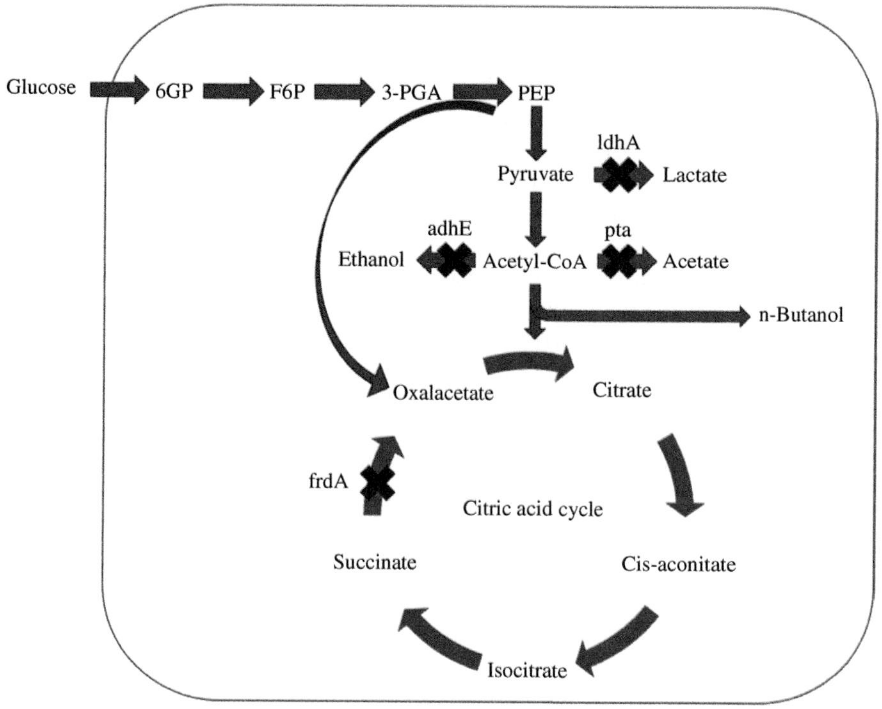

Figure 5.5 Genetically engineered metabolic pathway for the production of butanol in *E. coli*. *Source:* modified from Saini et al., 2016.

However, this compound is toxic to *E. coli*, which barely tolerates organic solvents at low concentrations, for example, the cell growth of this strain is completely inhibited in the presence of 1% (v/v) n-butanol, which has been one of the limitations in the development of strains producing this important compound (Si et al., 2016).

The overall market size in 2006 for this product was 2.8 million tons. Until 2011 the market demand and production was 2,300 kton/year (Dutta et al., 2016).

5.7 Malic Acid

Malic acid, or its ionized form, malate is one of the most abundant acids in nature. L-malic acid (L-hydroxy-butanedioic acid or L-hydroxy-succinic acid) is the naturally occurring isomer and is found in some fruits and vegetables. The raw material for fermentation with malic acid includes molasses, starch and cellulosic materials (Li et al., 2014b).

It is used mainly as a food additive or as an intermediate in the synthesis of fine chemicals (West et al., 2016), can be obtained as a racemic mixture of isomers, through the hydration of maleic or fumaric acid (Zou et al., 2013), from pyruvic acid through oxaloacetic acid by fermentation (Kumar et al., 2016) and conversion of fumaric acid under fumarase catalysis (Stojkovič et al., 2011). There are several studies in which the co-production of aflatoxins is associated with specific microorganisms (fungi) that are used in the production of malic acid (Zhang et al., 2011); On the other hand, large-scale production processes from these microorganisms have presented different complications in production, such as supplementation and diffusion of oxygen due to mycelial growth, carbon/nitrogen ratio in the medium, pH, Temperature, among others (Zhang et al., 2013, Zambanini et al., 2016).

To avoid such constrains, well-established (genetically modified) model organisms, such as *S. cerevisiae*, *A. niger*, and *E. coli*, were designed for the production of microbial malate (Figure 5.6) (Nakayama et al., 2012; Zhang et al., 2015). Dong et al., (2016) modified the *E. coli* strain for the production of L-malate by direct conversion of pyruvate in a single step, yielding L-malate yields of 3.62 g/L from 20.9 g/L pyruvate. This was obtained from the suppression of four genes directly involved in the catabolism of pyruvate (ldhA, pflB, poxB, and pta-ackA) and the reinforcement of the biosynthetic pathway by protein engineering (NAD+-sfcA, NAD+-ME2, NADP+-ME4, and NADP+-maeB). Of the three biosynthetic pathways, the former is preferred by industries (1 mol glucose/2 mol L-malic acid). However, it remains a problem to improve the yields of this product by metabolic engineering (Chi et al., 2016).

Currently, the annual production of malic acid is approximately 40,000 metric tons, with a price of US $1.80 to $2.00 per pound. Considering this price, the use

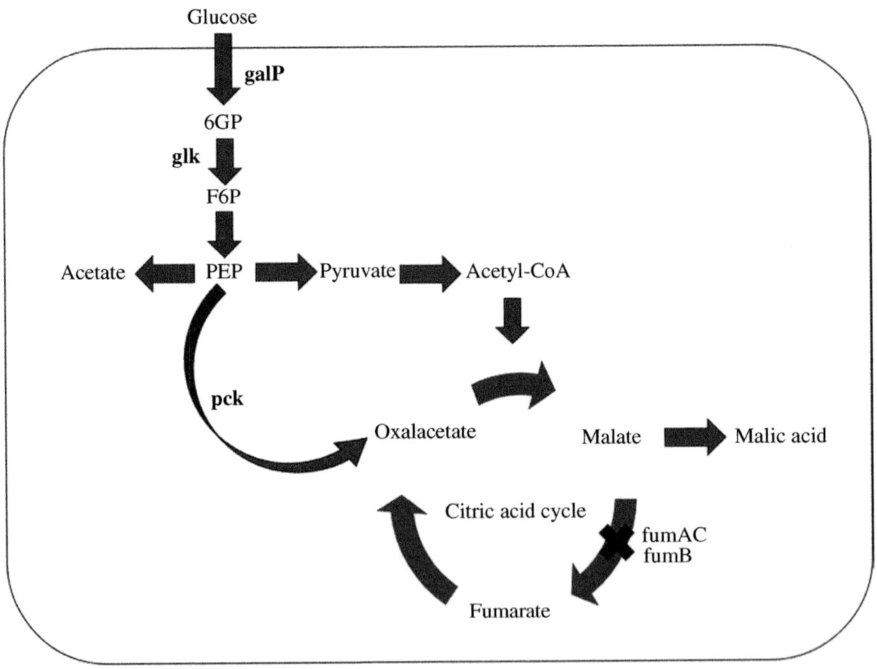

Figure 5.6 Genetically engineered metabolic pathway for the production of malic acid in *E. coli*. *Source:* modified from Zhang et al., 2011.

of a synthesis approach from low-value biomass could substantially reduce its cost of production (Mondala, 2015).

5.8 Gluconic Acid

Chemically, gluconic acid is a stereoisomer of 2,3,4,5,6-pentahydroxyhexanoic acid (Pal et al., 2016). It is used in the pharmaceutical, textile, construction and especially in the food industry (Cañete-Rodríguez et al., 2016). It is usually obtained by the partial oxidation of glucose, by biotransformation by microorganisms (fungi and bacteria mainly) (Mounir et al., 2016; Sainz et al., 2016), also obtained by chemical synthesis based on the oxidation (Zhang et al., 2016b) and by biochemical oxidation (use of enzymes), (Cui et al., 2016).

The raw materials for obtaining this compound are diverse: starch (Sun et al., 2012; An et al., 2013; Matsui et al., 2013), lignocellulosic biomass (Ajala et al., 2016; Zhang et al., 2016), sugarcane molasses (Sharma et al., 2008), cheese whey (Mukhopadhyay et al., 2005; Alonso et al., 2015), waste paper (Ikeda et al., 2006), and others; this makes it an excellent example for the optimal use of residues with high carbohydrate content (Ramachandran et al., 2017).

Among the different types of fungi used in the production of gluconic acid, *A. niger* has the highest productivity and is the microorganism currently used in industrial production (Wang et al., 2016). However, some of these processes are not viable due to the high cost in the commercial production from *A. niger*, due to interference of other endogenous enzymes (catalase, cellulase, and amylase) (Park et al., 2000). Considering the above, the enzyme glucose oxidase has been induced in genetically engineered microorganisms for the production of this compound and to improve the yields in the production (Figure 5.7) (Singh and Kumar, 2007).

The demand for this organic acid has increased over the last two decades, with production at 60,000 tonnes per year (Ahmed et al., 2015), its price ranges from the US$0.50 to 0.56 per pound) and $0.66 to $ 0. 70 for sodium gluconate (powder, Roger et al., 2013).

5.9 Aminovalaric Acid

5-Aminovalerate is a potential precursor compound for many value-added chemical derivatives and for the synthesis of polymers (Shin et al., 2016). It is considered as a potential chemical platform for the synthesis of glutarate, 5-hydroxyvalerate and 1,5 pentanediol (Adkins et al., 2013, Park et al., 2014). The direct fermentative production of 5-amino-valeric acid on an industrial scale has not yet been demonstrated (Pulicharla et al., 2016), because of this, *E. coli* has now been in the center of interest as a platform for the biosynthesis of 5–amino valerate (Rohles et al., 2016). The synthesis of this compound can be carried out by two methods (microbial and enzymatic), where L-lysine is used as the only substrate (Liu et al., 2014).

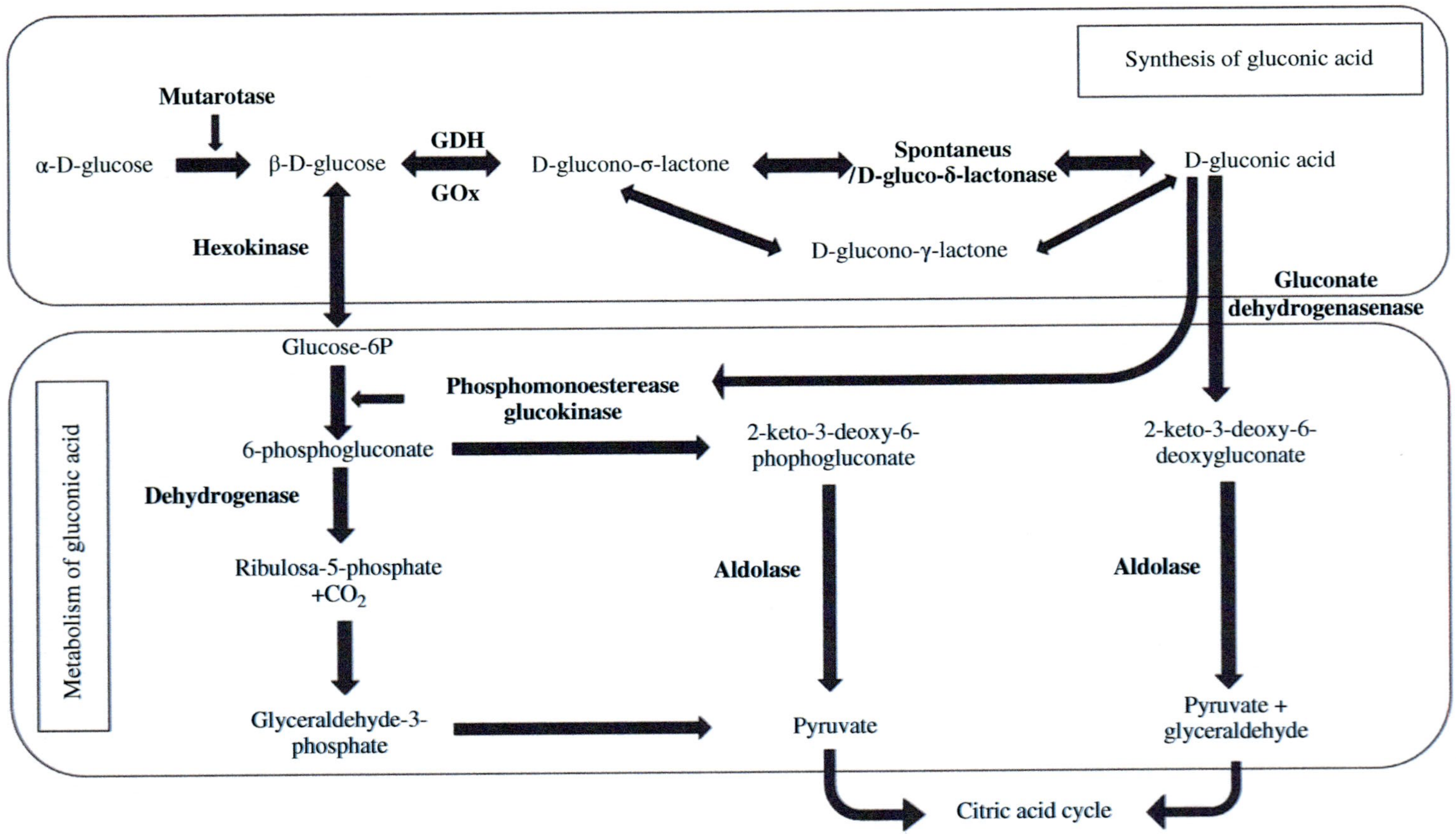

Figure 5.7 Genetically engineered metabolic pathway for the production of gluconic acid in *E. coli.*

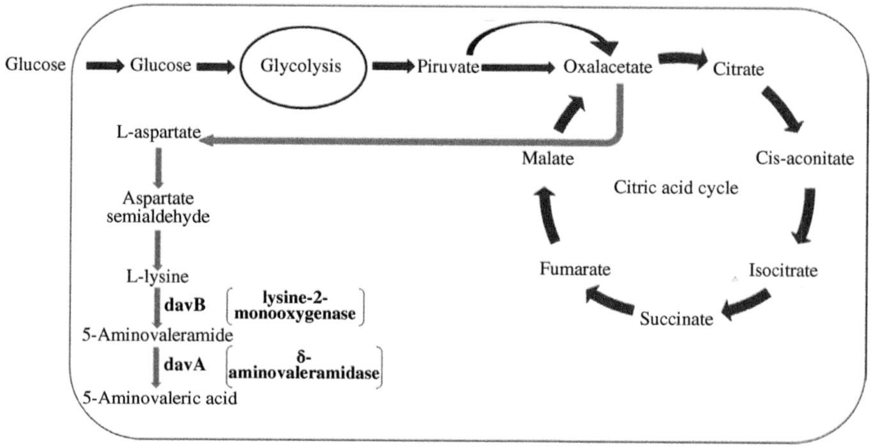

Figure 5.8 Genetically engineered metabolic pathway for the production of 5-aminovalerate in *E. coli*. *Source:* modified by Cho et al., 2015. (*See insert for color representation of this figure.*)

Li et al. (2016) modified the *E. coli* strain with two *P. putida* genes for the production of 5-aminovalerate from L-lysine, yielding 29.6 g/L in 48 h with a yield of 0.96 mol/mol (batch fermentation) and 63.2 g/L in 156 h, with a yield of 0.77 mol/mol (batch fermentation fed). The L-lysine biosynthesis process in this genetically modified strain consists of two stages (Figure 5.8):

1) The oxidation of L-lysine, (catalyzed by the enzyme lysine-2-monooxygenase); and
2) the hydrolysis of the intermediate compound reaction (catalyzed by the enzyme δ-aminovaleramidase).

However, in order to improve biosynthetic efficiency, optimum reaction conditions are essential (e.g. reaction temperature, metal ion additive and cell permeability). However, one of the factors that most affect the production is the presence and absence of oxygen, since under anaerobic conditions the yields are low compared to production in an aerobic system (3.2 g/L in and 7.5 g/L respectively) (Wang et al., 2016).

The production of 5-aminovalerate from L-lysine through a cell biocatalyst offers unique advantages, since the reaction mixture is simple and its production capacity highly efficient and profitable, therefore, this is a promising material for the production of this compound.

5.10 Glutamic Acid

Glutamic acid is commercially the most important amino acids produced (Abdenacer et al., 2012). In a conventional way, this amino acid is produced by microbiological fermentation (Pal et al., 2015), because in the chemical production a racemic mixture is formed (Pulicharla et al., 2016). Glucose is one of the most widely used carbon sources for the production of glutamic acid. Some researchers have used other sources of low-cost carbon, such as datil juice (Abdenacer et al., 2012), cassava starch (Jyothi et al., 2005) and cane bagasse (Nampoothiri and Pandey, 1996).

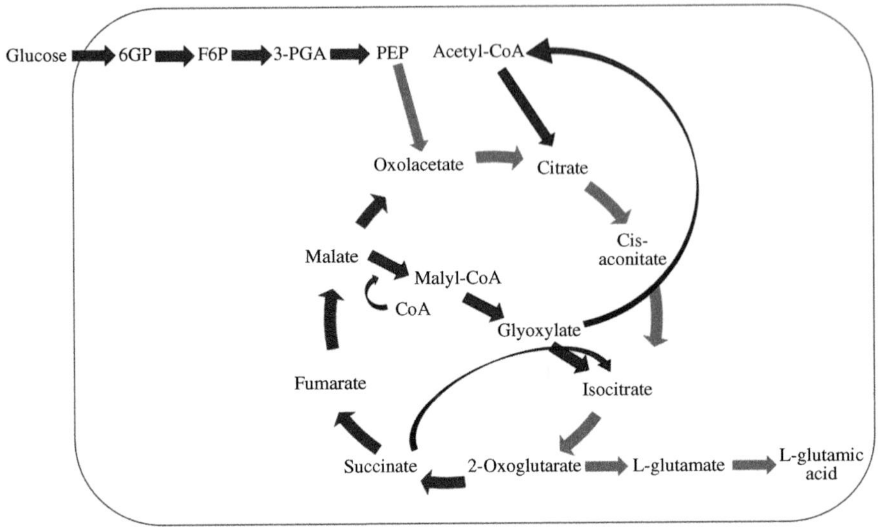

Figure 5.9 Genetically engineered metabolic pathway for the production of glutamic acid in *E. coli*. *Source:* modified from Vuoristo et al., 2016. (*See insert for color representation of this figure.*)

Currently, glutamic acid is commercially produced by fermentation, from native or genetically engineered strains (Kumar et al., 2014). An example of these strains is *Corynebacterium glutamicum*, an actinobacteria widely used for the industrial production of this amino acid (Sawada et al., 2010). Glutamic acid production by *C. glutamicum* can be induced by biotin limitation, the addition of Tween-40 and addition of penicillin (Vertes et al., 2013). For the biosynthesis of this compound, the intermediates of the citric acid cycle serve as precursor metabolites for the production of this amino acid (Figure 5.9) (Wada et al., 2016).

In order to design microorganisms producing amino acids, five strategies must be considered: 1) amplification of enzymes in the metabolic pathway, 2) reduction in the formation of co-products, 3) regulation of enzyme feedback, 4) reduction of flow of the citric acid cycle metabolites (TCA) and 5) the increase of reducers such as NADP + (Hirasawa and Shimizu, 2016).

This amino acid is widely used in human and animal feed, in pharmaceuticals, cosmetics, agrochemicals and other industrial derivatives (Li et al., 2015, Madura et al., 2015, Tahri et al., 2016). The demand for amino acids in the world market is of the magnitude of 106 tons/year, with sales prices of between US$9 and 10/kg (Pal et al., 2016).

5.11 Cadaverine (1,5-diaminopentane)

The cadaverine (1,5-diaminopentane) is a naturally occurring chemical substance of an animal decay (Li et al., 2014). It has application in agriculture (Kind and Wittmann, 2011; Regla-Marquez et al., 2016) and the production of polyamides (Matsushima et al., 2016). The main raw material for its production is lysine

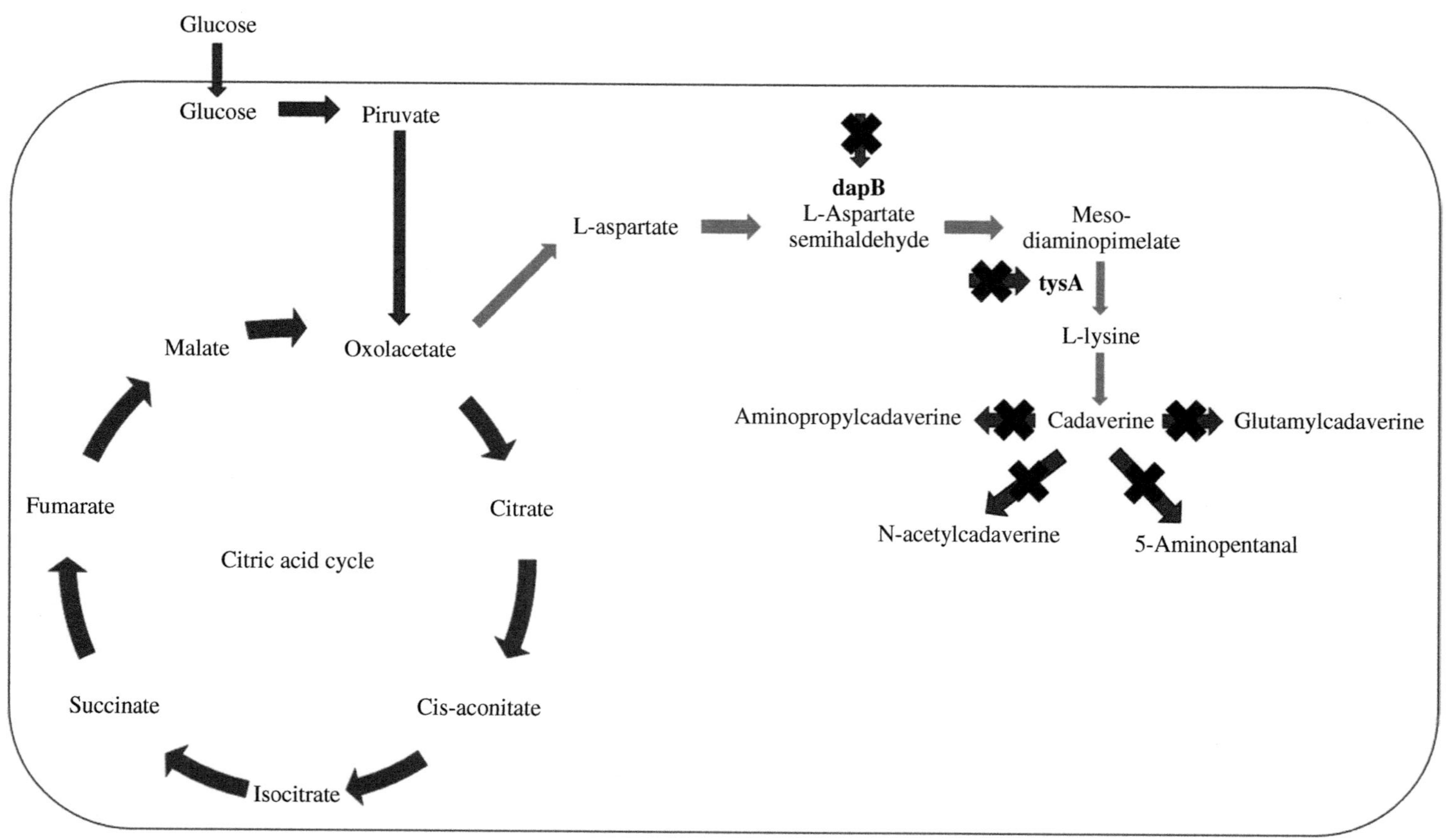

Figure 5.10 Genetically engineered metabolic pathway for the production of cadaverine in *E. coli. Source:* modified from Qian et al., 2011. (*See insert for color representation of this figure.*)

(Jeong et al., 2016), however, other substrates are also used for its production, for example, cellobiose (Ikeda et al., 2013), starch (Tateno et al., 2009), methanol (Leßmeier et al., 2015), among others. It can be produced biotechnologically by microbial fermentation from microorganisms such as *Corynebacterium glutamicum* and genetically modified *E. coli* (Ma et al., 2015), with which yields of 72–88 g/L have been obtained (Volkert et al., 2010; Kind et al., 2014).

The production of cadaverine in *E. coli* is biosynthesized from L-lysine by two L-lysine decarboxylase enzymes (Ikeda et al., 2013), which encode expression or silencing of various genes (cadA and dapA), which are directly involved in increasing production (Figure 5.10) (Chung et al., 2015).

Qian et al. (2011) metabolically modified a strain of *E. coli* that produces 1,5-diaminopentane in a medium with glucose and mineral salts, where they inactivated cadaverine degradation pathways and activated the enzyme L-lysine decarboxylase, which directly converts L-Lysine in cadaverine (amplified the cadA gene under a tac promoter), further increased expression of the dapA gene encoding dihydrodipicolinate synthase through the substitution of the native promoter with the trc promoter, yields of 9.61 g/L using a fed-batch culture was achieved.

5.12 Conclusion

The present review aimed to provide a concise summary of the cellular and biochemical control based on the *E. coli* strain as a biosynthetic platform for the production of various platform chemicals for different industrial applications. The chapter also briefly discussed the current state of the raw materials used to obtain them, the production, marketing and price aspects of these products. Advances in cellular and biochemical control from genetic engineering have markedly improved the yields of these compounds produced from microbial biosynthesis in organisms like *E. coli.*

Acknowledgment

Financial assistance by INRS-ETE has been thankfully acknowledged by K Hegde.

References

Abdenacer, M., Kahina, B.I., Aïcha, N. et al. (2012) Sequential optimization approach for enhanced production of glutamic acid from Corynebacterium glutamicum 2262 using date juice. *Biotechnology and Bioprocess Engineering*, 17, 795–803. doi:10.1007/s12257-011-0486-8

Adkins, J., Jordan, J., and Nielsen, D.R. (2013) Engineering Escherichia coli for Renewable Production of the 5-Carbon Polyamide Building-Blocks 5-Aminovalerate and Glutarate. *Biotechnology and Bioengineering*, 110, (6) 1726–1734.

Ahmed, A.S., Farag, S.S., Hassan, I.A., and Botros, H.W. (2015) Production of gluconic acid by using some irradiated microorganisms. *Journal of Radiation Research and Applied Sciences*, 8, 374–380. doi:10.1016/j.jrras.2015.02.006

Ahuja, M., Kaidery, N.A., Yang, L. et al. (2016) Distinct Nrf2 Signaling Mechanisms of Fumaric Acid Esters and Their Role in Neuroprotection against 1-Methyl-4-Phenyl-1, 2, 3, 6-Tetrahydropyridine-Induced Experimental Parkinson's-Like Disease. *The Journal of Neuroscience*, 36, 6332–6351.

Ajala, E.O., Ajala, M.A., Ogunniyi, D.S., and Sunmonu, M.O. (2016) Kinetics of gluconic acid production and cell growth in a batch bioreactor by *Aspergillus niger* using breadfruit hydrolysate: Kinetic of Gluconic Acid Production and Cell Growth. *Journal of Food Process Engineering.* doi:10.1111/jfpe.12461

Alonso, S., Rendueles, M., and Díaz, M. (2015) Simultaneous production of lactobionic and gluconic acid in cheese whey/glucose co-fermentation by Pseudomonas taetrolens. *Bioresource Technology*, 196, 314–323. doi:10.1016/j.biortech.2015.07.092

An, K., Hu, F., and Bao, J. (2013) Simultaneous Saccharification of Inulin and Starch Using Commercial Glucoamylase and the Subsequent Bioconversion to High Titer Sorbitol and Gluconic Acid. *Applied Biochemistry and Biotechnology*, 171, 2093–2104. doi:10.1007/s12010-013-0278-z

Bhandiwad, A., Shaw, A.J., Guss, A. et al. (2014) Metabolic engineering of Thermoanaerobacterium saccharolyticum for n-butanol production. *Metabolic Engineering*, 21, 17–25. doi:10.1016/j.ymben.2013.10.012

Bhatia, S.K., Bhatia, R.K., and Yang, Y.-H. (2016) Biosynthesis of polyesters and polyamide building blocks using microbial fermentation and biotransformation. *Reviews in Environmental Science and Bio/Technology*, 15, 639–663. doi:10.1007/s11157-016-9415-9

Bond-Watts, B.B., Bellerose, R.J., and Chang, M.C.Y. (2011) Enzyme mechanism as a kinetic control element for designing synthetic biofuel pathways. *Nature Chemical Biology*, 7, 222–227. doi:10.1038/nchembio.537

Bozell, J.J. and Petersen, G.R. (2010) Technology development for the production of biobased products from biorefinery carbohydrates—the US Department of Energy's "Top 10" revisited. *Green Chemistry*, 12, 539. doi:10.1039/b922014c

Branduardi, P. and Porro, D. (2016) n-butanol: challenges and solutions for shifting natural metabolic pathways into a viable microbial production. *FEMS, In press.*

Cañete-Rodríguez, A.M., Santos-Dueñas, I.M., Jiménez-Hornero, J.E. et al. (2016) Gluconic acid: Properties, production methods and applications—An excellent opportunity for agro-industrial by-products and waste bio-valorization. *Process Biochemistry*, 51, 1891–1903. doi:10.1016/j.procbio.2016.08.028

Chen, C.-T. and Liao, J.C. (2016) Frontiers in microbial 1-butanol and isobutanol production. *FEMS Microbiology Letters*, 363, fnw020. doi:10.1093/femsle/fnw020

Chen, Y., Zhou, T., Liu, D. et al. (2013) Production of butanol from glucose and xylose with immobilized cells of Clostridium acetobutylicum. *Biotechnology and Bioprocess Engineering*, 18, 234–241. doi:10.1007/s12257-012-0573-5

Chenyu Du, A.A. (2014) Fermentative Itaconic Acid Production. *Journal of Biodiversity, Bioprospecting and Development*, 01. doi:10.4172/2376-0214.1000119

Chi, Z., Wang, Z.-P., Wang, G.-Y. et al. (2016) Microbial biosynthesis and secretion of L -malic acid and its applications. *Critical Reviews in Biotechnology*, 36, 99–107. doi:10.3109/07388551.2014.924474

Chin, T., Sano, M., Takahashi, T. et al. (2015) Photosynthetic production of itaconic acid in Synechocystis sp. PCC6803. *Journal of Biotechnology*, 195, 43–45. doi:10.1016/j.jbiotec.2014.12.016

Cho, C., Choi, S.Y., Luo, Z.W., and Lee, S.Y. (2015) Recent advances in microbial production of fuels and chemicals using tools and strategies of systems metabolic engineering. *Biotechnology Advances*, 33, 1455–1466. doi:10.1016/j.biotechadv.2014.11.006

Choi, S., Song, C.W., Shin, J.H., and Lee, S.Y. (2015) Biorefineries for the production of top building block chemicals and their derivatives. *Metabolic Engineering*, 28, 223–239. doi:10.1016/j.ymben.2014.12.007

Chu, H.S., Kim, Y.S., Lee, C.M.et al. (2015) Metabolic engineering of 3-hydroxypropionic acid biosynthesis in Escherichia coli. *Biotechnology and Bioengineering*, 112, 356–364.

Chung, H., Yang, J.E., Ha, J.Y. et al. (2015) Bio-based production of monomers and polymers by metabolically engineered microorganisms. *Current Opinion in Biotechnology*, 36, 73–84. doi:10.1016/j.copbio.2015.07.003

Cui, C., Chen, H., Chen, B., and Tan, T. (2016) Genipin Cross-Linked Glucose Oxidase and Catalase Multi-enzyme for Gluconic Acid Synthesis. *Applied Biochemistry and Biotechnology.* doi:10.1007/s12010-016-2228-z

Das, R.K., Brar, S.K., and Verma, M. (2016a) Potential use of pulp and paper solid waste for the bio-production of fumaric acid through submerged and solid state fermentation. *Journal of Cleaner Production*, 112, 4435–4444. doi:10.1016/j.jclepro.2015.08.108

Das, R.K., Brar, S.K., and Verma, M. (2016b) Recent advances in the biomedical applications of fumaric acid and its ester derivatives: The multifaceted alternative therapeutics. *Pharmacological Reports*, 68, 404–414. doi:10.1016/j.pharep.2015.10.007

Das, R.K. and Brar, S.K. (2014) Enhanced Fumaric Acid Production from Brewery Wastewater and Insight into the Morphology of Rhizopus oryzae 1526. *Applied Biochemistry and Biotechnology*, 172, 2974–2988. doi:10.1007/s12010-014-0739-z

Das, R.K., Brar, S.K., and Verma, M., 2016. Fumaric Acid, In: *Platform Chemical Biorefinery*. Elsevier, pp. 133–157.

David, Y., Oh, Y.H., Baylon, M.G. et al. (2016) Microbial Production of 3-Hydroxypropionic Acid. *Industrial Biotechnology: Products and Processes.*

Day, D.F., Dequeiroz, G., Chung, C.H., and Kim, M. (2007) By-products from bagasse. Proc. ISSCT Cong. XXVI, Durban, South Africa 29.

de Jong, E., Higson, A., Walsh, P., and Wellisch, M. (2012) Bio-based chemicals value added products from biorefineries. IEA Bioenergy, Task42 Biorefinery.

Demain, A.L. and Martens, E. (2016) Production of valuable compounds by molds and yeasts. *The Journal of Antibiotics.* doi:10.1038/ja.2016.121

Ding, Y., Li, S., Dou, C. et al. (2011) Production of Fumaric Acid by Rhizopus oryzae: Role of Carbon–Nitrogen Ratio. *Applied Biochemistry and Biotechnology*, 164, 1461–1467. doi:10.1007/s12010-011-9226-y

Dong, X., Chen, X., Qian, Y. (2016) Metabolic engineering of *Escherichia coli* W3110 to produce L-malate: Metabolic Engineering of *Escherichia coli* W3110. *Biotechnology and Bioengineering.* doi:10.1002/bit.26190

Dürre, P. (2016) Butanol formation from gaseous substrates. *FEMS Microbiology Letters*, 363, fnw040. doi:10.1093/femsle/fnw040

Dutta, K., Mahanty, B., Daverey, A. et al. (2016) Biorefinery and Possible Negative Impacts on the Food Market. In: *Platform Chemical Biorefinery*. Elsevier, pp. 323–331.

El-Imam, A.M.A., Kazeem, M.O., Odebisi, M.B. et al. (2013) Production of Itaconic Acid from Jatropha curcas Seed Cake by Aspergillus terreus. *Notulae Scientia Biologicae*, 5, 57.

Felthouse, T.R., Burnett, J.C., Horrell, B. et al. (2001) Maleic anhydride, maleic acid, and fumaric acid. *Kirk-Othmer Encyclopedia of Chemical Technology.*

Fernández-Naveira, Á., Abubackar, H.N., Veiga, M.C., and Kennes, C. (2016) Efficient butanol-ethanol (B-E) production from carbon monoxide fermentation by Clostridium carboxidivorans. *Applied Microbiology and Biotechnology*, 100, 3361–3370. doi:10.1007/s00253-015-7238-1

Gooch, J.W. (Ed) (2011) *Encyclopedic Dictionary of Polymers*. Springer New York, New York, NY.

Gopinath, K.P., Sankaranarayanan, A.R., and Nivedhitha, L. (2016) Platform Chemical Biorefinery and Agroindustrial Waste Management. In: *Platform Chemical Biorefinery*. Elsevier, pp. 379–391.

Green, E.M. (2011) Fermentative production of butanol—the industrial perspective. *Current Opinion in Biotechnology*, 22, 337–343. doi:10.1016/j.copbio.2011.02.004

Gupta, A., Hicks, M.A., Manchester, S.P., and Prather, K.L.J. (2016) Porting the synthetic D-glucaric acid pathway from *Escherichia coli* to *Saccharomyces cerevisiae*. *Biotechnology Journal*, 11, 1201–1208. doi:10.1002/biot.201500563

Harder, B.-J., Bettenbrock, K., and Klamt, S. (2016) Model-based metabolic engineering enables high yield itaconic acid production by Escherichia coli. *Metabolic Engineering*, 38, 29–37. doi:10.1016/j.ymben.2016.05.008

Hegde, K., Prabhu, A., Sarma, S.J. (2016) Potential application of renewable itaconic acid for the synthesis of 3-methyltetrahydrofuran. In: Brar, S.K., Sarma, S.J. and Pakshirajan, K. (Eds) *Platform chemical biorefinery-future green industry*. Elsevier, Netherlands. pp. 181–200.

Hirasawa, T. and Shimizu, H. (2016) Recent advances in amino acid production by microbial cells. *Current Opinion in Biotechnology*, 42, 133–146. doi:10.1016/j.copbio.2016.04.017

Huang, L., Xiang, Y., Cai, J. et al. (2011) Effects of three main sugars in cane molasses on the production of butyric acid with Clostridium tyrobutyricum. *Korean J. Chem. Eng.*, 28 (12), 2312–2315.

Huang, X., Lu, X., Li, Y. et al. (2014) Improving itaconic acid production through genetic engineering of an industrial. *Microbial Cell Factories*,13:119. https://doi.org/10.1186/s12934-014-0119-y

Ikeda, N., Miyamoto, M., Adachi, N. et al. (2013) Direct cadaverine production from cellobiose using β-glucosidase displaying Escherichia coli. *AMB Express*, 3, 1.

Ikeda, Y., Park, E., and Okuda, N. (2006) Bioconversion of waste office paper to gluconic acid in a turbine blade reactor by the filamentous fungus Aspergillus niger. *Bioresource Technology*, 97, 1030–1035. doi:10.1016/j.biortech.2005.04.040

Jensen, N.B., Borodina, I., Chen, Y. et al. (2016) *Microbial production of 3-Hydroxypropionic acid*. Google Patents.

Jeon, H.-G., Cheong, D.-E., Han, Y. et al. (2016) Itaconic acid production from glycerol using Escherichia coli harboring a random synonymous codon-substituted 5′-coding region variant of the cadA gene. *Biotechnology and Bioengineering.*

Jeong, S., Yeon, Y.J., Choi, E.-G. et al. (2016) Alkaliphilic lysine decarboxylases for effective synthesis of cadaverine from L-lysine. *Korean Journal of Chemical Engineering*, 33, 1530–1533. doi:10.1007/s11814-016-0079-5

Jiang, L., Wang, J., Liang, S. (2011) Enhanced butyric acid tolerance and bioproduction by Clostridium tyrobutyricum immobilized in a fibrous bed bioreactor. *Biotechnology and Bioengineering*, 108, 31–40. doi:10.1002/bit.22927

Jimenez-Quero, A., Pollet, E., Zhao, M. (2016) Itaconic and Fumaric Acid Production from Biomass Hydrolysates by Aspergillus Strains. *Journal of Microbiology and Biotechnology*, 26, 1557–1565. doi:10.4014/jmb.1603.03073

Jo, J.-E., Mohan Raj, S., Rathnasingh, C. et al. (2008) Cloning, expression, and characterization of an aldehyde dehydrogenase from Escherichia coli K-12 that utilizes 3-Hydroxypropionaldehyde as a substrate. *Applied Microbiology and Biotechnology*, 81, 51–60. doi:10.1007/s00253-008-1608-x

Jones, D.T. and Woods, D.R. (1986) Acetone-butanol fermentation revisited. *Microbiological Reviews*, 50, 484.

Jung, W.S., Kang, J.H., Chu, H.S. et al. (2014) Elevated production of 3-hydroxypropionic acid by metabolic engineering of the glycerol metabolism in Escherichia coli. *Metabolic Engineering*, 23, 116–122. doi:10.1016/j.ymben.2014.03.001

Jyothi, A.N., Sasikiran, K., Nambisan, B., and Balagopalan, C. (2005) Optimisation of glutamic acid production from cassava starch factory residues using Brevibacterium divaricatum. *Process Biochemistry*, 40, 3576–3579. doi:10.1016/j.procbio.2005.03.046

Katsuya Tooyama, T.M. (2013) Simultaneous Saccharification of Corn Starch in Gluconic Acid Production by Aspergillus niger Immobilized on Nonwoven Fabric in a Pressurized Reactor. *Journal of Microbial & Biochemical Technology*, 05. doi:10.4172/1948-5948.1000106

Khan, I., Ullah, S., and Oh, D.-H. (2016) Chitosan grafted monomethyl fumaric acid as a potential food preservative. *Carbohydrate Polymers*, 152, 87–96. doi:10.1016/j.carbpol.2016.06.073

Kind, S., Neubauer, S., Becker, J. (2014) From zero to hero – Production of bio-based nylon from renewable resources using engineered Corynebacterium glutamicum. *Metabolic Engineering*, 25, 113–123. doi:10.1016/j.ymben.2014.05.007

Kind, S. and Wittmann, C. (2011) Bio-based production of the platform chemical 1,5-diaminopentane. *Applied Microbiology and Biotechnology*, 91, 1287–1296. doi:10.1007/s00253-011-3457-2

Klement, T. and Büchs, J. (2013) Itaconic acid – A biotechnological process in change. *Bioresource Technology*, 135, 422–431. doi:10.1016/j.biortech.2012.11.141

Kumar, M.B.A., Gao, Y., Shen, W., and He, L. (2015) Valorisation of protein waste: An enzymatic approach to make commodity chemicals. *Frontiers of Chemical Science and Engineering*, 9, 295–307. doi:10.1007/s11705-015-1532-4

Kumar, R., Vikramachakravarthi, D., and Pal, P. (2014) Production and purification of glutamic acid: A critical review towards process intensification. *Chemical Engineering and Processing: Process Intensification*, 81, 59–71. doi:10.1016/j.cep.2014.04.012

Kumar, V., Ashok, S., and Park, S. (2013) Recent advances in biological production of 3-hydroxypropionic acid. *Biotechnology Advances*, 31, 945–961. doi:10.1016/j.biotechadv.2013.02.008

Kwak, S., Park, Y.-C., and Seo, J.-H. (2013) Biosynthesis of 3-hydroxypropionic acid from glycerol in recombinant Escherichia coli expressing Lactobacillus brevis dhaB and dhaR gene clusters and E. coli K-12 aldH. *Bioresource Technology*, 135, 432–439. doi:10.1016/j.biortech.2012.11.063

Lan, E.I., Ro, S.Y., and Liao, J.C. (2013) Oxygen-tolerant coenzyme A-acylating aldehyde dehydrogenase facilitates efficient photosynthetic n-butanol biosynthesis in cyanobacteria. *Energy and Environmental Science*, 6, 2672. doi:10.1039/c3ee41405a

Lee, C.C., Kibblewhite, R.E., Paavola, C.D. et al. (2016) Production of Glucaric Acid from Hemicellulose Substrate by Rosettasome Enzyme Assemblies. *Molecular Biotechnology*, 58, 489–496. doi:10.1007/s12033-016-9945-y

Leßmeier, L., Pfeifenschneider, J., Carnicer, M. et al. (2015) Production of carbon-13-labeled cadaverine by engineered Corynebacterium glutamicum using carbon-13-labeled methanol as co-substrate. *Applied Microbiology and Biotechnology*, 99, 10163–10176. doi:10.1007/s00253-015-6906-5

Li, A., van Luijk, N., ter Beek, M. et al. (2011) A clone-based transcriptomics approach for the identification of genes relevant for itaconic acid production in Aspergillus. *Fungal Genetics and Biology*, 48, 602–611. doi:10.1016/j.fgb.2011.01.013

Li, M., Li, D., Huang, Y. et al. (2014) Improving the secretion of cadaverine in Corynebacterium glutamicum by cadaverine–lysine antiporter. *Journal of Industrial Microbiology and Biotechnology*, 41, 701–709. doi:10.1007/s10295-014-1409-4

Li, X., Liu, Y., Yang, Y. et al. (2014) High levels of malic acid production by the bioconversion of corn straw hydrolyte using an isolated Rhizopus delemar strain. *Biotechnology and Bioprocess Engineering*, 19, 478–492. doi:10.1007/s12257-014-0047-z

Li, X., Zhou, J., Ouyang, S. et al. (2016) Fumaric Acid Production from Alkali-Pretreated Corncob by Fed-Batch Simultaneous Saccharification and Fermentation Combined with Separated Hydrolysis and Fermentation at High Solids Loading. *Applied Biochemistry and Biotechnology.* doi:10.1007/s12010-016-2232-3

Li, Z., Xu, J., Jiang, T. et al. (2016) Overexpression of transport proteins improves the production of 5-aminovalerate from l-lysine in Escherichia coli. *Scientific Reports*, 6, 30884. doi:10.1038/srep30884

Li, Z., Zhang, C., and Xu, Q. (2015) Study on Green Crystallization Process for L-glutamic Acid Production. In: Zhang, T.-C., Nakajima, M. (Eds), *Advances in Applied Biotechnology*. Springer Berlin Heidelberg, Berlin, Heidelberg, pp. 547–555.

Lim, H.G., Noh, M.H., Jeong, J.H. et al. (2016) Optimum Rebalancing of the 3-Hydroxypropionic Acid Production Pathway from Glycerol in *Escherichia coli*. *ACS Synthetic Biology*. doi:10.1021/acssynbio.5b00303

Liu, H., Ma, J., Wang, M. et al. (2016) Food Waste Fermentation to Fumaric Acid by Rhizopus arrhizus RH7-13. *Applied Biochemistry and Biotechnology*, 180, 1524–1533. doi:10.1007/s12010-016-2184-7

Liu, P., Zhang, H., Lv, M. et al. (2014) Enzymatic production of 5-aminovalerate from l-lysine using l-lysine monooxygenase and 5-aminovaleramide amidohydrolase. *Scientific Reports* 4. doi:10.1038/srep05657

Liu, P., Zhu, X., Tan, Z. et al. (2015) Construction of Escherichia Coli Cell Factories for Production of Organic Acids and Alcohols. In: Ye, Q., Bao, J., and Zhong, J.-J. (Eds), *Bioreactor Engineering Research and Industrial Applications I*. Springer Berlin Heidelberg, Berlin, Heidelberg, pp. 107–140.

Liu, Y., Gong, X., Wang, C. et al. (2016) Production of glucaric acid from myo-inositol in engineered Pichia pastoris. *Enzyme and Microbial Technology*, 91, 8–16. doi:10.1016/j.enzmictec.2016.05.009

Luo, L.H., Seo, J.-W., Baek, J.-O. et al. (2011) Identification and characterization of the propanediol utilization protein PduP of Lactobacillus reuteri for 3-hydroxypropionic acid production from glycerol. *Applied Microbiology and Biotechnology*, 89, 697–703. doi:10.1007/s00253-010-2887-6

Ma, W., Cao, W., Zhang, H. et al. (2015) Enhanced cadaverine production from l-lysine using recombinant Escherichia coli co-overexpressing CadA and CadB. *Biotechnology Letters*, 37, 799–806. doi:10.1007/s10529-014-1753-5

Maghsoodlou, M.T., Karima, M., Lashkari, M. et al. (2016) Convenient approach for the one-pot, three-component synthesis of 1-(benzothiazolylamino) methyl-2-naphthol using fumaric acid as a green catalyst. *Bulgarian Chemical Communications*, 48, 369–372.

Mansur, A.R., Tango, C.N., Kim, G.-H., and Oh, D.-H. (2015) Combined effects of slightly acidic electrolyzed water and fumaric acid on the reduction of foodborne pathogens and shelf life extension of fresh pork. *Food Control*, 47, 277–284. doi:10.1016/j.foodcont.2014.07.019

Mao, Y.-M. (2016) Preparation of Gluconic Acid by Oxidation of Glucose with Hydrogen Peroxide: Preparation of Gluconic Acid. *Journal of Food Processing and Preservation*. doi:10.1111/jfpp.12742

Matsakas, L., Topakas, E., and Christakopoulos, P. (2014) New trends in microbial production of 3-hydroxypropionic acid. *Current Biochemical Engineering* 1, 141–154.

Matsushima, Y., Hirasawa, T., and Shimizu, H. (2016) Enhancement of 1,5-diaminopentane production in a recombinant strain of *Corynebacterium glutamicum* by Tween 40 addition. *The Journal of General and Applied Microbiology*, 62, 42–45. doi:10.2323/jgam.62.42

Moon, T.S., Yoon, S.-H., Lanza, A.M. et al. (2009) Production of Glucaric Acid from a Synthetic Pathway in Recombinant Escherichia coli. *Applied and Environmental Microbiology*, 75, 589–595. doi:10.1128/AEM.00973-08

Mounir, M., Shafiei, R., Zarmehrkhorshid, R. et al. (2016) Simultaneous production of acetic and gluconic acids by a thermotolerant Acetobacter strain during acetous fermentation in a bioreactor. *Journal of Bioscience and Bioengineering*, 121, 166–171. doi:10.1016/j.jbiosc.2015.06.005

Mukhopadhyay, R., Chatterjee, S., Chatterjee, B.P. et al. (2005) Production of gluconic acid from whey by free and immobilized Aspergillus niger. *International Dairy Journal*, 15, 299–303. doi:10.1016/j.idairyj.2004.07.010

Nagy, E. and Hegedüs, I. (2015) Second Generation Biofuels and Biorefinery Concepts Focusing on Central Europe. *Chemical Engineering*, 45.

Nakayama, S., Tabata, K., Oba, T. et al. (2012) Characteristics of the high malic acid production mechanism in Saccharomyces cerevisiae sake yeast strain No. 28. *Journal of Bioscience and Bioengineering*, 114, 281–285. doi:10.1016/j.jbiosc.2012.04.003

Nampoothiri, K.M. and Pandey, A. (1996) Solid state fermentation for L-glutamic acid production using Brevibacterium sp. *Biotechnology Letters*, 18, 199–204.

Niu, K., Xiong, T., Qin, H.-B. et al. (2016) 3-Hydroxypropionic acid production by recombinant *Escherichia coli* ZJU-3HP01 using glycerol-glucose dual-substrate fermentative strategy. *Biotechnology and Applied Biochemistry*. doi:10.1002/bab.1505

Okamoto, S., Chin, T., Nagata, K. et al. (2015) Production of itaconic acid in Escherichia coli expressing recombinant α-amylase using starch as substrate. *Journal of Bioscience and Bioengineering*, 119, 548–553. doi:10.1016/j.jbiosc.2014.10.021

Pachapur, V.L., Sarma, S.J., Brar, S.K., and Chaabouni, E. (2016) Platform Chemicals. In: *Platform Chemical Biorefinery*. Elsevier, pp. 1–20.

Pal, P., Dekonda, V.C., and Kumar, R. (2015) Fermentative production of glutamic acid from renewable carbon source: Process intensification through membrane-integrated hybrid bio-reactor system. *Chemical Engineering and Processing: Process Intensification*, 92, 7–17. doi:10.1016/j.cep.2015.03.022

Pal, P., Kumar, R., and Banerjee, S. (2016a) Manufacture of gluconic acid: A review towards process intensification for green production. *Chemical Engineering and Processing: Process Intensification*, 104, 160–171. doi:10.1016/j.cep.2016.03.009

Pal, P., Kumar, R., VikramaChakravarthi, D., and Chakrabortty, S. (2016b) Modeling and simulation of continuous production of L (+) glutamic acid in a membrane-integrated bioreactor. *Biochemical Engineering Journal*, 106, 68–86. doi:10.1016/j.bej.2015.11.008

Pan, X., Liu, H., Liu, J. et al. (2016) Omics-based approaches reveal phospholipids remodeling of Rhizopus oryzae responding to furfural stress for fumaric acid-production from xylose. *Bioresource Technology*, 222, 24–32. doi:10.1016/j.biortech.2016.09.101

Pang, Z.-W., Lu, W., Zhang, H. et al. (2016) Butanol production employing fed-batch fermentation by Clostridium acetobutylicum GX01 using alkali-pretreated sugarcane bagasse hydrolysed by enzymes from Thermoascus aurantiacus QS 7-2-4. *Bioresource Technology*, 212, 82–91. doi:10.1016/j.biortech.2016.04.013

Park, E.-H., Shin, Y.-M., Lim, Y.-Y. et al. (2000) Expression of glucose oxidase by using recombinant yeast. *Journal of Biotechnology*, 81, 35–44.

Park, S.J., Oh, Y.H., Noh, W. et al. (2014) High-level conversion of L-ysine into 5-aminovalerate that can be used for nylon 6,5 synthesis. *Biotechnology Journal*, 9, 1322–1328. doi:10.1002/biot.201400156

Pedroso, G.B., Montipó, S., Mario, D.A.N. et al. (2016) Building block itaconic acid from left-over biomass. *Biomass Conversion and Biorefinery.* doi:10.1007/s13399-016-0210-1

Pérez-Bibbins, B., Salgado, J.M., Torrado, A. et al. (2013) Culture parameters affecting xylitol production by Debaryomyces hansenii immobilized in alginate beads. *Process Biochemistry*, 48, 387–397. doi:10.1016/j.procbio.2013.01.006

Pulicharla, R., Lonappan, L., Brar, S.K., and Verma, M. (2016) Production of Renewable C5 Platform Chemicals and Potential Applications. In: *Platform Chemical Biorefinery*. Elsevier, pp. 201–216.

Qian, Z.-G., Xia, X.-X., and Lee, S.Y. (2011) Metabolic engineering of Escherichia coli for the production of cadaverine: A five carbon diamine. *Biotechnology and Bioengineering*, 108, 93–103. doi:10.1002/bit.22918

Qureshi, N., Cotta, M.A., and Saha, B.C. (2014) Bioconversion of barley straw and corn stover to butanol (a biofuel) in integrated fermentation and simultaneous product recovery bioreactors. *Food and Bioproducts Processing*, 92, 298–308. doi:10.1016/j.fbp.2013.11.005

Ramachandran, S., Nair, S., Larroche, C., and Pandey, A. (2017) Gluconic Acid. In: *Current Developments in Biotechnology and Bioengineering*. Elsevier, pp. 577–599.

Ramos, J.L., Udaondo, Z., Fernández, B. et al. (2016) First- and second-generation biochemicals from sugars: biosynthesis of itaconic acid: Biochemicals through fermentation. *Microbial Biotechnology*, 9, 8–10. doi:10.1111/1751-7915.12333

Rathnasingh, C., Raj, S.M., Jo, J.-E., and Park, S. (2009) Development and evaluation of efficient recombinant *Escherichia coli* strains for the production of 3-hydroxypropionic acid from glycerol. *Biotechnology and Bioengineering* n/a-n/a. doi:10.1002/bit.22429

Rathnasingh, C., Raj, S.M., Lee, Y. et al. (2012) Production of 3-hydroxypropionic acid via malonyl-CoA pathway using recombinant Escherichia coli strains. *Journal of Biotechnology*, 157, 633–640. doi:10.1016/j.jbiotec.2011.06.008

Regla-Márquez, C.F., Canto-Flick, A., Avilés-Viñas, S.A. et al. (2016) Cadaverine: a common polyamine in zygotic embryos and somatic embryos of the species Capsicum chinense Jacq. *Plant Cell, Tissue and Organ Culture* (PCTOC), 124, 253–264. doi:10.1007/s11240-015-0889-x

Reizman, I.M.B., Stenger, A.R., Reisch, C.R. (2015) Improvement of glucaric acid production in E. coli via dynamic control of metabolic fluxes. *Metabolic Engineering Communications*, 2, 109–116. doi:10.1016/j.meteno.2015.09.002

Rhodes, P.M. and Stanbury, P.F. (Eds), 1997. *Applied microbial physiology: a practical approach*, The practical approach series. IRL Press at Oxford University Press, Oxford; New York.

Roa Engel, C.A., Straathof, A.J.J., Zijlmans, T.W. et al. (2008) Fumaric acid production by fermentation. *Applied Microbiology and Biotechnology*, 78, 379–389. doi:10.1007/s00253-007-1341-x

Rogers, P., Chen, J.-S., and Zidwick, M.J. (2013) Organic Acid and Solvent Production: Acetic, Lactic, Gluconic, Succinic, and Polyhydroxyalkanoic Acids.

In: Rosenberg, E., DeLong, E.F., Lory, S. et al. (Eds), *The Prokaryotes*. Springer Berlin Heidelberg, Berlin, Heidelberg, pp. 3–75.

Rohles, C.M., Gießelmann, G., Kohlstedt, M. et al. (2016) Systems metabolic engineering of Corynebacterium glutamicum for the production of the carbon-5 platform chemicals 5-aminovalerate and glutarate. *Microbial Cell Factories*, 15. doi:10.1186/s12934-016-0553-0

Saha, B.C. (2016) Emerging biotechnologies for production of itaconic acid and its applications as a platform chemical. *Journal of Industrial Microbiology and Biotechnology.* doi:10.1007/s10295-016-1878-8

Saini, M., Li, S.-Y., Wang, Z.W. et al. (2016) Systematic engineering of the central metabolism in Escherichia coli for effective production of n-butanol. *Biotechnology for Biofuels*, 9. doi:10.1186/s13068-016-0467-4

Sainz, F., Navarro, D., Mateo, E. et al. (2016) Comparison of d-gluconic acid production in selected strains of acetic acid bacteria. *International Journal of Food Microbiology*, 222, 40–47. doi:10.1016/j.ijfoodmicro.2016.01.015

Sawada, K., Zen-in, S., Wada, M., and Yokota, A. (2010) Metabolic changes in a pyruvate kinase gene deletion mutant of Corynebacterium glutamicum ATCC 13032. *Metabolic Engineering*, 12, 401–407. doi:10.1016/j.ymben.2010.01.004

Schadeweg, V. and Boles, E. (2016) Increasing n-butanol production with Saccharomyces cerevisiae by optimizing acetyl-CoA synthesis, NADH levels and trans-2-enoyl-CoA reductase expression. *Biotechnology for Biofuels*, 9. doi:10.1186/s13068-016-0673-0

Sharma, A., Vivekanand, V., and Singh, R.P. (2008) Solid-state fermentation for gluconic acid production from sugarcane molasses by Aspergillus niger ARNU-4 employing tea waste as the novel solid support. *Bioresource Technology*, 99, 3444–3450. doi:10.1016/j.biortech.2007.08.006

Sheldon, R.A. (2014) Green and sustainable manufacture of chemicals from biomass: state of the art. *Green Chem.*, 16, 950–963. doi:10.1039/C3GC41935E

Shin, J.H., Park, S.H., Oh, Y.H. et al. (2016) Metabolic engineering of Corynebacterium glutamicum for enhanced production of 5-aminovaleric acid. *Microbial Cell Factories*, 15. doi:10.1186/s12934-016-0566-8

Shiue, E. and Prather, K.L.J. (2014) Improving d-glucaric acid production from myo-inositol in E. coli by increasing MIOX stability and myo-inositol transport. *Metabolic Engineering*, 22, 22–31. doi:10.1016/j.ymben.2013.12.002

Si, H.-M., Zhang, F., Wu, A.-N. et al. (2016) DNA microarray of global transcription factor mutant reveals membrane-related proteins involved in n-butanol tolerance in Escherichia coli. *Biotechnology for Biofuels*, 9. doi:10.1186/s13068-016-0527-9

Singh, O.V. and Kumar, R. (2007) Biotechnological production of gluconic acid: future implications. *Applied Microbiology and Biotechnology*, 75, 713–722. doi:10.1007/s00253-007-0851-x

Song, C.W., Kim, D.I., Choi, S. et al. (2013) Metabolic engineering of Escherichia coli for the production of fumaric acid. *Biotechnology and Bioengineering*, Vol. 110, (7), 2025–2034.

Steiger, M.G., Blumhoff, M.L., Mattanovich, D., and Sauer, M. (2013) Biochemistry of microbial itaconic acid production. *Frontiers in Microbiology*, 4. doi:10.3389/fmicb.2013.00023

Stojkovič, G., Plazl, I., and Žnidaršič-Plazl, P. (2011) l-Malic acid production within a microreactor with surface immobilised fumarase. *Microfluidics and Nanofluidics*, 10, 627–635. doi:10.1007/s10404-010-0696-y

Sun, C., Pang, Y., and Chen, R. (2015) Supramolecular assembly by co-crystallization of fumaric acid with 1H-benzotriazole: crystal structure and synthon cooperation. DOI: 10.2991/icmmcce-15.2015.167 Conference: Conference: 2015 4th International Conference on Mechatronics, Materials, Chemistry and Computer Engineering

Sun, W.-J., Zhou, Y.-Z., Zhou, Q. et al. (2012) Semi-continuous production of 2-keto-gluconic acid by Pseudomonas fluorescens AR4 from rice starch hydrolysate. *Bioresource Technology*, 110, 546–551. doi:10.1016/j.biortech.2012.01.040

Tahri, Y., Gagnière, E., Chabanon, E. et al. (2016) Investigation of the l-Glutamic acid polymorphism: Comparison between stirred and stagnant conditions. *Journal of Crystal Growth*, 435, 98–104. doi:10.1016/j.jcrysgro.2015.11.019

Tateno, T., Okada, Y., Tsuchidate, T. et al. (2009) Direct production of cadaverine from soluble starch using Corynebacterium glutamicum coexpressing α-amylase and lysine decarboxylase. *Applied Microbiology and Biotechnology*, 82, 115–121. doi:10.1007/s00253-008-1751-4

Tingirikari, J.M.R., Ahmed, S., and Yata, V.K. (2016) 3-Hydroxy-propionic Acid. In: *Platform Chemical Biorefinery*. Elsevier, pp. 101–117.

Tsai, Y.-C., Huang, M.-C., Lin, S.-F., and Su, Y.-C. (2001) *Method for the production of itaconic acid using Aspergillus terreus solid state fermentation*. Google Patents.

Valdehuesa, K.N.G., Liu, H., Nisola, G.M. et al. (2013) Recent advances in the metabolic engineering of microorganisms for the production of 3-hydroxypropionic acid as C3 platform chemical. *Applied Microbiology and Biotechnology*, 97, 3309–3321. doi:10.1007/s00253-013-4802-4

van der Straat, L., Tamayo-Ramos, J.A., Schonewille, T., and de Graaff, L.H. (2013) Overexpression of a modified 6-phosphofructo-1-kinase results in an increased itaconic acid productivity in Aspergillus niger. *AMB Express* 3, 1.

Van Nostrand's, (2006) Maleic anhydride, maleic acid, and fumaric acid. *Scientific Encyclopedia*. John Wiley & Sons, Inc.

Vassilev, N., Medina, A., Eichler-Löbermann, B. et al. (2012) Animal Bone Char Solubilization with Itaconic Acid Produced by Free and Immobilized Aspergillus terreus Grown on Glycerol-Based Medium. *Applied Biochemistry and Biotechnology*, 168, 1311–1318. doi:10.1007/s12010-012-9859-5

Vertès, A.A., Inui, M., and Yukawa, H. (2013) The Biotechnological Potential of Corynebacterium glutamicum, from Umami to Chemurgy. In: Yukawa, H. and Inui, M. (Ed.), *Corynebacterium Glutamicum*. Springer Berlin Heidelberg, Berlin, Heidelberg, pp. 1–49.

Vinoth Kumar, R., Pakshirajan, K., and Pugazhenthi, G. (2016) Malic and Succinic Acid. In: *Platform Chemical Biorefinery*. Elsevier, pp. 159–179.

Vuoristo, K.S., Mars, A.E., Sanders, J.P.M. et al. (2016) Metabolic Engineering of TCA Cycle for Production of Chemicals. *Trends in Biotechnology*, 34, 191–197. doi:10.1016/j.tibtech.2015.11.002

Wada, M., Sawada, K., Ogura, K. et al. (2016) Effects of phosphoenolpyruvate carboxylase desensitization on glutamic acid production in Corynebacterium glutamicum ATCC 13032. *Journal of Bioscience and Bioengineering*, 121, 172–177. doi:10.1016/j.jbiosc.2015.06.008

Wang, D., Wang, C., Wei, D. et al. (2016) Gluconic acid production by gad mutant of Klebsiella pneumoniae. *World Journal of Microbiology and Biotechnology*, 32. doi:10.1007/s11274-016-2080-x

Wang, X., Cai, P., Chen, K., and Ouyang, P. (2016) Efficient production of 5-aminovalerate from l-lysine by engineered Escherichia coli whole-cell biocatalysts. *Journal of Molecular Catalysis B: Enzymatic* 134, 115–121. doi:10.1016/j.molcatb.2016.10.008

Wei, D., Liu, X., and Yang, S.-T. (2013) Butyric acid production from sugarcane bagasse hydrolysate by Clostridium tyrobutyricum immobilized in a fibrous-bed bioreactor. *Bioresource Technology*, 129, 553–560. doi:10.1016/j.biortech.2012.11.065

Wen, R.C. and Shen, C.R. (2016) Self-regulated 1-butanol production in Escherichia coli based on the endogenous fermentative control. *Biotechnology for Biofuels*, 9. doi:10.1186/s13068-016-0680-1

Wendisch, V.F., Brito, L.F., Gil Lopez, M. et al. (2016) The flexible feedstock concept in Industrial Biotechnology: Metabolic engineering of Escherichia coli, Corynebacterium glutamicum, Pseudomonas, Bacillus and yeast strains for access to alternative carbon sources. *Journal of Biotechnology*, 234, 139–157. doi:10.1016/j.jbiotec.2016.07.022

West, T.P. (2016) Microbial Malic Acid Production: Exploring New Avenues of Synthesizing a Commercially-Valuable Chemical. *Journal of Microbial & Biochemical Technology*, 8. doi:10.4172/1948-5948.1000303

Xu, Q., Li, S., Fu, Y. et al. (2010) Two-stage utilization of corn straw by Rhizopus oryzae for fumaric acid production. *Bioresource Technology* 101, 6262–6264. doi:10.1016/j.biortech.2010.02.086

Yang, S.T., Zhang, K., Zhang, B., and Huang, H. (2011) 3.16-Fumaric acid. In Murray, M.Y. (ed.) *Comprehensive Biotechnology*, 2nd edn. Academic Press, Burlington 163–177.

Yu, S., Yao, P., Li, J. et al. (2016) Enzymatic synthesis of 3-hydroxypropionic acid at high productivity by using free or immobilized cells of recombinant Escherichia coli. *Journal of Molecular Catalysis B: Enzymatic*, 129, 37–42. doi:10.1016/j.molcatb.2016.03.011

Yuan, W., Ding, R.-H., Ge, H. et al. (2017) Solid-phase extraction of d-glucaric acid from aqueous solution. *Separation and Purification Technology*, 175, 352–357. doi:10.1016/j.seppur.2016.11.060

Zambanini, T., Sarikaya, E., Kleineberg, W. et al. (2016) Efficient malic acid production from glycerol with Ustilago trichophora TZ1. *Biotechnology for Biofuels*, 9. doi:10.1186/s13068-016-0483-4

Zhang, H., Li, N., Pan, X. et al. (2016) Oxidative conversion of glucose to gluconic acid by iron(III) chloride in water under mild conditions. *Green Chem.*, 18, 2308–2312. doi:10.1039/C5GC02614H

Zhang, H., Zhang, J., and Bao, J. (2016) High titer gluconic acid fermentation by Aspergillus niger from dry dilute acid pretreated corn stover without detoxification. *Bioresource Technology*, 203, 211–219. doi:10.1016/j.biortech.2015.12.042

Zhang, S., Qu, C., Huang, X. et al. (2016) Enhanced isopropanol and n-butanol production by supplying exogenous acetic acid via co-culturing two clostridium strains from cassava bagasse hydrolysate. *Journal of Industrial Microbiology and Biotechnology*, 43, 915–925. doi:10.1007/s10295-016-1775-1

Zhang, X., Wang, X., Shanmugam, K.T., and Ingram, L.O. (2015) *L-malate production by metabolically engineered Escherichia coli*. Google Patents.

Zhang, K., Zhang, B., and Yang, S.-T. (2013) Production of citric, itaconic, fumaric and malic acids in filamentous fungal fermentations. *Bioprocessing technologies in biorefinery for sustainable production of fuels, chemicals, and polymers*. John Wiley & Sons Inc, Hoboken 375–397.

Zhang, X., Wang, X., Shanmugam, K.T., and Ingram, L.O. (2011) L-Malate Production by Metabolically Engineered Escherichia coli. *Applied and Environmental Microbiology*, 77, 427–434. doi:10.1128/AEM.01971-10

Żółtaszek, R., Hanausek, M., Kiliańska, Z.M., and Walaszek, Z. (2016) Biologiczna rola kwasu D-glukarowego i jego pochodnych; potencjalne zastosowanie w medycynie The biological role of D-glucaric acid and its derivatives: Potential use in medicine. Journal cover 70.

Zou, X., Zhou, Y., and Yang, S.-T. (2013) Production of polymalic acid and malic acid by Aureobasidium pullulans fermentation and acid hydrolysis. *Biotechnology and Bioengineering*, 110, 2105–2113.

6

Pre-Treatment of Alternative Carbon Source: How Does it Make Sense to Microorganism at Cellular Level?

Joseph Sebastian[1†], Pratik Kumar[1†], Krishnamoorthy Hegde[1], Satinder Kaur Brar[3], Mausam Verma[1], and Ratul Kumar Das[2]

[1] *INRS-ETE, Université du Québec, Québec, Canada*
[2] *TERI-Deakin Nanobiotechnology Centre, Biotechnology and Management of Bioresources Division, The Energy and Resources Institute, Haryana, India*
[3] *Institut national de la recherche scientifique, Centre - Eau Terre Environnement, Québec, Canada*

6.1 Introduction

Biomass is a renewable carbon neutral source and its utilization as a sustainable feedstock for production of platform chemicals and liquid fuels, has drawn attention from academic and industry sectors over the last decade. This biobased economy is envisaged on biorefineries, where biofuels, commodity chemicals and novels materials like bioplastics are produced in an integrated manner at an industrial scale (Sheldon, 2014). Based on the biomass feedstock used, these biorefineries can be categorized as:

- First generation biorefineries: This approach uses the edible parts of corn, sugar cane, cassava, wheat, barley, rye, sorghum, sugar beet and soybean as feedstock. The sugars obtained from these edible cultivars are converted to fuels or chemicals by means of biological or catalytic transformation;
- second generation biorefineries: To overcome the drawbacks of first generation biorefineries, where edible feedstocks were used, second generation biorefineries utilize non-edible feedstocks, such as agricultural residues, forest residues, biodegradable fraction of municipal solid waste and industrial waste, for production of chemicals and fuels;
- third generation biorefineries: Utilize algal biomass for the production of the chemicals and fuels, such as biodiesel and bioethanol. The main drawback being the requirement of dewatering techniques; and
- fourth generation biorefineries: These biorefineries are based on the utilization of synthetic biology techniques for the development of novel photosynthetic organisms that produce biofuels and chemicals.

[†] *Authors equally contributed to the chapter*

Microbial Sensing in Fermentation, First Edition. Edited by Satinder Kaur Brar, Ratul Kumar Das, and Saurabh Jyoti Sarma.

Pretreatment of biomass is a key primary stage in the valorisation of biomass for the production of biofuels and chemicals in a biorefining process. It is estimated that approximately 10^{11} tonnes of biomass is produced per annum and carbohydrate and lignin accounts for approximately 75% and 20% respectively. The lignocellulose, present in the cell wall of plants, is available in large quantities. It is made up of 20% lignin, 40-50% cellulose and 20-40% hemicellulose. Cellulose is a linear polymer made up of (1,4)-D-glucopyranose units, linked by alpha 1-4 linkages, whereas hemicelluloses are a complex of polysaccharides, such as glucose, xylose, mannose, arabinose, along with galacturonic acid and methylglucoronic acid, and exists in association with cellulose in plant cell wall. Lignin is a highly branched polymer of polyphenolic aromatic compounds. Hence to release sugars, from cellulose and hemicellulose, the biomass has to be depolymerized either by thermochemical (also known as biomass-to-liquid (BTL) conversion) or biochemical techniques (Saxena et al., 2007; Antoni et al., 2007; Sheldon, 2014; Sindhu et al., 2016). The strategies for conversion of lignocellulose are represented in Figure 6.1.

During, thermochemical process of pretreatment the biomass is subjected to either pyrolysis to obtain pyrolysis oil, which can be used for energy production, or gasification to produce synthesis gas or syn gas (carbon monoxide and hydrogen), which can then be converted to fuels by means of catalytic or biological conversion. The biochemical or hydrolytic means of conversion utilizes enzymes or chemicals for pretreatment to release the sugars from cellulose and hemicellulose. These treatments are milder when compared to thermochemical but may require additional steps, such as steam or ammonia explosion. The additional steps are required to improve accessibility and to expose the sites of action, especially during enzymatic treatment, but may be subjected to inhibition by compounds produced during thermal and chemical treatment (Saxena et al., 2007; Balan, 2014; Sheldon, 2014). In recent years the advent of biological treatment, using brown, white and soft rot fungi, has shown to significantly improve the efficiency of hydrolysis as well as reduce energy consumption and is environment friendly. Moreover, the chances of producing inhibitory compounds that prevent further conversion of the sugars is low and higher degree of delignification is observed, which in turn improves enzymatic saccharification rate (Sindhu et al., 2016). However, the process is slow and requires sterile conditions, thereby

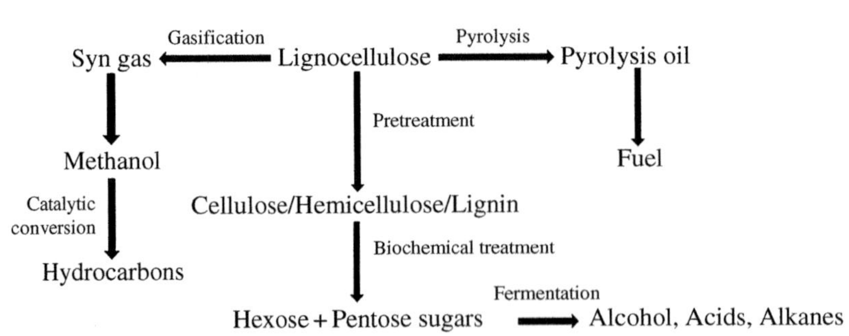

Figure 6.1 Conversion of lignocellulose to fuels and chemicals.

increasing operational costs and hence has not been recommended for use at industrial scale (Chaturvedi and Verma, 2013).

The sugars obtained after the pretreatment of biomass can then be converted to fuels and chemicals either by fermentation or by established chemocatalytic processes. In this chapter, fermentation mediated conversion of sugars to different biofuels and chemicals, and the metabolic pathways involved will be discussed. In addition to sugars, pretreatment also produces different inhibitory compounds. The impact of the presence of these inhibitors, their mode of action and the strategies to prevent these inhibitory effects are discussed as well in the sections to follow.

6.2 Pre-Treated Carbon Source and Microbial Assimilation: Cellular and Biochemical Aspects

The primary objective of biomass pretreatment is to convert the complex polysaccharides present in the biomass into simple sugars, such as glucose and xylose. These simple sugars are then converted into fuel or commodity chemicals either by microbial action or chemical processes. The biological means of conversion of the sugars to desired products and the metabolic pathways involved will be discussed in the following sections and an overview of the various products that can be obtained, via fermentation, is represented in Figure 6.2.

The metabolic conversion of glucose happens in two stages. Glycolysis, wherein glucose is converted to pyruvate, in the cytoplasm, followed by tricarboxylic acid (TCA) or Krebs cycle, under aerobic conditions in the mitochondria, wherein pyruvate is initially converted to acetyl-CoA and then into a sequence of organic acids. The glycolytic pathway and TCA cycle, represented in Figures 6.3 and 6.4

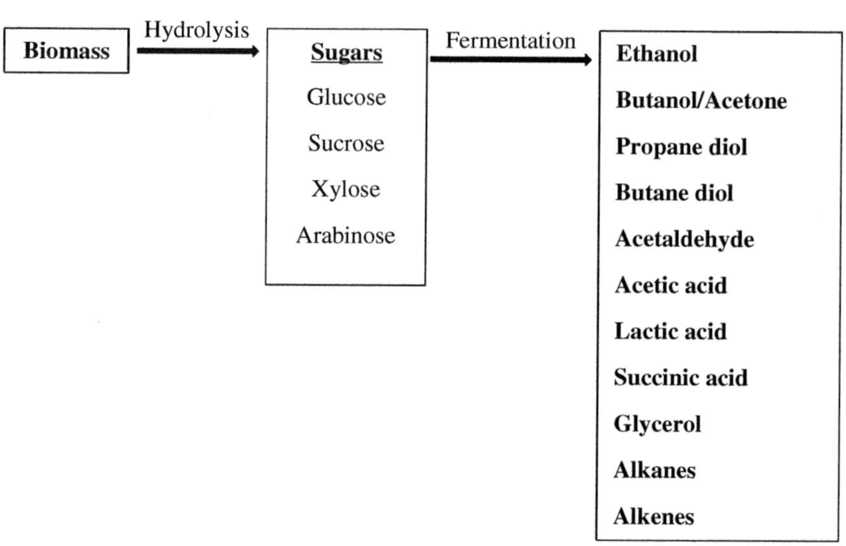

Figure 6.2 Production chemicals from lignocellulosic biomass.

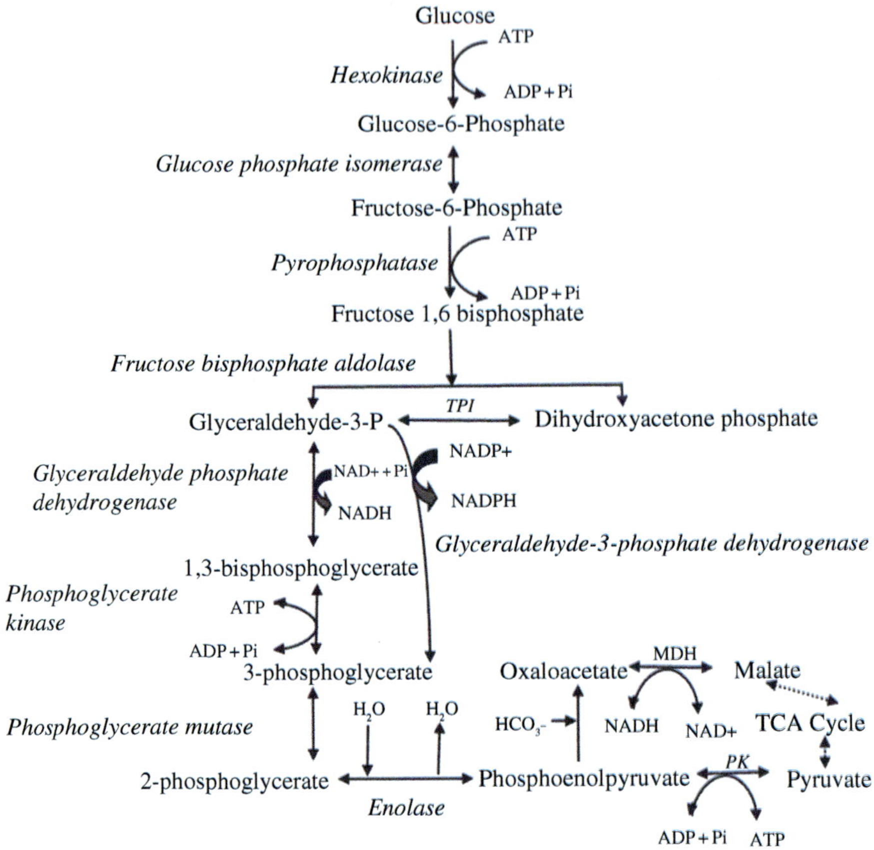

Figure 6.3 Glycolytic or Embden-Meyerhof-Parnas pathway.

respectively, are used to generate adenosine triphosphate (ATP), nicotinamide adenine dinucleotide (NADH) and several biosynthetic precursors (Spector, 2009; Pereto, 2011).

In addition to glucose, pentose, xylose and arabinose can be obtained by pretreatment of biomass which are converted to pyruvate by two metabolic pathways, as represented in Figure 6.5. The glygeraldehyde-3-phosphate (GAP) generated from the pentose phosphoketolase pathway (PPK) and pentose phosphate pathway (PPP), is converted to pyruvate by glycolytic pathway generating ATP and NADH (Tanaka et al., 2002; Spector, 2009; Okano et al., 2010; Balan, 2014).

Among the various metabolites produced during utilization of hexose and pentose sugars, the metabolites, pyruvate, acetyl-CoA, glyceraldehyde-3-phosphate and phosphoenolpyruvate, act as precursor for the synthesis of a wide variety of products, such as ethanol, acetic acid, lactic acid, etc. Most of the strategies for achieving high production rates of chemicals most of often target to alter metabolic pathways involved. These modifications induce excess production and efficient conversion of these precursors. The metabolic pathways involved and the strategies that have

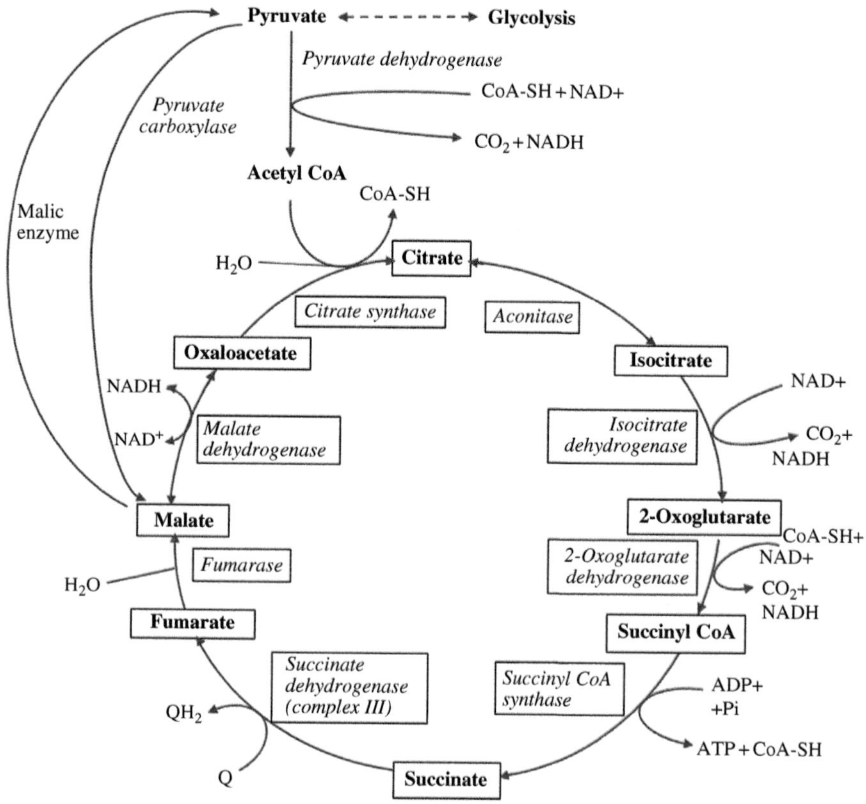

Figure 6.4 TCA or Krebs cycle.

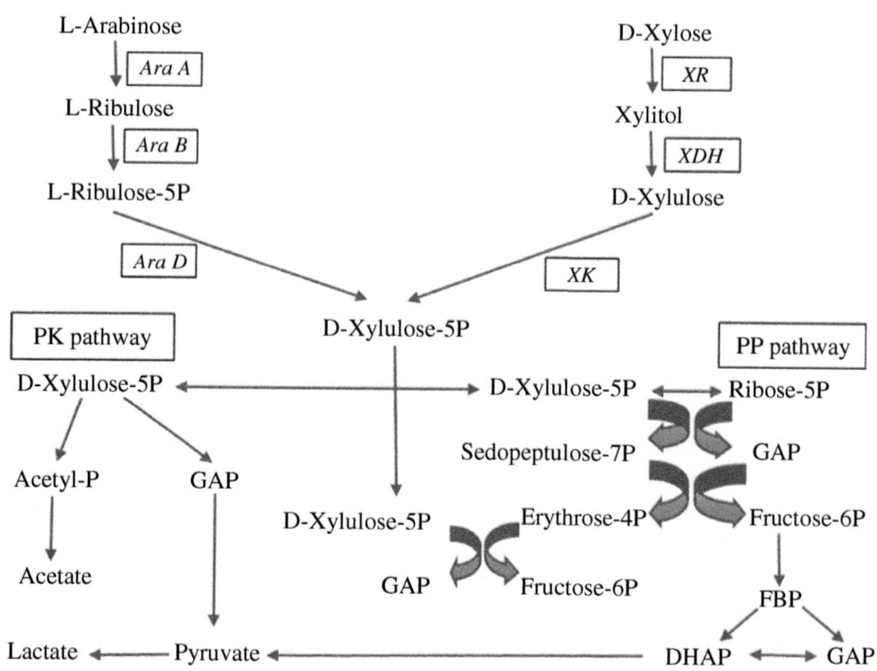

Figure 6.5 Pathway of pentose sugar utilization.

been used to induce higher production rate of some of the desired products, such as fuels and chemicals, are discussed in the sections to be followed.

6.2.1 Alcohols

Alcohol, specifically ethanol, was the primary fuel for the automotive industry before the advent of cheap gasoline. The first prototype for the modern internal combustion engine developed by Nikolaus August Otto used ethanol as fuel and was sponsored by the sugar factory. Moreover, the first affordable automobile, Henry Ford's Model T, was designed to run on 100% ethanol but the advent of cheap gasoline rendered the use of ethanol as an automotive fuel unfeasible. The interest in use of ethanol as fuel was rekindled in the 1970s in Brazil but this ethanol was mainly sourced from first generation biorefineries (Antoni et al., 2007).

The number of oxygen atoms per carbon atom has to be reduced to extract as much energy as possible from the biomass. This is referred to as disproportionisation reaction and can be achieved by anaerobic conversion of glucose to ethanol, as represented in Equation 6.1 below (Zeikus, 1980). Ethanol, with its low oxygen to carbon plus hydrogen ratio, compared to glucose, retains most of the energy content of glucose upon combustion. The use of anaerobic conditions additionally induces greater degree of substrate utlisation by microorganisms, as glycolysis produces far lesser amount of ATP when compared to aerobic respiration, as well as, avoid the need for energy intensive aeration techniques during industrial production (Antoni et al., 2007).

$$\underset{\text{Glucose}}{C_6H_{12}O_6} \xrightarrow[\text{Redox}]{} \underset{\text{Ethanol}}{2C_2H_6O} + 2CO_2 \tag{6.1}$$

6.2.1.1 Bioethanol

Bioethanol production is the largest microbial process and currently almost all fuel ethanol is produced by fermentation of glucose obtained from enzymatic hydrolysis of either corn starch or sugar cane molasses, which forms first generation bioethanol. The push for the switch to second generation ethanol production from non-food plants and agricultural residues, such as corn stover, wheat and rice straw, forestry, paper mill residues and paper portion of municipal waste, has necessitated the use of pretreatment strategies that can be used for effective conversion of cellulose and hemicellulose to glucose and xylose sugars. It has been observed that the enzymatic conversion of cellulose is slower than the rate of ethanol production by yeast and this issue has to be addressed to achieve simultaneous saccharification and fermentation (SSF) and achieve effective conversion biomass to bioethanol (Gray et al., 2006; Hahn-Hagerdahl et al., 2007).

Glucose and xylose sugars are obtained as a result of hydrolysis of biomass and the proportions these sugars can vary based on plant source used as the substrate. However, the yeast strains currently available are not capable to utilizing xylose component of the hydrolysate completely and novel yeast strains are being developed that are able to utilize both the sugars thereby leading to efficient ethanol production (Hahn-Hagerdahl et al., 2007). The enzymes of interest that play a

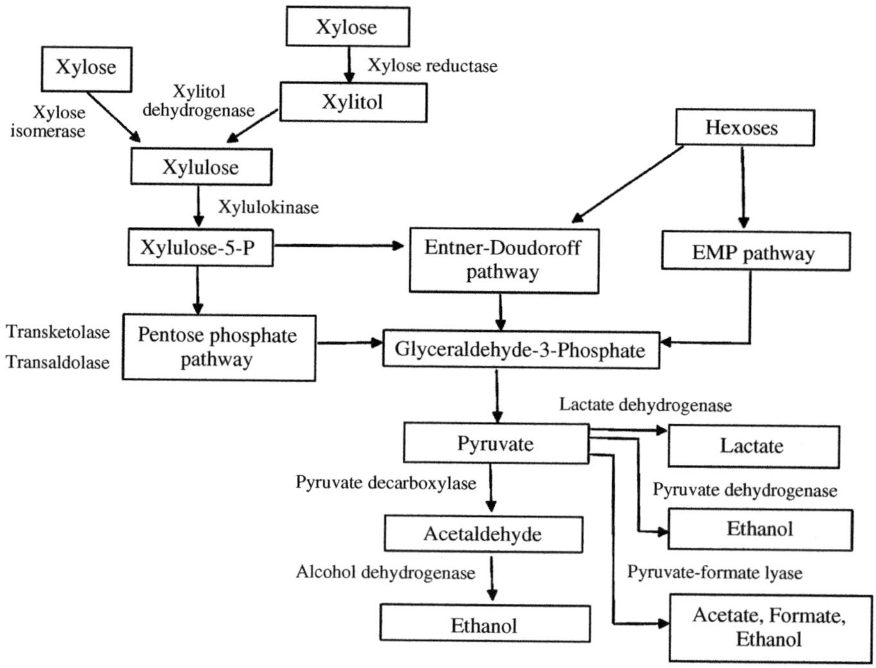

Figure 6.6 Target enzymes for increasing bioethanol production (Danner and Braun, 1999).

key role in ethanol fermentation are represented in Figure 6.6 and these have become a target for the development novel strains that are able to produce significantly high concentrations of ethanol. Genetically modified yeast strains capable to utilizing xylose, as observed in *Pichia stipitis*, is an attractive alternate for yeast mediated ethanol production (Danner and Braun, 1999).

It has been observed that the bacterial strain *Zymomonas mobilis* is able to produce ethanol and considered as the workhorse for bacterial ethanol production as well as the source of genes for genetic modification and development of novel bacterial strains (Alterthum and Ingram, 1989). The strain, *Z. mobilis*, in addition to ethanol produces large quantities of other metabolic products such as sorbitol, glycerol, acetoin, acetic acid and extracellular polysaccharide levan, and was used for continuous production of ethanol in a fluidized bed bioreactor to obtain a maximum ethanol yield of 50 g/L/h (Weuster-Botz, 1993).

To overcome the difficulties associated with the use of *Z. mobilis* and yeast, *E. coli* has been subjected to genetic modification. The modification involved introduction of genes for pyruvate decarboxylase and alcohol dehydrogenase from *Z. mobilis* and removal of genes responsible for side product formation. Ethanol concentrations of up to 9% were obtained with fermentations performed with this *E. coli strain, E. coli LY165* (Antoni et al., 2007). Another similar ethanologenic modified *E. coli* strain (KO11) with additional inactivation of succinic acid production was able to convert glucose and xylose to ethanol with yield equaling theoretical maxima (Gonzalez et al., 2003; Lin and Tanaka, 2006).

These modified bacterial strains are able to produce high concentrations of ethanol but require neutral pH for growth. This causes fermentation to be more susceptible to contamination unlike yeast mediated fermentations which are performed at acidic pH range (Lin and Tanaka, 2006).

In addition to the ability to utilize glucose and xylose efficiently, process parameters during fermentation also play an important role in efficient ethanol fermentation. The influence of aeration on ethanol production by xylose fermenting *Pichia stipitis* was studied in wheat straw hydrolysate. It was observed that the optimization of the aeration conditions led to an increase of 29.1% in ethanol production and a considerable reduction of 42.9% in duration of fermentation when compared to non-aerated fermentation (Bellido et al., 2013). Aeration and mixing play key roles during bioethanol production. It is essential to have enough biomass in the production medium for efficient conversion of sugars to ethanol and low dissolved oxygen levels can lead to sluggish or stuck fermentations. Excess dissolved oxygen might also lead to non-availability of NADH for alcohol dehydrogenase due to increased activity of NADH-oxidase, a strategy that has been identified useful for acetaldehyde production (Tanaka et al., 1990).

6.2.1.2 Butanol and Acetone

Butanol was produced in large quantities, during ABE (Acetone:Butanol:Ethanol::3:6:1) fermentation process, developed by chemist C. Weizmann. Rice and maize starch were used as substrate for the *C. acetobutylicum* mediated fermentation. The butanol produced during fermentation did not hold any value during the time of first world war. ABE fermentation was the second largest bioindustry after ethanol production. The availability of acetone and butanol from fossil fuel led to the winding up of fermentation mediated production of butanol and acetone in most parts of the world (Antoni et al., 2007). Interest in butanol production was rekindled due to the demand for alternate renewable sources of fuel and chemicals. Butanol has a higher energy density and lower volatility than ethanol and can be transformed, by dehydration, to butene, which can then be used for the synthesis of propylene, in combination with ethylene obtained by dehydration of ethanol (Sheldon, 2014).

Clostridium sp. has shown to have the ability to use hexose and pentose sugars as well as CO_2, H_2 and CO for butanol production. Hence, the microbe has been proposed for the production of butanol from lignocellulosic hydrolysates however the issues pertaining to recovery of butanol, sluggish growth of the microbe, substrate inhibition and low solvent tolerance needs to be addressed before a feasible industrial scale production can be achieved (Jang et al., 2012; Tracy et al., 2012). The metabolic pathway and the enzymes involved in the production of butanol are represented in Figure 6.7.

As mentioned above a wide variety of lignocellulosic waste can be used as substrate for the production of butanol. The use of domestic organic waste in one of the study with *C. acetobutylicum*, yielded butanol concentrations in the range of 3-4.2 g/L whereas the use of wheat straw by *C. beijerinckii* P260 strain gave butanol yield of 12 g/L. Other lignocellulosic materials that have been investigated for use as substrate were dried distiller's grains and soluble (DDGS),

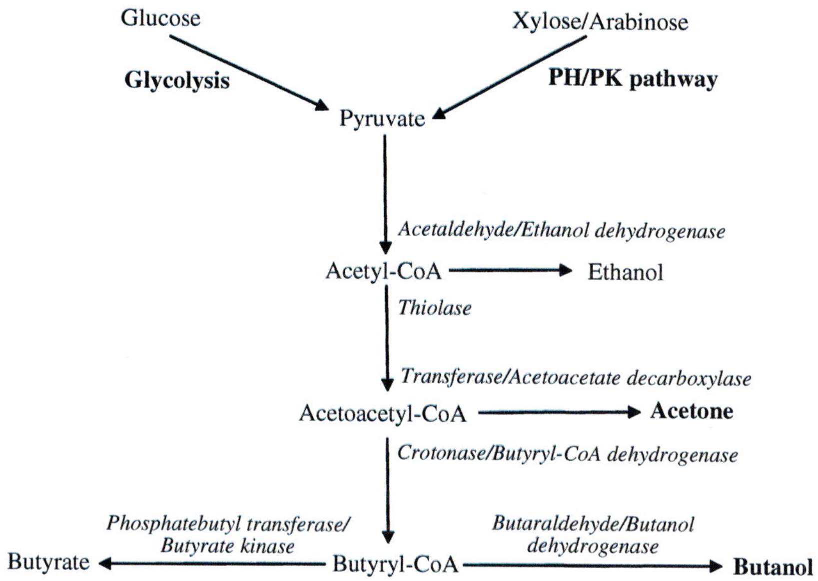

Figure 6.7 Metabolic pathway of butanol synthesis.

rice bran, corn fiber, and cassava bagasse. Table 6.1 provides a summary of the various lignocellulosic materials that have been investigated and the corresponding butanol yield (Jang et al., 2012).

To establish a competitive and feasible butanol fermentation system, attempts have been made to develop innovative fermentation strategies to overcome the issues of butanol production by ABE fermentation. One such strategy that has been found to be effective and improve ethanol yield is the use of co-culture batch fermentation. *B. subtilis* WD 161 was cultured along with *Clostridium butylicum* TISTR 1032 and a tenfold increase in amylase activity and 5.4–6.5 times increase in butanol production was achieved (Tran et al., 2010). These strategies, in addition to fed-batch fermentation, have not been able to achieve the level of butanol production as achieved with continuous fermentation strategy. The continuous plug flow biofilm reactor, inoculated with *C. beijerinckii* BA101 cells immobilized on clay bricks, was able to achieve a maximum solvents productivity of 16.2 g/L/h (Lienhardt et al., 2002). Another continuous fermentation strategy that has shown promising results is the use of membrane cell recycling bioreactors (MCRB) in which a hollow fiber ultramembrane filter was used to separate and recycle the cells into the bioreactor, thereby achieving high cell density. Maximum ABE production rate of 11.0 g/L/h was achieved during the fermentation and a productivity of 7.2 g/L/h during continuous culture with cell recycling of *C. saccharoperbutylacetonicum* N1–4 (Tashiro et al., 2005).

Another major limitation of fermentation mediated butanol production is butanol toxicity, which in turn prevents complete utilization of carbon substrate. In order to overcome this issue, several in situ product recovery strategies are under investigation which include adsorption, liquid–liquid extraction, perstraction (a combination of membrane permeation and extraction by contacting the fermentation broth with an extracting solvent), reverse osmosis, pervaporation,

Table 6.1 Various lignocellulosic waste investigated for butanol production and the corresponding concentration of butanol produced and duration of fermentation (Adapted from Jang et al., 2012).

Substrate	Microorganism	Fermentation duration (Hours)	Butanol Concentration (g/L)
10% domestic organic waste (DOW)	*C. acetobutylicum* ATCC 824	120	3.0
10% DOW hydrolysate	*C. acetobutylicum* ATCC 824	48	4.2
86 g/L wheat straw hydrolysate	*C. beijerinckii* P260	42	12.0
Dilute acid pretreated dried distiller's grains and soluble (DDGS)	*C. saccharobutylicum* 262	120	7.3
Liquid hot water pretreated DDGS	*C. butylicum* 592	72	7.0
AFEX pretreated DDGS	*C. butylicum* 592	72	7.0
Wheat bran hydrolysate	*C. beijerinckii* ATCC 55025	72	8.0
Rice bran	*C. saccharoperbutyl-acetonicum* N1–4	128	7.7
Dilute sulfuric acid treated corn fiber	*C. beijerinckii* BA101	88	6.4
Cassava bagasse	*C. acetobutylicum* JB200	40	9.71
Lime treated barley straw	*C. beijerinckii* P260	68	18
Lime treated switchgrass hydrolysates	*C. beijerinckii* P260	85	14.5

and gas stripping (Jang et al., 2012). Among these techniques gas stripping has been observed as the most efficient and low-cost technique for in situ recovery. It was observed that gas stripping was able to improve butanol productivity and yield by 200% and 118%, respectively (Ezeji et al., 2003). Another study that used concentrated cassava bagasse hydrolysate for butanol production by *C. acetobutylicum* JB200 in fed-batch mode. Gas stripping in this fed-batch fermentation with product recovery resulted in 76.4 g/L of butanol and volumetric productivity of 0.29 g/L/h (Lu et al., 2012). Additionally, attempts are being made to produce butanol from lignocellulosic biomass either using genetically modified *S. cerevisiae* or *Escherichia coli* strains (Balan, 2014).

6.2.2 Hydrogen

Hydrogen is considered as an ideal fuel for future transportation requirement, the proof being that automobile manufacturers are investing extensively in developing and launching hydrogen fuel cell cars. It can be used to produce electrical energy in fuel cells or burned to be converted into mechanical energy with no carbon dioxide. Hydrogen is usually produced by thermal/ chemical means, which is energy intensive and hence cannot be regarded as a renewable energy. Biological means of hydrogen production provides an attractive alternate for renewable and eco-friendly technique of hydrogen production (Antoni et al., 2007; Saxena et al., 2009). Biological hydrogen production can be performed by algal and cyanobacterial bio-photolysis of water or by photo-fermentation of organic substrates by photosynthetic bacteria. Moreover, anaerobic microorganisms, such as acidogenic bacteria, are capable of producing hydrogen by dark fermentation. Hydrogen is a common by product of anaerobic bacterial fermentation, as observed during industrial production of butanol. 40×10^6 m^3 of hydrogen was produced annually during biobutanol plant as by-product (Zverlov et al., 2006). It must be noted that all processes of biological hydrogen production depend on the presence of hydrogen-producing enzymes, such as nitrogenase, Fe-hydrogenase and NiFe hydrogenase (Saxena et al., 2009). The two main strategies of biohydrogen production are discussed briefly below:

a) Fermentative hydrogen production:

During dark fermentation, anaerobic bacteria and some microalgae, growing on carbohydrate rich substrates, produce hydrogen. Bacterial species belonging to the genus *Enterobacter*, *Bacillus* and *Clostridium*, have shown to produce hydrogen. Carbohydrates, especially glucose, is the preferred organic carbon source for hydrogen production, as represented in the Equation 6.2 below and the metabolic pathway represented in figure 6.8. An amount of 4 mol of hydrogen is produced from a single molecule of glucose when acetic acid is produced, as shown below. This observation provides us with the opportunity for the utilization of lignocellulosic biomass for the production of hydrogen (Saxena et al., 2009).

$$C_6H_{12}O_6 + 2H_2O \longrightarrow 2CH_3COOH + 2CO_2 + 4H_2 \tag{6.2}$$

It has been observed that anaerobic fermentative hydrogen production is cost-effective when compared to photosynthetic approach. Moreover, the use of lignocellulosic biomass further adds to the attractiveness of bio-hydrogen production. Dilute acid hydrolysis has been observed to be an efficient and ideal pretreatment strategy for the production of hydrogen. Biomass such as rice straw, cornstalk, grass, poplar leaves, barley straw, sugarcane bagasse and sun flower stalk has been subjected to the dilute acid hydrolysis pretreatment. Biohydrogen concentrations of 33.45 ml H_2/g and 72.21 ml H_2/g have been obtained from poplar leaves and grass, subject to dilute acid hydrolysis, respectively. The use of more drastic acid hydrolysis treatment has been predicted to lead to reduction of biohydrogen production due presence of higher concentrations of inhibitory compounds (Monlau et al., 2013).

Acetic and butyric acid production favors hydrogen production with optimum H_2 yield observed when acetic acid is the end product of fermentation (Hawkes et al., 2002; Nath and Das 2004; Antonopoulou et al., 2008). The production of hydrogen is highly influenced by process conditions, such as pH, hydraulic

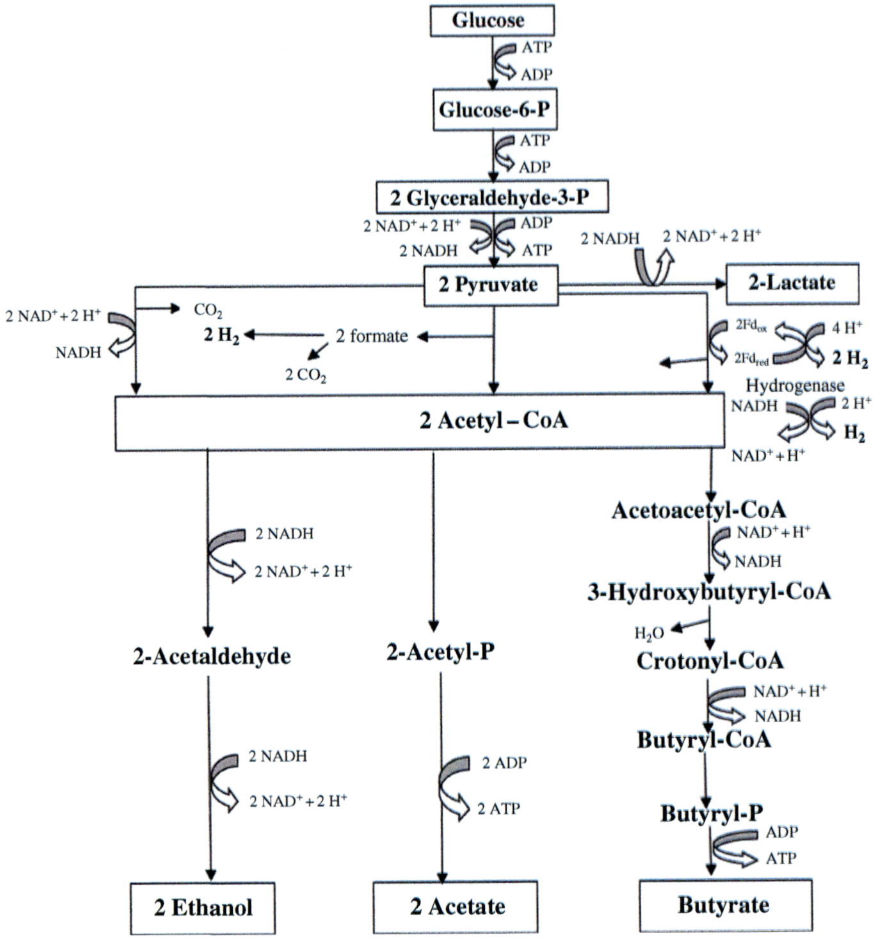

Figure 6.8 Metabolic pathway of glucose during dark fermentation (Xia et al., 2015).

retention time (HRT) and gas partial pressure. These parameters influence microbial metabolic balance and hence, fermentation end products. High hydrogen partial pressure prevents re-oxidisation of ferredoxin and hydrogen-carrying coenzymes. Hence, reduction of dissolved hydrogen concentration by sparging with nitrogen and agitation will aid to improve hydrogen production. HRT and pH are two important parameters that influence hydrogen production. Both these parameters can be varied in tandem or isolation to prevent methanogenesis, which significantly influence hydrogen yield. Since, hydrogen is a key metabolic intermediate it can be utilized methanogens efficiently thereby reducing the yield. pH range of 5-6 and HRT of 8-12 hours have been identified for successful hydrogen production (Hawkes et al., 2002; Nath and Das, 2004; Guo et al., 2014); and

b) biological water gas shift reaction

The biological water-gas shift reaction is a new route for hydrogen production. Some photo-heterotrophic bacteria, such as *Rhodospirulum rubrum* and *Rubrivivax gelatinosus*, are able to produce hydrogen under ambient conditions as represented in the Equation 6.3 below. They are able to utilize carbon monoxide as the sole carbon source, which can be obtained by the thermo-chemical conversion (Ni et al., 2006; Saxena et al., 2009).

$$\text{Biomass} \longrightarrow CO + H_2O \longrightarrow CO_2 + 2H_2 \tag{6.3}$$

High degree of utilization of carbon monoxide and near stoichiometric amount of hydrogen production was observed when purple non-sulphur bacteria was used to perform CO-water gas shift reaction in the dark. Studies performed using these bacteria has indicated that 1 kg of cells can produce 1 kg of hydrogen per day. Moreover, the cost of producing hydrogen by using these microbes, capable of biological water gas shift, is drastically lower compared to other methods (Saxena et al., 2009). The key step in this process is the gasification of biomass. To this effect, pyrolysis can be performed at a temperature of 650-800 K at 0.1-0.5 MPa in the absence of air. This pretreatment converts biomass to charcoal, gaseous compounds and liquid oils. Pyrolysis can be further classified into fast and slow. Fast pyrolysis, at higher temperature, is preferred for hydrogen production as it produces lesser amount charcoal (Ni et al., 2006).

6.2.3 Methane/biogas

Biogas plants provides us with an environmental friendly and sustainable means of producing methane gas from organic domestic or industrial residues, to be used as fuel, especially for public transportation. The science behind biogas production is understood, to an extent, and have resulted in successful adoption of flexible and easy-to-manage domestic scale biogas plants and industrial scale units. Another main advantage of biogas production from plant biomass is that it enables complete recycling of minerals and nutrients back to soil, thereby playing vital role in maintaining soil's vitality for future plantations. The production of biogas from plant biomass consists of a three-stage process involving different anaerobic or facultative anaerobic microorganisms during each stage, explained below and depicted in figure 6.9 (Antoni et al., 2007; Demirel and Scherer 2008).

Stage 1: Hydrolysis of polysaccharides present in plant biomass, such as cellulose and hemicellulose, into sugars or oligosaccharides, fatty acids and glycerol. This is followed by the production of mainly organic acids, such as acetic acid, propionic acid and butyric acid, hydrogen and CO_2. This process is known as acidogenesis.

Stage 2: Acetogenesis is the stage where acetic acid and CO_2 is generated and this stage is generally considered as the limiting process due to the long generation time. Acetogenic bacteria play a crucial role during this stage.

Stage 3: Upto 70% (v/v) of methane during this stage, referred to as methanogenesis, by the action of archaea bacteria that are sensitive to acidification, ammonia accumulation and low oxygen concentration. Hydrogenotrophic methanogens and aceticlastic methanogens play a crucial role during this stage.

The composition of microbial community involved in biogas production has been observed to vary with the substrate, the type of fermenter and the process, mesophilic or thermophilic hydrolytic (O'Sullivan et al., 2005; Cirne et al., 2007;

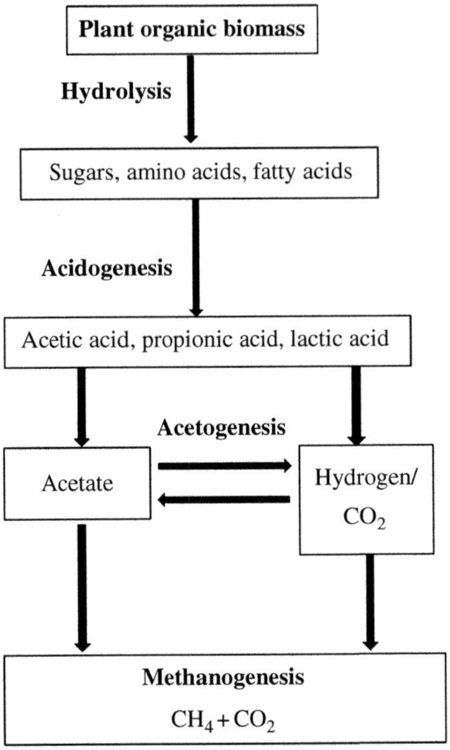

Figure 6.9 Conversion of biomass to methane.

Klocke et al., 2007). Hence a comprehensive understanding of the microbes involved in the process will aid in improving the production efficiency. Currently only 70% of the organic matter is converted to methane and the adoption of a multistage stage strategy, where hydrolysis stage is separated from acetogenesis and methanogenesis, has shown to significantly improve the efficiency of the process. To this effect the separate pretreatment of biomass holds special significance and the adoption of thermophilic processes for biomass hydrolysis has shown to further speed up reactions (Demirel and Scherer, 2008). Moreover, it was found that a stable anaerobic digestion is accomplished only in the presence of microbial communities belonging to four major metabolic groups: hydrolytic fermentative bacteria, proton-reducing acetogenic bacteria, hydrogenotrophic methanogens, and aceticlastic methanogens (Zinder et al., 1984).

The performance of the biogas plant is primarily linked to the microbial community present in the digester. Acetotrophic methanogens has a crucial role in the conversion of acetate to methane and the optimization of the conditions to aid maximum performance are of paramount importance (Demirel and Scherer, 2008). The two main species of acetotrophic methanogens that have been isolated and identified from biogas plants are *Methanosaeta* spp. and *Methanosarcina* spp., with higher concentrations of *Methanosarcina* spp. observed in digesters with high acetate concentrations and in lower acetate concentration *Methanosaeta* spp. was observed to be dominant (Stroot et al., 2001; McMahon et al., 2001).

The environmental and physical conditions have a profound impact on the performance of acetotrophic methanogens in digesters (Ghangrekar et al., 2005). Environmental parameters like temperature and pH have a profound impact on biogas production. Significant reduction in daily biogas production was observed when temperature dropped from 36 °C to 24 °C (Rastogi et al., 2007). It was found that the optimum pH range for acetotrophic methanogens was between 6.6 and 7.3 and that they were inhibited strongly below a pH of 6.2. Additionally, free ammonia (FA) concentrations were also found to have inhibitory effect at pH levels above 7.4 (Siegrist et al., 2002). Another important observation made was that the concentration of ammonia and volatile fatty acids (VFA) had an impact on the diversity of methanogens in the bioreactor. It was observed that at high levels of NH_3 and VFA, *Methanosarcinaceae* spp. was dominant in manure digesters, while *Methanosaetaceae* spp. was dominant in sewage sludge digesters with low levels of NH_3 and VFA (Karakashev et al., 2005). The other factors that have been suggested to have profound impact on the microbial community present in the digester are hydraulic retention time (HRT), seed inoculum used, hydrogen partial pressure and the substrate used for biogas production. The presence and activity of hydrogenotrophic methanogens, such as *Methanomicrobiales*, in addition to acetogenic methanogens, is crucial for a stable and efficient process performance (Demirel and Scherer, 2008).

6.2.4 Organic Acids

The direct conversion sugars obtained after lignocellulosic pre-treatment can be transformed to platform chemicals using biotechnological and fermentation mediated processes. A wide variety of carboxylic acids can be produced from carbohydrates by fermentation. Carboxylic acids such as lactic acid, citric acid, succinic acid, acetic acid, maleic acid, are some of the examples of carboxylic acids that are currently commercially available. Most of the acids produced during fermentation are either intermediates or end product of metabolic activity. Pyruvate and acetyl-CoA act as a precursor for many acids, as represented in figures 6.10 and 6.11, as well as the intermediates of TCA cycle are carboxylic acids. The over production and accumulation of these acids can be achieved by genetic modification and optimization of fermentation conditions (Okano et al., 2010). The bioproduction strategies for a few of the commercially available and novel carboxylic acids are discussed briefly below.

Lactic acid is a first generation biobased commodity chemical, with many applications. The rapidly growing use of polylactate as bioplastic, has necessitated the development of industrial scale efficient production strategies, especially from lignocellulosic biomass hydrolysate, to meet the future requirements and remain cost effective. Lactic acid bacteria (LAB) are mainly used for the production of the acid but due to inherent limitation of LAB, the use of genetically modified improved strains of LAB and other organisms such as *E. coli*, *Corynebacterium glutamicum* and yeast, is being considered for use in commercial production and these genetically modified organisms, capable of utilizing

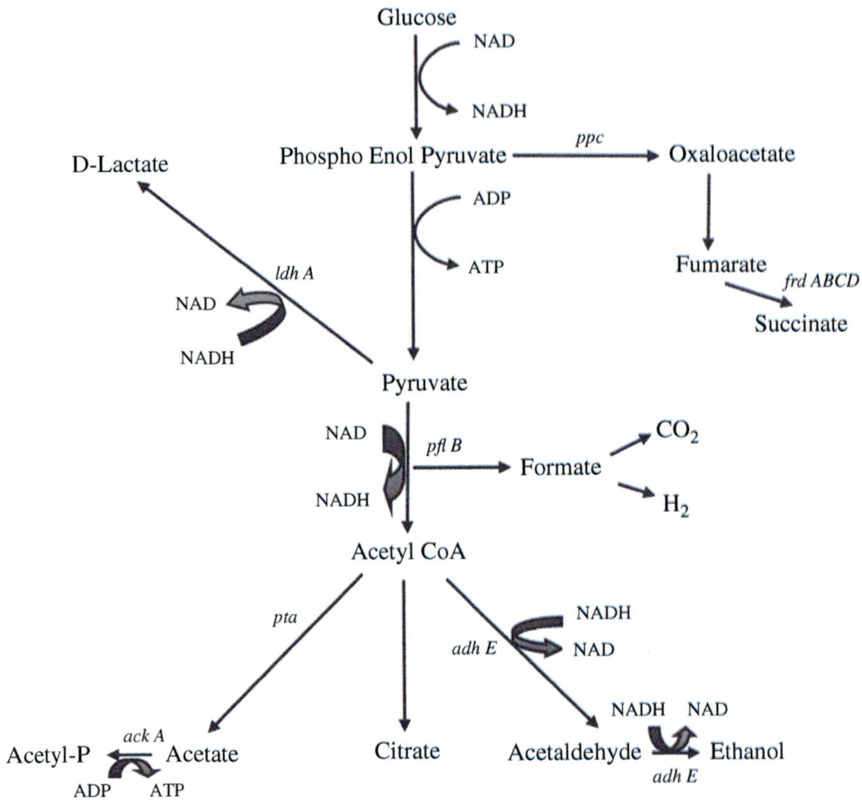

Figure 6.10 Illustration of glycolysis in E. coli.

both hexose and pentose sugars efficiently, can be used for lactic acid production on hydrolysates of cellulosic biomass (Danner and Braun, 1999; Okano et al., 2010; Sheldon, 2014).

Citric acid and acetic acid are examples of other commodity chemicals that can be produced by fermentation of sugars obtained after pretreatment of lignocellulosic biomass (Sheldon, 2014). Currently the demands for these acids are met by synthetic means of production. The fermentation mediated synthesis of acetic acid is performed by *Acetobacter* that converts ethanol to acetic acid and that of citric acid by growing *Aspergillus niger* and *Candida* sp. (Danner and Braun, 1999; Vandenberghe, 1999). Agro-industrial residues, such as apple pomace, cassava bagasse, coffee husk, wheat straw, pineapple waste, etc., have been investigated for citric acid production, using solid state fermentation techniques, and have been found to be ideal for use in the production (Pandey and Soccol, 1998; Pandey et al., 1999; Vandenberghe et al., 1999). Hence, the sugar rich hydrolysates of cellulosic biomass provide an attractive for the production these acids. One such strategy for utilization of pyrolysis product of lignocellulosic biomass was studied and involved the production of levoglucosan by pyrolysis and subsequent production of citric acid by sequential fermentation initially by

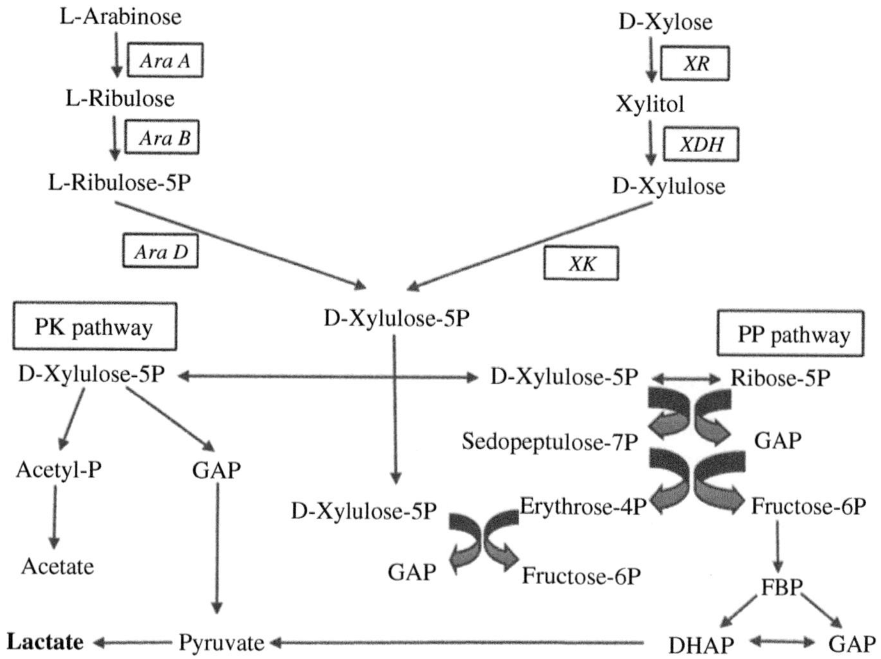

Figure 6.11 Pentose sugar utilization and lactic acid synthesis in lactic acid bacteria.

Phanerochaete chrysosporium, to remove inhibitory compounds, followed by fermentation to produce citric acid using *A. niger* CBX-209 (Yang et al., 2014).

Succinic acid, due to its high-volume applications in developing polyesters and polyamides, has led to considerable advances in microbial production of succinic acid (Sheldon, 2014). Biosuccinium™, developed by Reverdia (a joint venture of DSM and Roquette), has used a proprietary low pH yeast technology for the production of succinic acid (http://www.reverdia.com/). Similar technology is also being used by BIOAMBER for the production of succinic acid. This technology enables the production of succinic acid directly instead of a salt, thereby aiding in considerable reduction in the amount of chloride and sulphate waste generated (Sheldon, 2014; https://www.bio-amber.com). Currently these companies utilize corn derived sugar for commercial production but would have to switch biomass derived sugars for sustainable production in the future.

Another important carboxylic acid that is being investigated for commercial production from renewable sources is 3-hydroxypropionic acid. It has garnered interest, as it can be chemically converted into acrylic acid, acrylic esters and amides and enable sustainable production of superabsorbent polymers, plastics, paints, etc... 7.5 g/L of 3-HPA was obtained by the culture of a genetically modified yeast grown on xylose sugar (Kildegaard et al., 2015). BASF, Cargill and Novozyme, in 2013, was able to demonstrate a successful bioproduction of 3-HPA at pilot scale and in 2014 was able to achieve conversion of 3-HPA to glacial acrylic acid and superabsorbent polymers (https://www.novozymes.com/).

6.3 Challenges of Inhibitory Hydrolysis Products and Strategic Solution

Lignocellulosic biomasses produce variety of inhibitory compounds during pretreatments. These inhibitors need to be removed or transformed into non-inhibitory products to achieve optimum bioproduction of desired target products. The following sections discuss the various inhibitory compounds produced during pretreatment, their mode of inhibition and strategies to alleviate the inhibitory effect of such compounds.

6.3.1 Inhibitory Products: Pretreatment Metabolites or By-products

Lignocellulosic biomass has shown intense utility in terms of biochemical conversion in enhancing the available useful compound such as glucose, xylose, and other sugars by uncovering their natural state into the more available form (Yang et al., 2008). However, the recalcitrant nature of native biomass becomes a physical and chemical barrier in making cellulose amenable to enzymatic hydrolysis which serves as a key step in the biochemical conversion of lignocellulosic material with context to the sugar-based platform (Ko et al., 2015). Thus, the concept of pre-treatment becomes very handy in availing more of the cellulose surface area to undergo further biochemical steps with larger efficacy in converting useful products such as biogas, lipids or ethanol. However, pre-treatment method involves various side chain reactions producing distinct inhibitory products along with the amenable cellulose compound (which undergoes enzymatic hydrolysis). These reactions are accompanied by solubilization of hemicellulose and presence of other recalcitrant compounds such as lignin which coats the native cellulose structure (Jönsson et al., 2016).

Formation of these inhibitory compounds, their nature and concentration vary according to the kind of pre-treatment and the feedstock being employed. Various pre-treatment methods, such as acid-based methods, hydrothermal methods, alkaline methods, oxidative methods (wet oxidation and use of peroxides) or use of alternative solvents produces various kinds of inhibitory compounds (aliphatic carboxylic acids, phenolic compounds, furans, furans aldehydes, etc.). These compounds affect the microbial growth and cellulolytic enzymes production that can be broadly categorized based on the chemical structure specificity such as a) aliphatic compounds; b) furan aldehyde; c) aromatic compounds; and d) inorganic compounds and other inhibitors.

6.3.1.1 Aliphatic Compounds

These are the most common inhibitors formed during the pre-treatment process. Compounds such as acetic acid (formed due to the hydrolysis of acetyl groups of hemicellulose), levulinic acid (degradation of HMF compound), formic acid (formed due to acid-catalyzed process: degradation product of furfural and HMF), etc. are major ingredients of the lignocellulosic hydrolysates. They exhibit relatively lower toxicity as compared to other inhibitors but can be present at a higher concentration in a final mixture after pre-treatment process.

These aliphatic compounds have been shown to inhibit the ethanol production by affecting the biomass growth during fermentation (due to intracellular accumulation of H^+ ions). The general trend for the toxicity effect on the microorganisms, such as *S. cerevisiae* follows the order: acetic acid < levulinic acid < formic acid (Taherzadeh et al., 2011). Acetic acid is not only the hydrolysis by-product but also a well-known fermentation by-product and their concentration depends a lot on the residence time and temperature condition inside the reactor that affects the degree of hydrolysis of acetyl groups. The major reason accounting for the degree of toxicity is based on the undissociated ions of these compounds remaining in solution. Also, the degree of un-dissociation of these compounds in solution depends on the kind of feedstock too, i.e. whether pre-treatment is processed on softwoods or hardwoods or spruce, stalks, etc. For example, feedstock derived from agricultural residues and hardwood gave a higher concentration of aliphatic compounds than softwood (Jönsson et al., 2016). In some case, these concentrations were demarcated to obtain efficient ethanol production with a set value of 100 mM (hardwood: aliphatic compounds >100 mM affected ethanol production but the concentration of <100 mM from softwood feedstock alleviated these effects).

The more the undissociated ions in the solution more will be the possibility of their entrance inside the cell membrane via diffusion. After their entry into the membrane, they are dissociated due to the change in a cytosolic environment having a pH of around 7, eventually leading to cell death. This may be further accounted due to the accumulation of H^+ ions, thereby lowering the intracellular pH and disrupting the balance between ATP and ADP. Some studies have also mentioned that due to this phenomenon, ethanol productivity rate (not yield) decreases at the expense of biomass production as these cells must pump out more H^+ ions to maintain the natural neutral environment (Verduyn et al., 1990 and Viegas et al., 1991). However, the tolerance level of baker's yeast in presence of acetic acid concentration, even 5 g/L has been found impressive under undissociated state (Taherzadeh et al., 1997). Also, acetic acid at 4.4 g/L in the culture medium inhibited 15% growth of *C.saccharolyticus.* Panagiotopoulos et al. (2009) found effective hydrogen production from fermentation of wheat and barley biomass-derived sugars with acetic acid concentration reaching up to 1 g/L and no inhibition till 3 g/L.

6.3.1.2 Aromatic Compounds

These inhibitors mainly include the phenolic compounds, such as phenolic aromatic carboxylic acids (ferulic acid and 4-hydroxybenzoic acid), and non-phenolic aromatic carboxylic acids too such as cinnamic acid. They are generally formed due to acid-catalyzed hydrolysis. However, carboxylic acids can be categorized under both aliphatic and aromatic. To note, the inhibitory effect of aromatic carboxylic acid is generally higher as compared to the aliphatic ones (though former is inhibitory at low concentration). For example, Larsson et al. (2000) studied that lower concentration of ferulic acid (1.0 mM) showed an inhibitory effect to *S. cerevisiae* as compared to two-fold concentration of common aliphatic compounds, such as acetic acid, levulinic acid, and formic acid. Even for different lignocellulosic biomass studied, such as sugarcane bagasse

(Martin et al., 2002) and corn stover (Du et al., 2010), small ferulic acid concentration (1.1 mM and 0.033 mM respectively) showed inhibitory effects to *S. cerevisiae* that was used for ethanol production. These inhibitory compounds (phenolics and other aromatic compounds) inhibit microbial growth and affect cellular yield too. However, they are variable in this regard and depends on the functional group. In fact, the inhibitory property is affected by the position of the substituents (hydroxyl or methoxyl) in a compound too. For example, Larsson et al. (2000) observed that during ethanolic fermentation, the oxidized and reduced form of diphenol/quinone showed a significant difference in inhibition property towards *Saccharomyces cerevisiae* (oxidized form: more inhibitory) in oxygen-limited assays. Even the inhibitory effects of the phenylpropane derivatives were found to be dependent on the presence and position of the unsaturated bonds in the aliphatic side chain.

Like aliphatic compounds, aromatic derivatives production too depends on the reaction time and experimental conditions such as acidity/alkalinity, temperature, pressure, etc. Samuel et al. (2013) studied that the increase in temperature and time of autohydrolysis pre-treatment decreases the lignin hydroxyl groups (related to an increasing of phenolic hydroxyl groups) which are associated with the aliphatic side chain. A possible mechanism was examined for the above study, which showed auto-catalyzed elimination of water from lignin group forming phenol due to cleavage of β-O-4 aryl ether linkages. This suggested that apart from the whole lignocellulosic biomass pre-treatment, autohydrolysis resulted in depolymerization rather than general condensation. In many studies, the toxicity mechanism was not examined and elucidated. Early research by Keweloh et al. (1990) showed that the phenolics interfere with the cell membrane thereby changing its functions and protein-to-lipid ratio. This could be attributed to an increase in the cell respiratory activity which is induced by the change in property of such phenolic compounds and also their formation during the fermentation. For example, *S. cerevisiae* was shown to decrease the inhibitory effect of phenolic compounds (coniferyl aldehyde is reduced to coniferyl alcohol) (Larsson et al., 2000). Recent studies have also suggested that the biomass sugar themselves can undergo transformation into the pseudo-lignin compoundwhen exposed to high temperature (pretreatment, such as acid hydrolysis). Even hydrothermal pre-treatment of lignocellulosic biomass can undergo such transformation of monosaccharides to form pseudo-inhibitory compounds generated from an aldehyde, ketones and other aromatic compounds (Rasmussen et al., 2014).

6.3.1.3 Furan Aldehydes

These are the most explored and commonly found inhibitors formed during the pretreatment of lignocellulosic biomass. These include furfural, HMF (commonly found in lignocellulosic hydrolysates) which mainly occurs due to dehydration of common sugars such as pentose, hexose, and others. They affect the metabolism of microorganism along with the product yield (commonly, ethanol). However, under anaerobic condition, these compounds cause an increase in product yield by converting furfural into furfuryl alcohol and HMF to 2,5-bis-hydroxymethylfuran. A study by Wahlbom et al. (2002) showed that the xylose-utilizing *S. cerevisiae*, formed a less recalcitrant compound, i.e. xylitol which transformed the process to undergo more ethanol (product) yield. Table 1 shows some of the studies on pretreatment of lignocellulosic biomass where furfural and HMF were shown as the major inhibitors studied by the researchers over the years.

Understanding the mechanism of Furan aldehyde inhibitors: The reduction of HMF is linked to the consumption of the co-factor, NADPH whereas reduction of furfural is associated with the co-factor NADH (Hanly et al., 2014). Generally, furfural is more toxic and shows more inhibitory effect than HMF. Also, furfural is found in the hydrolysates and easily formed as compared to HMF during acid hydrolysis (or high-temperature process). It is also reported that furfural strongly inhibits the ethanol fermentation than other inhibitors (Guo et al., 2008). Even the most sensitive enzymes, pyruvate and aldehyde dehydrogenases depicted lower inhibitory effect than furfural. Also, HMF increased the lag phase of the growing microorganisms and hence affects the conversion rate of the final product (four times lower as compared to the furfural presence) (Gad-kariem et al., 2009). Hence, HMF makes the microbial process to last relatively longer (Ask et al., 2013).

6.3.2 Strategies to Control Inhibitory Effects

Inhibitory compounds as discussed earlier affect the growth of microorganisms and the product yield. Even within the same pretreatment method, variables such as temperature and fermentation period affect the overall output of available glucose (cellulose), thereby modifying the product conversion rate and quantity. These inhibitory compounds need to be detoxified to extract the maximum product output from the available lignocellulosic biomass. However, their removal from the system varies with the physical and chemical parameters of the experimental conditions and hence become a tedious task.

The most common pretreatment methods employed these days are dilute acid hydrothermal method with or without steam explosion, liquid hot water treatment, and alkaline methods. It has been found that within the same pretreatment method, variables such as time and temperature of reaction produces different amount of inhibitor compounds. However, the optimization of the employed pretreatment methods can also minimize/control the inhibitors concentration to a significant extent. For example, Rajan et al. (2014) founded that three variables employed in dilute acid pretreatment viz. temperature (140 °C–160 °C), reaction time (10 min, 20 min, 30 min, 45 min and 60 min) and acid volume (5, 10 and 20 $dm^3 m^{-3}$) were responsible for incurring different degree of formed inhibitors (formic acid, furfural, acetic acid and 5-hydroxymethylfurfural). It was found that at 140 °C, 30 min of reaction time and at 10 $dm^3 m^{-3}$ concentration of dilute sulfuric acid, the maximum glucose was yielded (at 89%) followed by maximum decrease of 37% in inhibitors concentration (reported concentrations of furfural, formic acid, 5-hydroxymethylfurfural and acetic acid were 12.08 ± 1.69, 32.37 ± 4.91, and 1.14 ± 0.22, 7.98 ± 1.02 $g\,kg^{-1}$, respectively). However, not every time, the optimization process works satisfactorily as it takes a lot of experimental time and hence industry avoids doing the same. Usually, two strategies are employed for controlling the inhibitory compounds: a) detoxification of the inhibitory compounds and; b) adaptation of the microorganisms to the reaction process. These detoxification methods will be discussed, under the category of biological detoxification and physical and chemical detoxification methods, respectively in the coming sub-sections.

6.3.2.1 Biological Detoxification Strategy for the Inhibitors

Detoxification step is not economical. For example, physical detoxification methods such as membrane method, ion exchange resin are quite costly to operate. Thus, sometimes, the logical strategy is to allow the microorganisms to adapt to the native environmental situation which facilitates them to bring out the best economic strategies in controlling the inhibitory effect and maximizing the product yield (Behera et al., 2014). *In-situ* detoxification of the inhibitory compounds plays a vital role where bacteria are actively involved in processing the metabolic degradation pathway to break these compounds into useful or harmless end products. For instance, Guo et al. (2013) studied the butanol production by non-detoxified hemicellulosic hydrolysate from corn fiber (pretreatment with dilute acid), which showed higher intolerance of inhibitory products with *Clostridium beijerinckii* mutant RT66. This microorganism was obtained by continuous culture method which helps them to adapt to the situation and effectively deal with the inhibitors thereby enhancing the product output.

In fact, an adaptation of microorganism technique benefits industries also, due to the fact that there is no detoxification step involved during the fermentation, where ethanol yield is improved naturally without the input of other chemicals and hence more side chain reactions are avoided (Parawira et al., 2010). The bioreduction capability of fermenting yeasts or other types of microorganisms has been explored in order to detoxify the inhibitory effects of the compounds (removal of inhibitors before fermentation process) formed due to the pretreatment process. These microbial and enzymatic pathways of detoxification are in fact very specific to the inhibitory compounds and their action. Even some bacteria, such as ethanologenic bacteria have been shown to adapt to a molecular mechanism in response to lignocellulosic-derived inhibitory compounds (Ibraheem et al., 2013). These microorganisms were shown to act positive under stress response and perform via tolerance mechanism which allows them to maintain the integrity of the cell membrane (which gets affected due to protonation/pH disturbance as discussed in the mechanism section of aliphatic compounds 6.2.1.1). Silva and Roberto (2001) adapted *Candida guilliermondii* to rice straw hemicellulose hydrolysate and studied the xylitol production for alleviating the inhibitory effects produced by xylose.

6.3.2.2 Understanding the Mechanism of Microorganism Adaptation for The Detoxification of Inhibitory Compounds

6.3.2.2.1 Homeostasis

pH homeostasis plays a key role during microorganism adaptation to counter inhibitory products formation during the pretreatment step. As an example, an increase in ammonia production or additional component can counter the pH imbalance as ammonia reacts with excess H^+ ions present in the cell which during pretreatment process forms NH_4^+ ions and raises intracellular pH. In fact, some enzymes, such as ureases (urea into ammonia and CO_2) and arginine deaminases produce ammonia (into NH_3 and L-citruline).

Activation of amino acid decarboxylase too showed expulsion of $2H^+$ (studied with *E. coli* where glutamate-dependent decarboxylation was found to be more robust than lysine-dependent decarboxylation) which led to an increase in intracellular pH (Lin et al., 1996). Decarboxylation mechanism generates a proton motive force which drives the ATP activity and helps in maintaining the metabolic activity within the cells (Hara et al., 2015). However, this acid resistant mechanism depends on the strain, type of media used, growth phase and the length of acid stress (Warnecke et al., 2005). As -discussed earlier, glutamate-dependent decarboxylation is more robust than lysine or arginine since glutamate amino acid decarboxylase can handle even a pH stress of 4 as compared to optimal pH level of 5 and 5.7 for arginine and lysine counterpart respectively. However, above stated statement is a general notion as a study by Álvarez-Ordóñez et al, (2010) showed that lysine and arginine counterpart is better in handling acid stress than glutamate for *Salmonella typhimurium*.

6.3.2.2.2 Enzymatic Detoxification

Inhibitors degradation are also being practiced to remove the inhibitory effects of the above-discussed products during pretreatment of lignocellulosic biomass. The role of protein (such as enzymes and co-enzymes) synthesis and activity can alleviate the negative impact of inhibitors by following specific degradation pathway (Ibraheem et al., 2013). Many different bacterial species had been studied so far for the degradation mechanism of inhibitory products mainly phenolic compounds and weak acids. Even some bacterial species such as *Caldicellulosiruptor owensensis* has been found capable of converting lignocellulosic biomass into the useful product even without pretreatment (Peng et al., 2015). Various gram-positive bacteria, such as *Arthrobacter, Lactobacillus, Rhodococcus,* etc. and gram-negative bacteria, such as *Enterobacter, E.coli, Pseudomonas,* etc., had shown active degradation of the phenolic inhibitory compounds with ability to degrade ferulic acid (ferulic acid decarboxylase enzyme responsible), vanillic acid (vanillate demethylase is responsible), catechol (catechol-1, 2-dioxygenase) via series of enzymatic.Even laccase, a multicopper oxidase enzyme was shown to largely reduce the lignocellulosic inhibitory compounds (HMF, catechol, vanillic acid) on fermentation (Jönsson et al., 1998). Laccase is produced by *Bacillus subti*lis, *Azospirillum lipoferum, Streptomyces cyaneus,* etc. which helps in ring cleavage of the aromatic inhibitors.

These techniques of biological treatment of inhibitors have an advantage over physical and chemical detoxification, discussed in a later section) because of a little waste generation with the added benefit of performing the reaction in the same reactor itself (Hou-Rui et al., 2009). Though this process is environmentally friendly, it undergoes fewer side reactions; still, it takes longer process time. These bio-abatement processes have drawn attention to some researchers lately. For example, Nichols et al. (2008) studied the hydrolysate treatment with fungus, *C. ligniaria* which resulted in significant improvement of pentose with a recombinant bacterial strain (*E.coli FBR5*). This strain was found to have specific potential in ethanol production with furfural inhibitor converted into furfuryl alcohol. In another study by Cao et al. (2013), the formation of inhibitors during the pretreatment of lignocellulosic slurries was detoxified, resulted in an increase of cellulose conversion (1.2-1.5 fold). These bio-abatement microbes selectively induce the mechanism to metabolize the inhibitory compound and remove them by adapting to the strains followed during the fermentation (Nichols et al., 2010). Some other biological detoxification can be studied from Table 6.2.

6.3.2.3 Physical and Chemical Detoxification Strategy for Inhibitors

Other detoxification methods apart from the biological method include Evaporation, Neutralization, Ion-exchange resins, use of peroxidases, solvent extraction, activated charcoal, etc. These treatments can basically be broadly categorized into two forms, viz. physical, physicochemical treatment:

- Physical treatment mainly includes evaporation and membrane techniques. Evaporation process has been shown to reduce the volatile inhibitor compounds such as acetic acid, furfural, and vanillin. However, the major disadvantage of using this method is that, at the end, it increases the non-volatile compound concentration and hence increases the toxicity behavior of the final product. However, some studies in the past revealed good possibilities of employing this method successfully. Larsson et al. (1999) achieved over 90% removal of furfural from wood hemicellulosic hydrolysates. Palmqvist et al. (1996) successfully studied removal of most of the volatile fraction from the willow hemicellulosic hydrolysates through roto-evaporation technique.
  - Other physical methods, such as membrane technique helped to remove the inhibitory compounds by isolating solvent groups from the hydrolysates and internal pores were functionalized to adsorb these inhibitor groups, such as HMF, furfural, and acetic acids to retain over the membrane surface. Some studies reported membrane technique performing better than the ion-exchange resin method in removing acetic acid from the biomass hydrolysates of corn stover (Grzenia et al., 2008, 2010). In fact, the detoxification efficiency was higher in the continued extraction with octanol, ethyl alcohol (in organic phase). Some other studies are tabulated under Table 6.2 for the physical method of detoxification.
- Physico-chemical treatment includes ion-exchange resin, neutralization, overliming, activated charcoal and solvent extraction method. A study by Larsson et al. (1999) compared different methods such as anion exchange, alkali treatment, evaporation with sulfite treatment and enzymatic treatment for the detoxification of aliphatic acids, furan derivatives, and phenolic compounds from spruce biomass fermented via *Saccharomyces cerevisiae.* They found that anion exchange at pH 10 was overall the best condition as compared to other processes (evaporation with sulfite treatment was least effective and laccase (enzymatic treatment) only detoxified phenolic compounds).

Neutralization, on the other hand, became an unavoidable step before the fermentation step as the hydrolysates after pretreatment is mainly acidic in nature and thus base material, such as calcium hydroxide and sodium hydroxide is used which eventually precipitates common inhibitors such as furfural and other

Table 6.2 Various pre-treatment effects on inhibitors produced by different lignocellulosic biomass and discussion on possible solution/conclusion.

Lignocellulose material	Process details/ pretreatment method	Experimental details	End product target	Inhibitors	Concentration/ effects of inhibitors	Microorganism	Possible solution/ inferences	References
Wheat straw and wet slurry	Steam explosion	30 °C and 175 rpm for 168 h; not strictly anaerobic	Ethanol	Acetic acid, furfural, and HMF	Acetic acid reduced ethanol productivity with increasing concentration; 3.5 g/L. Furfural produced a delay on sugar consumption rates and HMF did not have a significant effect on the concentration range studied.	*P. stipitis*	Solid fraction and wet slurry were mixed for slightly better production of sugar (xylose and glucose)	(Bellido et al., 2011)
Wheat straw	Wet oxidation, hydrothermal process (oxygen pressure and carbonate addition)	wheat straw mixed with Na_2CO_3 and water; Anaerobic medium prepared with N_2/CO2 (4:1)	Ethanol	Furfural and 5-hydroxymethyl-furfural	Without oxygen and carbonate least, biomass production (Total sugar: 8 g/L) and 0.2 g/L ethanol production [Oxygen has low effect on inhibitor production, With oxygen and base/no base: lowest ethanol production (0.2 g/L)	Strain A3 and Thermoanaerobacterium *thermosaccharolyticum HG8*	Without carbonate but with oxygen, maximum biomass production (Total sugar: 11 g/L) Carbonate but no oxygen: 1.2 g/L ethanol	(Ahring et al., 1996)
Wheat straw	Combined Wet Oxidation and Alkaline Hydrolysis	20 g/L straw, 170 °C, 5 to 10 min 85% yield of glucose	Enzymatic convertible cellulose to glucose	Furfural and hydroxymethyl-furfural,	No pretreatment with base (only 22% w/w convertible cellulose was obtained	*Aspergillus niger*	Addition of Na_2CO_3 (base) increased the convertible cellulose from 22% w/w to around 80 % w/w Oxygen pressure improved filterability but no effect on ECC	(Bjerre et al., 2000)

(Continued)

Table 6.2 (Continued)

Lignocellulose material	Process details/ pretreatment method	Experimental details	End product target	Inhibitors	Concentration/ effects of inhibitors	Microorganism	Possible solution/ inferences	References
Wheat straw	Dilute sulphuric acid pretreatment	50 °C ;pH = 10	Lipids	Acetic acid, furfural, and HMF	Biomass and lipid content in C. curvatus was 10.9 g/L and 37.2% (control) respectively. HMF >3 g/L low lipid production. At <3 g/L, biomass and lipid content dropped only by 5.7% and 7.9%, respectively. At 1 g/L, furfural inhibited biomass and lipid production by 72.0% and 62.0%, respectively. In the presence of 0.7 g/L (7.3 mM) furfural, biomass production decreased by 44.2%	*C. curvatus* (ATCC 20509),	Good among other reporter literature (The high tolerance of *C. curvatus* to the inhibitors furfural and HMF also suggested that detoxification of the hydrolysate is not necessary for optimal yeast growth and lipid production.	(Yu et al., 2011)

Rice straw	Electron beam irradiation	Rice straw irradiated by cellulase using an electron beam at a current of 0.03–0.24 mA, a dose of 7.6–90 kGy and 1–2 MeV.	Glucose	Acetic acid, furfural, and HMF	Enzymatic hydrolysis of untreated rice straw with 60 FPU of cellulase and 30 CBU of b-glucosidase, resulted in a yield of 5.1% and 22.6% of theoretical maximum after hydrolysis for 24 and 132 h, respectively; Glucose yield of 43.1% and 52.1% after hydrolysis for 24 and 132 h, respectively	*Aspergillus niger*	Crystallinity and enzyme digestibility pretreated by EBI under different conditions showed some correlation with a correlation coefficient of 0.5107	(Bak et al., 2009)
Switch grass	Lime pretreatment	50 °C, 24 h, 0.10 g $Ca(OH)_2$/g raw biomass; wash intensity of 100 ml water/g raw biomass	Ethanol	Lignin compounds	Resulted in limited lignin reduction, ranging from 16.7% to 35.5%,	Cellulase from *Trichoderma reesei* supplemented with cellobiase from *Aspergillus niger*	porosity improvement of the biomass pretreated at various pretreatment severities. Yields of glucose, xylose, and total reducing sugars reached 239.6, 127.2, and 433.4 mg/g raw biomass, which were respectively 3.15, 5.78, and 3.61 times those of untreated biomass	(Xu et al., 2010)

(*Continued*)

Table 6.2 (Continued)

Lignocellulose material	Process details/ pretreatment method	Experimental details	End product target	Inhibitors	Concentration/ effects of inhibitors	Microorganism	Possible solution/ inferences	References
Softwood (spruce)	Dilute acid treatment (DA) and Sulfite pretreatment (SPORL)	180 °C for 30 min with sulfuric acid; liquor to wood ratio =5	Ethanol	Furfural, 5-hydroxymethylfurfural, and formic, acetic and levulinic acids	Inhibitory products were only 35% of what was formed in DA pretreatment. SPORL pretreatment dissolved approximately 32% of the lignin as lignosulfonate, Cellulose-to-glucose conversion yields were 91% at 24 h for the SPORL substrate and 55% at 48 h for the DA substrate, respectively.	Commercial enzymes, Cellulase and b-glucosidase	Complete enzymatic hydrolysis in SPORL (>90%) within 24 h SPORL removed the recalcitrance not only by dissolving hemicellulose and depolymerizing cellulose but also by partially (32%) dissolving lignin and sulfonating the residual lignin in substrate higher sugar recovery in SPORL and produced much lower levels of traditional fermentation inhibitors than dilute acid.	(Shuai et al., 201)
Saccharum spontaneum, a weedy material	Pretreated with H_2SO_4, NaOH, and Aqueous ammonia (AA)	dilute sulfuric acid (0.75%, 1.5%); Substrate pretreated at 160 C for varying time periods (15–60 min)	Ethanol	Furfurals Phenolics	Detoxification of S. spontaneum hydrolysate (160 C, 15 min, and 1.5% sulfuric acid) by calcium hydroxide over-liming; 41.75 % (furfural removal) 33.21% (furfural removal)	*Pichia stipitis* NCIM3498, and m *A. oryzae* MTCC1846	AA treatment showed the maximum release of sugars after enzymatic hydrolysis (93.3% HE) hydrolytic eff., and eventually showed the maximum ethanol production with *P. stipitis* NCIM3498	(Chandel et al., 2011)

phenolics to some extent. Also, the product yield was not greatly enhanced by neutralization as compared to other physicochemical treatment methods.

For example, Chandel et al. (2007) studied four different detoxification methods viz. ion exchange treated hydrolysate, activated charcoal, laccase treatment, overliming and neutralized hydrolysate. It was found that the ethanol yield was in the order: neutralized hydrolysate (0.22 g/g) < overliming (0.30 g/g) < activated charcoal (0.42 g/g) < ion exchange treatment (0.48 g/g). However, when compared with the activated carbon treatment method, overliming performed better in terms of generating more fermentable hydrolysates whereas the former was not producing it at all (Barbosa et al., 1992). Further, the main mechanism governing overliming detoxification was metal cation removal as compared to the apparent decrease in HMF or other inhibitors that are expected to occur (as cation removal changes the inorganic ion content which might increase the fermentability).

Activated charcoal method performs better in removing inhibitor compounds as compared to overliming. For instance, Hodge et al. (2009) showed that overliming achieved higher fermentable yield with HMF removal between 48%–72% while for activated charcoal, it was >85% of treatment. These results also suggested that optimum detoxification may be dependent on the hydrolysates and the conditions under which they are generated. As an example, Carvalheiro et al. (2005) found that the best detoxification method for hydrolysates generated from hot water (pretreated brewery spent grain) was activated charcoal as compared to overliming or ion exchange resin. One important inference can be gathered from above studies that under hot water pretreatment, there will be less inorganic hydrolysates as compared to when dilut sulphuric acid pretreatment is performed and the removal of inhibitors will be enhanced. It is because the primary inhibitor then comprises of phenolics and furans which can later be easily removed by the activated charcoal method. Again, it depends on the pretreatment method being employed, as for dilute sulphuric acid; the function of the inhibitory compound present is due to both organic and inorganic material. Hence, the activity of inhibitor removal via any discussed process depends on many parameters mainly including the type of lignocellulosic biomass, kind of pre-treatment employed and some experimental parameters.

In industry, activated charcoal and overliming are generally used for the removal of the inhibitor compounds formed during the pre-treatment process. Both activated charcoal method and overliming have been proved to be economical than other discussed methods (especially for the furan derivatives). Researchers also laid stress on the improvement factors, such as the ratio of charcoal and hydrolysates, temperature, pH, contact time, etc. that need to be investigated from time to time to lay an optimized protocol for the industrial scale production.

Another kind of physicochemical treatment which is very common is the use of extraction solvents. Mainly, ethyl acetate, chloroform, and trichloroethylene are being used for this purpose (Cantarella et al., 2004). However, other extraction solvents, such as octanol and kerosene are also found to be of equally potential. A recent study by Zhu et al. (2011) achieved elimination of over 70% of acetic acid, >45% of HMF and 100% of furfural from prehydrolysate corn stover using trialkylamine extraction (30% trialkylamine-50% n-octanol -20% kerosene). However, ethyl acetate is the most common extraction solvent that had been employed over the years. The increase in fermentation inhibitors has been found to increase from 0% to 93% using *P. stipitis* as compared to the reference fermentation (Wilson et al., 1998).

A study by Cruz et al. (1999) showed that the removal of acetic acid, (56%) and other phenolic compounds (>84%) occured when eucalyptus wood hemicelluloses hydrolysates were used and extracted via ethyl acetate or diethyl ether. It has been found lately that the low molecular weight compound (phenolic) were the most inhibiting compounds extracted via ethyl acetate extraction (Zhuang et al., 2009). Also, the detoxification effect combined with overliming (calcium hydroxide) in conjugation with ethyl acetate showed an increase in fermentation efficiency (>90%) and higher yield (0.45 g/g) and productivity (0.67 g/l/h) using thermotolerant *Saccharomyces cerevisiae VS3 strain* (Pasha et al., 2007). Some further physical and chemical detoxification can be comprehended from Table 6.3.

6.3.3 Correlation (Synergistic Effects) of Inhibitory Compounds and their Detoxification

The mixture of different inhibitors exhibits a synergistic effect and affects the bacterial metabolism in alleviating the inhibitory property. For example, Zhao et al. (2012) investigated the effect of some inhibitors on the growth of lipid-producing microorganisms (oleaginous): *Rhodosporidium toruloides* and found that when formic acid, acetic acid, furfural, and vanillin were present at isolated concentration in medium at 2, 5, 0.5 and 1.5 g/L respectively, they pose inhibitory effects on the growth of cell biomass. However, the presence of all such compounds in the medium had a synergistic effect which found to dramatically reduce the minimum critical inhibition concentration and led to decrease in lipid production. When concentration of acetic acid, furfural, and vanillin were 2, 0.25, 0.25 g/L respectively and 2, 0 and 0.125 g/L respectively, the glucose and xylose utilization even decrease to zero and biomass growth reduced down to 0.38 g/L all the way from 11.59 g/L when compared to control experiment (no inhibitory compounds presence). The exact mechanism of these synergistic effects of the inhibitors needs to be evaluated in future as it will help to design the detoxification steps in industrial level more easily.

For instance, detoxification should not be the only prime focus for the removal/minimization of the inhibitory compound but focus should also be on the eradication of synergistic effects which hampers biomass growth. By removing the synergistic effects of inhibitory compounds, one can evenly optimize the detoxification steps which will lead to a reduction in cost and a better economy for products obtained at industrial scale. For example, furfural is known to be the strongest inhibitor formed during the lignocellulosic pretreatment procedure. Its small concentration, if left unremoved, can lead to higher inhibition of the product yield. Thus, even though other inhibitors are efficiently removed following the detoxification process, it can lead to inhibition at the end/downstream process due to persistent nature of furfural which has a

Table 6.3 Detoxification studies on various lignocellulosic biomass and their effect on inhibitor removal formed during pre-treatment step.

Detoxification	Pretreatment used	Microorganism/ enzyme used	Inhibitors targeted/removed	Product and lignocellulosic biomass used	Results	References
Bio-detoxification	Steam explosion	*Amorphotheca resinae* ZN1	Acetic acid, Levulinic acid, HMF, Furfural Acetic acid, Levulinic acid, HMF and furfural removed from 2.5 g/L to 0.6 g/l, 3 g/L t 0.6 g/L, 0.6 g/L to 0, 0.9 g/L to 0 respectively.	Ethanol fermentation using corn stover material; Rice straw, cotton stalk Wheat straw and rape straw	32.9 ± 1.2 highest glucose yield control (21.4 ± 0.8)	(Zhang et al., 2010)
Alkaline peroxide treatment (AP; chemical treatment)	steam-explosion	*Clostridium acetobutylicum* ATCC 824	soluble lignin compounds (SLC)	Acetone–butanol–ethanol (ABE) production from corn stover	1.77 g/L SLC in control 1.12 ± 0.08 g/L after AP treatment ABE yield of 0.30 and a productivity of 0.17 g/(L h) yield of 0.23; productivity of 0.15 g/(L h) in the control medium containing 50 g/L mixed sugars.	(Wang et al., 2011)
Drying and water-washing	Steam explosion (SE)	*Trichoderma reesei*	Furfural and phenol compounds	Ethanol from corn straw	Drying and Water-washing removed 81% of the furfural and 85% of the phenol compounds from SECS.	(Li et al., 2008)
Soluble polyelectrolyte adsorption; polyelectrolyte polyethyleneimine (PEI)	Dilute acid pretreated	Saccharomyces cerevisiae	Acetic acid, furfural, and 5-hydroxymethylfurfural (HMF)	Ethanol from Ponderosa pine slurry	Maximum furfural and HMF separation: 88.3% of furfural and 66.4% of HMF could be removed. On the other hand, only 23.1% and 13.4% of furfural and HMF, respectively, were removed from a post-enzymatic hydrolysis sample	(Carter et al., 2011)

(*Continued*)

Table 6.3 (Continued)

Detoxification	Pretreatment used	Microorganism/ enzyme used	Inhibitors targeted/removed	Product and lignocellulosic biomass used	Results	References
Optimization of overliming treatments	Dilute sulfuric acid treatment	Recombinant Escherichia coli LY01 as the biocatalyst	Acetic, formic, levulinic, Furfural, hydroxymethylfurfural	Ethanol from sugar cane bagasse hydrolysates (primarily pentose sugars)	Organic acids (acetic, formic, levulinic) were not affected; revealed a substantial reduction in furfural, hydroxymethylfurfural, bagasse hydrolysate was rapidly and efficiently fermented (>90% yield) by LY01. reduced total furan (furfural + HMF) content by more than half, from 1333 mg/L to 590 mg/L	(Martinez et al., 2000)
Two-stage hydrothermal pretreatments with different severities combined with physical refining (PFI mill beating)	Hydrothermal pretreatment	Hydrolysis of commercial enzymes Cellic CTec2 at an enzyme loading of 6 FPU/g organic matter	Formic acid, furfural, acetic acid, HMF and furfural	wheat straw	highest total sugars recovery of 66 %; Acetic acid and formic acid: not removed significantly for furfural and HMF:while generated a relative low level of inhibitors (ND) for low temp (100 °C) first stage treatment and relatively higher when high temp was used (140 °C).	(Min et al., 2015)

Zeolites treatment	Acid hydrolysis with 0.2 M sulfuric acid	*Escherichia coli* FBR5	5-hydroxymethylfurfural (HMF), furfural and vanillin	ethanol	Untreated model media contained 3 g/L of HMF, 2.57 g/L of furfural, and 2.7 g/L of vanillin; zeolite treated media contained 0.09 g/L of HMF, 0.0 g/L of furfural, and 0.07 g/L of vanillin, respectively. The untreated hydrolyzate contained 0.15 g/L of HMF, 1.3 g/L of furfural, and 0.04 g/L of vanillin while the zeolite treated hydrolyzate contained no detectable amount of inhibitors.	(Ranjan et al., 2009)
Cation exchange resin (CS16GC), neutral polymer adsorbent (XAD-16), and granulated activated carbon (GAC)	Adsorbates were dissolved in 20 wt.% H_2SO_4(aq).	NA	furfural, HMF, and acetic acid	Bioethanol from wood-based biomass	GAC has the highest adsorption capacity XAD-16 and CS16GC had lower adsorption capacities but also shorter cycle times due to easier regeneration. Productivity increased when regenerating with 50 wt.% EtOH(aq) solution.	(Sainio et al., 2011)

(Continued)

Table 6.3 (Continued)

Detoxification	Pretreatment used	Microorganism/ enzyme used	Inhibitors targeted/removed	Product and lignocellulosic biomass used	Results	References
Bio abatement	Dilute acid hydrolysate	*Coniochaeta ligniaria* NRRL30616 for bioabatement. For fermentation: *Saccharomyces* sp. LNH-ST	Furfural and 5-hydroxymethylfurfural (HMF)	corn stover	Improved metabolism of pentoses by a recombinant bacterial strain, *E. coli* FBR5; furfural to furfuryl alcohol	(Nichols et al., 2008)
Supplemented cellulase: xylanase and β-xylosidase	Pretreated with ammonia fiber expansion (AFEX) and dilute acid	NA	xylo-oligomer and xylan inhibition	corn stover	glucan conversions increased by 27% and 8% for corn stover pretreated with ammonia fiber expansion (AFEX) and dilute acid, respectively.	(Qing et al., 2011)
Optimizing pretreatment	dilute acid treatment	NA; cellulase cocktail, hydrolyzed with commercial cellulase enzymes	Formic acid, furfural, acetic acid and 5-hydroxymethylfurfural	wheat straw	27% increase in monosaccharide yield Rinsing with deionized water effective in removing inhibitors, such as 86% of furfural; Optimizing pretreatment conditions reduced inhibitors by 37%.	(Rajan et al., 2014)

Bioabatement	dilute acid or liquid hot water at 10% (w/v) solids	*C. ligniaria* NRRL30616 ; Bio-abatement of liquors was followed by enzymatic hydrolysis of cellulose	furan aldehydes (furfural and HMF), acetate, and phenolic compounds	corn stover liquors	By 48 h, nearly all acetate was removed, and the concentrations of HMF, furfural, and phenolic compounds were decreased by more than 50%. For bio-abated LHW liquors, a substantial amount of acetate removed (>95%), as well as HMF, furfural, and phenolic compounds (>65%) from liquor bio-abated for 48 h ; Bio-abatement improved yield; cellulose conversion was increased between 1.2-1.5 fold as compared to non-abated process.	(Cao et al., 2013)
Biological abatement	Dilute acid-pretreated hydrolysates	*Coniochaeta ligniaria;* bioabatement with strain NRRL strain 30616 (fermentation by *Saccharomyces cerevisiae*)	Acetate; Furan aldehydes (furfural and HMF)	Ethanol from switchgrass, reed canarygrass, and alfalfa stems	Bioabatement eliminated the extended fermentation lag times associated with inhibitory compounds; The concentration of acetic acid was reduced 32 ± 23% by bioabatement with *C. ligniaria* ; Bioabatement reduced the total furan concentration by 33–100%, while overliming reduced total furans by 13–36%.	(Nichols et al., 2010)

(*Continued*)

Table 6.3 (Continued)

Detoxification	Pretreatment used	Microorganism/ enzyme used	Inhibitors targeted/removed	Product and lignocellulosic biomass used	Results	References
Overliming treatment; Sulfite evaporation; Anion exchange; enzymatic detoxification with laccase	Dilute-acid	For fermentation: *Saccharomyces cerevisiae* detoxification with *Trichoderma reesei*	Aliphatic acids, furan derivatives, and phenolic compounds	Spruce	Ion exchange treatment with calcium hydroxide, and treatment with *T. reesei* were the most effective detoxification methods. Evaporation treatment with 0.1% sulfite were the least efficient detoxification methods. Treatment with laccase specifically removed only one group of the inhibitors(phenolic compounds). Anion exchange: most efficient method in removing all three inhibitors	(Larsson et al., 1999)
Trialkylamine extraction(30% trialkylamine-50% n-octanol -20% kerosene)	Sulfuric acid at moderate temperature was applied	*Pichia stipitis* CBS 5776	Acetic acid, 5-hydroxymethylfurfural furfural	Ethanol from corn stover	Detoxification results indicated that 73.3% of acetic acid, 45.7% of 5-hydroxymethylfurfural and 100% of furfural could be removed. The sugar utilization ratio was 93.2%; the ethanol concentration reached its peak value of 21.8 g/l, which was corresponding to 82.3% of the theoretical value.	(Zhu et al., 2011)

Ion exchange treated hydrolysate, activated charcoal, laccase overliming and neutralized hydrolysate	Acid hydrolysis	*Candida shehatae* NCIM 3501	Total phenolics, furans	sugarcane bagasse hydrolysate	Candida shehatae NCIM 3501 showed maximum ethanol yield (0.48 g/g) from ion exchange treated hydrolysate, followed by activated charcoal (0.42 g/g), laccase (0.37 g/g), overliming (0.30 g/g) and neutralized hydrolysate (0.22 g/g).	(Chandel et al., 2007)
Overliming by calcium hydroxide	Enzymatic hydrolysis of S. spontaneum pretreated with H2SO4, NaOH, and Aqueous ammonia (AA)	*Pichia stipitis* NCIM3498, and m A. oryzae MTCC1846	Furfurals Phenolics 41.75 % (furfural removal) 33.21% (furfural removal)	Ethanol	AA treatment shown maximum release of sugars after enzymatic hydrolysis (93.3% hydrolytic eff), and the maximum ethanol production too; 41.75 % (furfural removal) 33.21% (furfural removal) with P. stipitis NCIM3498	(Chandel et al., 2011)
Supercritical fluid extraction (SFE): methanol and subsequent CO2 treatmtent	Dilute-acid	*Saccharomyces cerevisiae*	Phenolics and furan derivatives	Ethanol from Spruce	Ethanol yield: 0.30 ± 0.03 (control) 0.43 ± 0.01 (SPE); Concentration of MeOH extraction higher for HMF (20.2 g/L), furfural (35 g/L), acetic acid (16.5 g/L), levulinic acid (5.6 g/L), vanillin (8.2 g/L) as compared to normal extraction in order: 4.6 g/L, 0.054 g/L, 2.1 g/L, 1.6 g/L, 3 g/L	Persson et al., 2002)

tendency to show synergistic effects with other residual (if any) inhibitors. Thus, the focus (designing the detoxification steps) should rather be on the removal of furfural (in this example, where vacuum evaporation method has found to be quite influential). These steps can then be followed by activated charcoal treatment based on the amount of other lignin or phenolic compounds left to be detoxified.

6.4 Conclusion

The main challenge in bioproduction is the production of inhibitory compounds during pretreatment of biomass. These inhibitors formed during the pre-treatment of lignocellulose biomass undergoing fermentation leads to less availability of fermentable cellulose by decreasing their surface area. This affects the product yield and productivity (also decrease in biomass growth) as these inhibitors affect the microorganism's cellular activity by creating an unfriendly environment where the enzymatic reaction is significantly disrupted. Acid pre-treatment has been found to be the most commonly employed technique for obtaining fermentable biomass substrate. However, some modification/optimization or overhaul of existing pre-treatment process can reduce the inhibitory effects and maintain the cellular activity by enhancing the homeostasis of the process and enzymatic activity of the cells. Some inhibitory compounds may still persist in the solution and their removal becomes a necessary step (detoxification).

Some successful detoxification process especially bio-abatement method of removing inhibitors has been lately found as an economical, viable and specific technique. The advantages, such as producing less side chain reaction, less waste material/products, and less energy input, has made it more attractive over the usual physical and chemical detoxification methods. The bio-abatement process involves the culture of another microorganism apart from the one used for the fermentation process. These organisms do not hamper the existing activity and in fact, enhances the product yield by detoxifying the inhibitors formed during the pre-treatment step. In addition, synergistic effects due to the presence of multiple inhibitors and other pretreatment by-products need to investigate. This might be essential to improve the economical assessment of industrial-scale production.

The need for sustainable development, depletion of fossil fuels and the increasing interest in green chemistry has necessitated the use of renewable biomass for the production of commodity chemicals and fuels. Recent advances in molecular biology, as well as chemical and physical processes, have enabled the development of novel and efficient techniques for both pre-treatments of biomass and production of commodity chemicals. Commercial production utilizing some of these novel processes has commenced and others have proven their feasibility at pilot scale. More of these commercial scale bioproduction using renewable biomass as feedstock is essential to replace the dependence on fossil fuels, paving the way for a sustainable biobased economy.

Acknowledgments

The authors are sincerely thankful to the Natural Sciences and Engineering Research Council of Canada (Discovery Grant 355254 and NSERC Strategic Grant) and Ministère des Relations Internationales du Québec (coopération Québec-Catalanya 2012-2014; project 07.302). R K Das sincerely acknowledges the financial support by the TERI Deakin Nanobiotechnology Centre, Biotechnology and Management of Bioresources Division, The Energy and Resources Institute, New Delhi.

References

Ahring, B. K., Jensen, K., Nielsen, P. et al. (1996) Pretreatment of wheat straw and conversion of xylose and xylan to ethanol by thermophilic anaerobic bacteria. *Bioresour Technol*, 58(2), 107–113. doi:10.1016/s0960-8524(96)00090-9

Álvarez-Ordóñez, A., Fernández, A., Bernardo, A., and López, M. (2010) Arginine and lysine decarboxylases and the Acid Tolerance Response of Salmonella Typhimurium. *Int. J. Food Microbiol.*, 136(3), 278–282. doi:10.1016/j.ijfoodmicro.2009.09.024

Antoni, D., Zverlov, V. V., and Schwarz, W. H. (2007) Biofuels from microbes. *Appl. Microbiol. Biotechnol.*, 77(1), 23–35. doi:10.1007/s00253-007-1163-x

Antonopoulou, G., Gavala, H. N., Skiadas, I. V. et al. (2008) Biofuels generation from sweet sorghum: Fermentative hydrogen production and anaerobic digestion of the remaining biomass. *Bioresource Technology*, 99(1), 110–119. doi:10.1016/j.biortech.2006.11.048

Ask, M., Bettiga, M., Mapelli, V., and Olsson, L. (2013) The influence of HMF and furfural on redox-balance and energy-state of xylose-utilizing Saccharomyces cerevisiae. *Biotechnol. Biofuels.*, 6(1), 22. doi:10.1186/1754-6834-6-22

Bak, J. S., Ko, J. K., Han, Y. H. et al. (2009) Improved enzymatic hydrolysis yield of rice straw using electron beam irradiation pretreatment. *Bioresour. Technol.*, 100(3), 1285–1290. doi:10.1016/j.biortech.2008.09.010

Balan, V. (2014) Current Challenges in Commercially Producing Biofuels from Lignocellulosic Biomass. *ISRN Biotechnology*, 2014, 1–31. doi:10.1155/2014/463074

Balat, H. and Kırtay, E. (2010) Hydrogen from biomass – Present scenario and future prospects. *International Journal of Hydrogen Energy*, 35(14), 7416–7426. doi:10.1016/j.ijhydene.2010.04.137

Behera, S., Arora, R., Nandhagopal, N., and Kumar, S. (2014) Importance of chemical pretreatment for bioconversion of lignocellulosic biomass. *Renewable and Sustainable Energy Reviews*, 36, 91–106. doi:10.1016/j.rser.2014.04.047

Bellido, C., Bolado, S., Coca, M. et al. (2011) Effect of inhibitors formed during wheat straw pretreatment on ethanol fermentation by Pichia stipitis. *Bioresour. Technol.*, 102(23), 10868–10874. doi:10.1016/j.biortech.2011.08.128

Bjerre, A. B., Olesen, A. B., Fernqvist, T. et al. (2000) Pretreatment of wheat straw using combined wet oxidation and alkaline hydrolysis resulting in convertible cellulose and hemicellulose. *Biotechnol. Bioeng.*, 49(5), 568–577. doi:10.1002/(sici)1097-0290(19960305)49:5 < 568::aid-bit10 > 3.0.co;2-6

Canilha, L., Chandel, A. K., Suzane dos Santos Milessi, T. et al. (2012) Bioconversion of Sugarcane Biomass into Ethanol: An Overview about Composition, Pretreatment Methods, Detoxification of Hydrolysates, Enzymatic Saccharification, and Ethanol Fermentation. *Journal of Biomedicine and Biotechnology*, 2012, 1–15. doi:10.1155/2012/989572

Cantarella, M., Cantarella, L., Gallifuoco, A. et al. (2004) Comparison of different detoxification methods for steam-exploded poplar wood as a substrate for the bioproduction of ethanol in SHF and SSF. *Process Biochemistry*, 39(11), 1533–1542. doi:10.1016/s0032-9592(03)00285-1

Cao, G., Ximenes, E., Nichols, N. N. et al. (2013) Biological abatement of cellulase inhibitors. *Bioresour. Technol.*, 146, 604–610. doi:10.1016/j.biortech.2013.07.112

Carter, B., Squillace, P., Gilcrease, P. C., and Menkhaus, T. J. (2011) Detoxification of a lignocellulosic biomass slurry by soluble polyelectrolyte adsorption for improved fermentation efficiency. *Biotechnol. Bioeng.*, 108(9), 2053–2060. doi:10.1002/bit.23152

Carvalheiro, F., Duarte, L. C., Lopes, S. et al. (2006) Supplementation requirements of brewery's spent grain hydrolysate for biomass and xylitol production by Debaryomyces hansenii CCMI 941. *J. Ind. Microbiol. Biotechnol.*, 33(8), 646–654. doi:10.1007/s10295-006-0101-8

Carvalheiro, F., Duarte, L. C., Lopes, S. et al. (2005) Evaluation of the detoxification of brewery's spent grain hydrolysate for xylitol production by Debaryomyces hansenii CCMI 941. *Process Biochemistry*, 40(3-4), 1215–1223. doi:10.1016/j.procbio.2004.04.015

Chandel, A. K., Kapoor, R. K., Singh, A., and Kuhad, R. C. (2007) Detoxification of sugarcane bagasse hydrolysate improves ethanol production by Candida shehatae NCIM 3501. *Bioresour. Technol.*, 98(10), 1947–1950. doi:10.1016/j.biortech.2006.07.047

Chandel, A. K., Singh, O. V., Venkateswar Rao, L. et al. (2011) Bioconversion of novel substrate Saccharum spontaneum, a weedy material, into ethanol by Pichia stipitis NCIM3498. *Bioresour. Technol.*, 102(2), 1709–1714. doi:10.1016/j.biortech.2010.08.016

Chaturvedi, V. and Verma, P. (2013) An overview of key pretreatment processes employed for bioconversion of lignocellulosic biomass into biofuels and value added products. *3 Biotech.*, 3(5), 415–431. doi:10.1007/s13205-013-0167-8

Cirne, D. G., Lehtomäki, A., Björnsson, L., and Blackall, L. L. (2007) Hydrolysis and microbial community analyses in two-stage anaerobic digestion of energy crops. *J. Appl. Microbiol.*, 103(3), 516–527. doi:10.1111/j.1365-2672.2006.03270.x

Cruz, J. M., Domínguez, J. M., Domínguez, H., and Parajó, J. C. (1999). Solvent extraction of hemicellulosic wood hydrolysates: a procedure useful for obtaining both detoxified fermentation media and polyphenols with antioxidant activity. *Food Chemistry*, 67(2), 147–153. doi:10.1016/s0308-8146(99)00106-5

Danner, H. and Braun, R. (1999). Biotechnology for the production of commodity chemicals from biomass. *Chemical Society Reviews*, 28(6), 395–405. doi:10.1039/a806968i

Demirel, B. and Scherer, P. (2008). The roles of acetotrophic and hydrogenotrophic methanogens during anaerobic conversion of biomass to methane: a review. *Reviews in Environmental Science and Bio/Technology*, 7(2), 173–190. doi:10.1007/s11157-008-9131-1

Du, B., Sharma, L. N., Becker, C. et al. (2010a) Effect of varying feedstock-pretreatment chemistry combinations on the formation and accumulation of potentially inhibitory degradation products in biomass hydrolysates. *Biotechnol Bioeng*, 107(3), 430–440. doi:10.1002/bit.22829

Du, B., Sharma, L. N., Becker, C. et al. (2010b) Effect of varying feedstock-pretreatment chemistry combinations on the formation and accumulation of potentially inhibitory degradation products in biomass hydrolysates. *Biotechnol. Bioeng.*, 107(3), 430–440. doi:10.1002/bit.22829

Ezeji, T. C., Groberg, M., Qureshi, N., and Blaschek, H. P. (2003) Continuous Production of Butanol from Starch-Based Packing Peanuts. *Appl. Biochem. Biotechnol.*, 106(1-3), 375–382. doi:10.1385/abab:106:1-3:375

Gad kariem, E. A., Alabdeen M. S. Z., Idrees Tah, M. et al. (2009) Identification and Quantification of 5-Hydroxymethyl Furfural HMF in Some Sugar-Containing Food Products by HPLC. *Pakistan Journal of Nutrition*, 8(9), 1391–1396. doi:10.3923/pjn.2009.1391.1396

Ghangrekar, M. M., Asolekar, S. R., and Joshi, S. G. (2005) Characteristics of sludge developed under different loading conditions during UASB reactor start-up and granulation. *Water Research*, 39(6), 1123–1133. doi:10.1016/j.watres.2004.12.018

Gonzalez, R., Tao, H., Purvis, J. E. et al. (2003) Gene Array-Based Identification of Changes That Contribute to Ethanol Tolerance in Ethanologenic Escherichia coli: Comparison of KO11 (Parent) to LY01 (Resistant Mutant). *Biotechnol. Prog.*, 19(2), 612–623. doi:10.1021/bp025658q

Gray, K. A., Zhao, L., and Emptage, M. (2006) Bioethanol. *Curr Opin Chem Biol*, 10(2), 141–146. doi:10.1016/j.cbpa.2006.02.035

Grzenia, D. L., Schell, D. J., and Ranil Wickramasinghe, S. (2012). Membrane extraction for detoxification of biomass hydrolysates. *Bioresour Technol*, 111, 248–254. doi:10.1016/j.biortech.2012.01.169

Grzenia, D. L., Schell, D. J., and Wickramasinghe, S. R. (2008) Membrane extraction for removal of acetic acid from biomass hydrolysates. *Journal of Membrane Science*, 322(1), 189–195. doi:10.1016/j.memsci.2008.05.030

Guo, G., Chen, W., Chen, W. et al. (2008) Characterization of dilute acid pretreatment of silvergrass for ethanol production. *Bioresour. Technol.*, 99(14), 6046–6053. doi:10.1016/j.biortech.2007.12.047

Guo, T., He, A.-Y., Du, T.-F. et al. (2013a) Butanol production from hemicellulosic hydrolysate of corn fiber by a Clostridium beijerinckii mutant with high inhibitor-tolerance. *Bioresour. Technol.*, 135, 379–385. doi:10.1016/j.biortech.2012.08.029

Guo, X., Cavka, A., Jönsson, L. J., and Hong, F. (2013b) Comparison of methods for detoxification of spruce hydrolysate for bacterial cellulose production. *Microb. Cell Fact.*, 12(1), 93. doi:10.1186/1475-2859-12-93

Guo, X. M., Trably, E., Latrille, E. et al. (2014) Predictive and explicative models of fermentative hydrogen production from solid organic waste: Role of butyrate and lactate pathways. *International Journal of Hydrogen Energy*, 39(14), 7476–7485. doi:10.1016/j.ijhydene.2013.08.079

Hahn-Hägerdal, B., Karhumaa, K., Fonseca, C. (2007) Towards industrial pentose-fermenting yeast strains. *Appl Microbiol Biotechnol.*, 74(5), 937–953. doi:10.1007/s00253-006-0827-2

Hanly, T. J. and Henson, M. A. (2014). Dynamic model-based analysis of furfural and HMF detoxification by pure and mixed batch cultures of S. cerevisiae and S. stipitis. *Biotechnol. Bioeng.*, 111(2), 272–284. doi:10.1002/bit.25101

Hara, K. Y. and Kondo, A. (2015) ATP regulation in bioproduction. *Microb. Cell Fact.*, 14(1). doi:10.1186/s12934-015-0390-6

Hawkes, F. (2002) Sustainable fermentative hydrogen production: challenges for process optimisation. *International Journal of Hydrogen Energy*, 27(11-12), 1339–1347. doi:10.1016/s0360-3199(02)00090-3

Hodge, D. B., Andersson, C., Berglund, K. A., and Rova, U. (2009) Detoxification requirements for bioconversion of softwood dilute acid hydrolyzates to succinic acid. *Enzyme and Microbial Technology*, 44(5), 309–316. doi:10.1016/j.enzmictec.2008.11.007

Hu, C., Zhao, X., Zhao, J. et al. (2009) Effects of biomass hydrolysis by-products on oleaginous yeast Rhodosporidium toruloides. *Bioresour. Technol.*, 100(20), 4843–4847. doi:10.1016/j.biortech.2009.04.041

Ibraheem, O. and Ndimba, B. K. (2013) Molecular Adaptation Mechanisms Employed by Ethanologenic Bacteria in Response to Lignocellulose-derived Inhibitory Compounds. *Int. J. Biol. Sci.*, 9(6), 598–612. doi:10.7150/ijbs.6091

Jang, Y.-S., Malaviya, A., Cho, C. et al. (2012) Butanol production from renewable biomass by clostridia. *Bioresour. Technol.*, 123, 653–663. doi:10.1016/j.biortech.2012.07.104

Jönsson, L. J. and Martín, C. (2016) Pretreatment of lignocellulose: Formation of inhibitory by-products and strategies for minimizing their effects. *Bioresour. Technol.*, 199, 103–112. doi:10.1016/j.biortech.2015.10.009

Jönsson, L. J., Palmqvist, E., Nilvebrant, N. O., and Hahn-Hägerdal, B. (1998). Detoxification of wood hydrolysates with laccase and peroxidase from the white-rot fungus Trametes versicolor. *Appl. Microbiol. Biotechnol.*, 49(6), 691–697. doi:10.1007/s002530051233

Karakashev, D., Batstone, D. J., and Angelidaki, I. (2005) Influence of Environmental Conditions on Methanogenic Compositions in Anaerobic Biogas Reactors. *Appl. Environ. Microbiol.*, 71(1), 331–338. doi:10.1128/aem.71.1.331-338.2005

Keweloh, H., Weyrauch, G., and Rehm, H. J. (1990) Phenol-induced membrane changes in free and immobilized Escherichia coli. *Appl. Microbiol. Biotechnol.*, 33(1), 66–71.

Kildegaard, K. R., Wang, Z., Chen, Y. et al. (2015) Production of 3-hydroxypropionic acid from glucose and xylose by metabolically engineered Saccharomyces cerevisiae. *Metabolic Engineering Communications*, 2, 132–136. doi:10.1016/j.meteno.2015.10.001

Klocke, M., Mähnert, P., Mundt, K. et al. (2007) Microbial community analysis of a biogas-producing completely stirred tank reactor fed continuously with fodder beet silage as mono-substrate. *Syst. Appl. Microbiol.*, 30(2), 139–151. doi:10.1016/j.syapm.2006.03.007

Ko, J. K., Um, Y., Park, Y.-C. et al. (2015) Compounds inhibiting the bioconversion of hydrothermally pretreated lignocellulose. *Appl. Microbiol. Biotechnol.*, 99(10), 4201–4212. doi:10.1007/s00253-015-6595-0

Larsson, S., Quintana-Sainz, A., Reimann, A. et al. (2000a) Influence of lignocellulose-derived aromatic compounds on oxygen-limited growth and ethanolic fermentation by Saccharomyces cerevisiae. *Appl. Biochem. Biotechnol.*, 84-86, 617–632.

Larsson, S., Quintana-Sáinz, A., Reimann, A. et al. (2000b) Influence of Lignocellulose-Derived Aromatic Compounds on Oxygen-Limited Growth and Ethanolic Fermentation by Saccharomyces cerevisiae. 617–632. doi:10.1007/978-1-4612-1392-5_47

Larsson, S., Reimann, A., Nilvebrant, N.-O., and Jönsson, L. J. (1999). Comparison of Different Methods for the Detoxification of Lignocellulose Hydrolyzates of Spruce. *Appl. Biochem. Biotechnol.*, 77(1-3), 91–104. doi:10.1385/abab:77:1-3:91

Li, H. and Chen, H. (2008) Detoxification of steam-exploded corn straw produced by an industrial-scale reactor. *Process Biochemistry*, 43(12), 1447–1451. doi:10.1016/j.procbio.2008.05.003

Lienhardt, J., Schripsema, J., Qureshi, N., and Blaschek, H. P. (2002) Butanol Production by Clostridium beijerinckii BA101 in an Immobilized Cell Biofilm Reactor. *Appl. Biochem. Biotechnol.*, 98-100(1-9), 591–598. doi:10.1385/abab:98-100:1-9:591

Lin, J., Smith, M. P., Chapin, K. C. et al. (1996) Mechanisms of acid resistance in enterohemorrhagic Escherichia coli. *Appl. Environ. Microbiol.*, 62(9), 3094–3100.

Lin, Y. and Tanaka, S. (2005) Ethanol fermentation from biomass resources: current state and prospects. *Appl. Microbiol. Biotechnol.*, 69(6), 627–642. doi:10.1007/s00253-005-0229-x

Lu, C., Zhao, J., Yang, S.-T., and Wei, D. (2012) Fed-batch fermentation for n-butanol production from cassava bagasse hydrolysate in a fibrous bed bioreactor with continuous gas stripping. *Bioresour Technol*, 104, 380–387. doi:10.1016/j.biortech.2011.10.089

Martin, C., Galbe, M., Nilvebrant, N. O., and Jonsson, L. J. (2002) Comparison of the fermentability of enzymatic hydrolyzates of sugarcane bagasse pretreated by steam explosion using different impregnating agents. *Appl. Biochem. Biotechnol.*, 98-100, 699–716.

Martinez, A., Rodriguez, M. E., York, S. W. et al. (2000) Effects of Ca(OH)2 treatments ("overliming") on the composition and toxicity of bagasse hemicellulose hydrolysates. *Biotechnol. Bioeng.*, 69(5), 526–536. doi:10.1002/1097-0290(20000905)69:5 < 526::aid-bit7 > 3.0.co;2-e

Masai, E., Katayama, Y., and Fukuda, M. (2007) Genetic and biochemical investigations on bacterial catabolic pathways for lignin-derived aromatic compounds. *Biosci. Biotechnol. Biochem.*, 71(1), 1–15.

McMahon, K. (2001) Anaerobic codigestion of municipal solid waste and biosolids under various mixing conditions—II: microbial population dynamics. *Water Research*, 35(7), 1817–1827. doi:10.1016/s0043-1354(00)00438-3

Min, D.-Y., Xu, R.-S., Hou, Z., Lv, J.-Q., Huang, C.-X., Jin, Y.-C., & Yong, Q. (2015) Minimizing inhibitors during pretreatment while maximizing sugar production in enzymatic hydrolysis through a two-stage hydrothermal pretreatment. *Cellulose*, 22(2), 1253–1261. doi:10.1007/s10570-015-0552-z

Nichols, N. N., Dien, B. S., and Cotta, M. A. (2010) Fermentation of bioenergy crops into ethanol using biological abatement for removal of inhibitors☆. *Bioresour. Technol.*, 101(19), 7545–7550. doi:10.1016/j.biortech.2010.04.097

Nichols, N. N., Sharma, L. N., Mowery, R. A. et al. (2008) Fungal metabolism of fermentation inhibitors present in corn stover dilute acid hydrolysate. *Enzyme and Microbiol. Technology*, 42(7), 624–630. doi:10.1016/j.enzmictec.2008.02.008

Monlau, F., Aemig, Q., Trably, E. et al. (2013) Specific inhibition of biohydrogen-producing Clostridium sp. after dilute-acid pretreatment of sunflower stalks. *International Journal of Hydrogen Energy*, 38(28), 12273–12282. doi:10.1016/j.ijhydene.2013.07.018

Okano, K., Tanaka, T., Ogino, C., et al. (2009) Biotechnological production of enantiomeric pure lactic acid from renewable resources: recent achievements, perspectives, and limits. *Appl. Microbiol. Biotechnol.*, 85(3), 413–423. doi:10.1007/s00253-009-2280-5

O'Sullivan, C. A., Burrell, P. C., Clarke, W. P., and Blackall, L. L. (2005) Structure of a cellulose degrading bacterial community during anaerobic digestion. *Biotechnol. Bioeng.*, 92(7), 871–878. doi:10.1002/bit.20669

Nath, K. and Das, D. (2004). Improvement of fermentative hydrogen production: various approaches. *Applied Microbiology and Biotechnology*, 65(5). doi:10.1007/s00253-004-1644-0

Ni, M., Leung, D. Y. C., Leung, M. K. H., and Sumathy, K. (2006) An overview of hydrogen production from biomass. *Fuel Processing Technology*, 87(5), 461–472. doi:10.1016/j.fuproc.2005.11.003

Palmqvist, E., Hahn-Hägerdal, B., Galbe, M., and Zacchi, G. (1996) The effect of water-soluble inhibitors from steam-pretreated willow on enzymatic hydrolysis and ethanol fermentation. *Enzyme and Microbial Technology*, 19(6), 470–476. doi:10.1016/s0141-0229(95)00234-0

Panagiotopoulos, I. A., Bakker, R. R., Budde, M. A. W. et al. (2009) Fermentative hydrogen production from pretreated biomass: A comparative study. *Bioresour. Technol.*, 100(24), 6331–6338. doi:10.1016/j.biortech.2009.07.011

Pandey, A., Soccol, C. R., Nigam, P., and Soccol, V. T. (2000) Biotechnological potential of agro-industrial residues. I: sugarcane bagasse. *Bioresour. Technol.*, 74(1), 69–80. doi:10.1016/s0960-8524(99)00142-x

Parawira, W. and Tekere, M. (2011) Biotechnological strategies to overcome inhibitors in lignocellulose hydrolysates for ethanol production: review. *Crit. Rev. Biotechnol.*, 31(1), 20–31. doi:10.3109/07388551003757816

Pasha, C., Kuhad, R. C., and Rao, L. V. (2007) Strain improvement of thermotolerant Saccharomyces cerevisiae VS3 strain for better utilization of lignocellulosic substrates. *J. Appl. Microbiol.*, 103(5), 1480–1489. doi:10.1111/j.1365-2672.2007.03375.x

Peng, X., Qiao, W., Mi, S. et al. (2015) Characterization of hemicellulase and cellulase from the extremely thermophilic bacterium Caldicellulosiruptor owensensis and their potential application for bioconversion of lignocellulosic biomass without pretreatment. *Biotechnol. Biofuels.*, 8(1). doi:10.1186/s13068-015-0313-0

Persson, P., Larsson, S., Jönsson, L. J. et al. (2002) Supercritical fluid extraction of a lignocellulosic hydrolysate of spruce for detoxification and to facilitate analysis of inhibitors. *Biotechnol. Bioeng.*, 79(6), 694–700. doi:10.1002/bit.10324

Qing, Q. and Wyman, C. E. (2011) Supplementation with xylanase and β-xylosidase to reduce xylo-oligomer and xylan inhibition of enzymatic hydrolysis of cellulose and pretreated corn stover. *Biotechnol. Biofuels*, 4(1), 18. doi:10.1186/1754-6834-4-18

Rajan, K. and Carrier, D. J. (2014) Effect of dilute acid pretreatment conditions and washing on the production of inhibitors and on recovery of sugars during wheat straw enzymatic hydrolysis. *Biomass and Bioenergy*, 62, 222–227. doi:10.1016/j.biombioe.2014.01.013

Ranjan, R., Thust, S., Gounaris, C. E. et al. (2009) Adsorption of fermentation inhibitors from lignocellulosic biomass hydrolyzates for improved ethanol yield and value-added product recovery. *Microporous and Mesoporous Materials*, 122(1-3), 143–148. doi:10.1016/j.micromeso.2009.02.025

Rasmussen, H., Sørensen, H. R., and Meyer, A. S. (2014) Formation of degradation compounds from lignocellulosic biomass in the biorefinery: sugar reaction mechanisms. *Carbohydrate Research*, 385, 45–57. doi:10.1016/j.carres.2013.08.029

Rastogi, G., Ranade, D. R., Yeole, T. Y. et al. (2008) Investigation of methanogen population structure in biogas reactor by molecular characterization of methyl-coenzyme M reductase A (mcrA) genes. *Bioresour. Technol.*, 99(13), 5317–5326. doi:10.1016/j.biortech.2007.11.024

Rohland, B., Nitsch, J., and Wendt, H. (1992) Hydrogen and fuel cells — the clean energy system. *Journal of Power Sources*, 37(1-2), 271–277. doi:10.1016/0378-7753(92)80084-o

Sainio, T., Turku, I., and Heinonen, J. (2011) Adsorptive removal of fermentation inhibitors from concentrated acid hydrolyzates of lignocellulosic biomass. *Bioresour. Technol.*, 102(10), 6048–6057. doi:10.1016/j.biortech.2011.02.107

Saxena, R. C., Adhikari, D. K., and Goyal, H. B. (2009) Biomass-based energy fuel through biochemical routes: A review. *Renewable and Sustainable Energy Reviews*, 13(1), 167–178. doi:10.1016/j.rser.2007.07.011

Sheldon, R. A. (2014) Green and sustainable manufacture of chemicals from biomass: state of the art. *Green Chem.*, 16(3), 950–963. doi:10.1039/c3gc41935e

Shuai, L., Yang, Q., Zhu, J. Y. et al. (2010) Comparative study of SPORL and dilute-acid pretreatments of spruce for cellulosic ethanol production. *Bioresour. Technol.*, 101(9), 3106–3114. doi:10.1016/j.biortech.2009.12.044

Siegrist, H., Vogt, D., Garcia-Heras, J. L., and Gujer, W. (2002) Mathematical Model for Meso- and Thermophilic Anaerobic Sewage Sludge Digestion. *Environ. Sci. Technol.*, 36(5), 1113–1123. doi:10.1021/es010139p

Silva, C. J. S. M. and Roberto, I. C. (2001) Improvement of xylitol production by Candida guilliermondii FTI 20037 previously adapted to rice straw hemicellulosic hydrolysate. *Lett. Appl. Microbiol.*, 32(4), 248–252. doi:10.1046/j.1472-765X.2001.00899.x

Sindhu, R., Binod, P., and Pandey, A. (2016) Biological pretreatment of lignocellulosic biomass – An overview. *Bioresour. Technol.*, 199, 76–82. doi:10.1016/j.biortech.2015.08.030

Stroot, P. (2001) Anaerobic codigestion of municipal solid waste and biosolids under various mixing conditions—I. digester performance. *Water Research*, 35(7), 1804–1816. doi:10.1016/s0043-1354(00)00439-5

Taherzadeh, M. J., Niklasson, C., and Lidén, G. (1997) Acetic acid—friend or foe in anaerobic batch conversion of glucose to ethanol by Saccharomyces cerevisiae? *Chemical Engineering Science*, 52(15), 2653–2659. doi:10.1016/s0009-2509(97)00080-8

Tanaka, H., Ishikawa, H., Osuga, K., and Takagi, Y. (1990) Fermentative ability of Zymomonas mobilis under various oxygen supply conditions in batch culture. *Journal of Fermentation and Bioengineering*, 69(4), 234–239. doi:10.1016/0922-338x(90)90219-m

Tanaka, K., Komiyama, A., Sonomoto, K.(2002) Two different pathways for D - xylose metabolism and the effect of xylose concentration on the yield coefficient of L -lactate in mixed-acid fermentation by the lactic acid bacterium Lactococcus lactis IO-1. *Appl. Microbiol. Biotechnol.*, 60(1-2), 160–167. doi:10.1007/s00253-002-1078-5

Tashiro, Y., Takeda, K., Kobayashi, G., and Sonomoto, K. (2005) High production of acetone–butanol–ethanol with high cell density culture by cell-recycling and bleeding. *J. Biotechnol.*, 120(2), 197–206. doi:10.1016/j.jbiotec.2005.05.031

Tracy, B. P., Jones, S. W., Fast, A. G. (2012) Clostridia: the importance of their exceptional substrate and metabolite diversity for biofuel and biorefinery applications. *Current Opinion in Biotechnology*, 23(3), 364–381. doi:10.1016/j.copbio.2011.10.008

Tran, H. T. M., Cheirsilp, B., Hodgson, B., and Umsakul, K. (2010). Potential use of Bacillus subtilis in a co-culture with Clostridium butylicum for acetone–butanol–ethanol production from cassava starch. *Biochemical Engineering Journal*, 48(2), 260–267. doi:10.1016/j.bej.2009.11.001

Vandenberghe, L. P. S., Soccol, C. R., Pandey, A., and Lebeault, J.-M. (1999). Microbial production of citric acid. *Brazilian Archives of Biology and Technology*, 42(3), 263–276. doi:10.1590/s1516-89131999000300001

Verduyn, C., Postma, E., Scheffers, W. A., and van Dijken, J. P. (1990) Physiology of Saccharomyces Cerevisiae in Anaerobic Glucose-Limited Chemostat Culturesx. *Journal of General Microbiology*, 136(3), 395–403. doi:10.1099/00221287-136-3-395

Viegas, C. A. and Sa-Correia, I. (1991) Activation of plasma membrane ATPase of Saccharomyces cerevisiae by octanoic acid. *Journal of General Microbiology*, 137(3), 645–651. doi:10.1099/00221287-137-3-645

Wahlbom, C. F. and Hahn-Hägerdal, B. (2002) Furfural, 5-hydroxymethyl furfural, and acetoin act as external electron acceptors during anaerobic fermentation of xylose in recombinant Saccharomyces cerevisiae. *Biotechnol. Bioeng.*, 78(2), 172–178. doi:10.1002/bit.10188

Wang, L. and Chen, H. (2011). Increased fermentability of enzymatically hydrolyzed steam-exploded corn stover for butanol production by removal of fermentation inhibitors. *Process Biochemistry*, 46(2), 604–607. doi:10.1016/j.procbio.2010.09.027

Warnecke, T. and Gill, R. T. (2005) *Microb. Cell Fact.*, 4(1), 25. doi:10.1186/1475-2859-4-25

Weuster-Botz, D. (1993) Continuous ethanol production by Zymomonas mobilis in a fluidized bed reactor. Part I. Kinetic studies of immobilization in macroporous glass beads. *Appl. Microbiol. Biotechnol.*, 39(6), 679–684. doi:10.1007/bf00164449

Wilson, J. J., Deschatelets, L., and Nishikawa, N. K. (1989) Comparative fermentability of enzymatic and acid hydrolysates of steam-pretreated aspenwood hemicellulose by Pichia stipitis CBS 5776. *Appl. Microbiol. Biotechnol.*, 31-31(5-6), 592–596. doi:10.1007/bf00270801

Xia, A., Cheng, J., Song, W. et al. (2015) Fermentative hydrogen production using algal biomass as feedstock. *Renewable and Sustainable Energy Reviews*, 51, 209–230. doi:10.1016/j.rser.2015.05.076

Xu, J., Cheng, J. J., Sharma-Shivappa, R. R., and Burns, J. C. (2010) Lime pretreatment of switchgrass at mild temperatures for ethanol production. *Bioresour. Technol.*, 101(8), 2900–2903. doi:10.1016/j.biortech.2009.12.015

Yang, B. Wyman, C. E. (2008) Pretreatment: the key to unlocking low-cost cellulosic ethanol. *Biofuels, Bioproducts and Biorefining*, 2(1), 26–40. doi:10.1002/bbb.49

Yang, Z., Bai, Z., Sun, H., et al. (2014) Biomass pyrolysis liquid to citric acid via 2-step bioconversion. *Microb. Cell Fact.*, 13(1). doi:10.1186/s12934-014-0182-4

Yu, X., Zheng, Y., Dorgan, K. M., and Chen, S. (2011) Oil production by oleaginous yeasts using the hydrolysate from pretreatment of wheat straw with dilute sulfuric acid. *Bioresour. Technol.*, 102(10), 6134–6140. doi:10.1016/j.biortech.2011.02.081

Zeikus, J. G. (1980) Chemical and Fuel Production by Anaerobic Bacteria. *Annual Review of Microbiology*, 34(1), 423–464. doi:10.1146/annurev.mi.34.100180.002231

Zhang, J., Zhu, Z., Wang, X. et al. (2010) Biodetoxification of toxins generated from lignocellulose pretreatment using a newly isolated fungus, Amorphotheca resinae ZN1, and the consequent ethanol fermentation. *Biotechnol. Biofuels*, 3(1), 26. doi:10.1186/1754-6834-3-26

Zhao, X., Peng, F., Du, W., et al. (2012) Effects of some inhibitors on the growth and lipid accumulation of oleaginous yeast Rhodosporidium toruloides and preparation of biodiesel by enzymatic transesterification of the lipid. *Bioprocess. Biosyst. Eng.*, 35(6), 993–1004. doi:10.1007/s00449-012-0684-6

Zhu, J., Yong, Q., Xu, Y., and Yu, S. (2011) Detoxification of corn stover prehydrolyzate by trialkylamine extraction to improve the ethanol production with Pichia stipitis CBS 5776. *Bioresour. Technol.*, 102(2), 1663–1668. doi:10.1016/j.biortech.2010.09.083

Zverlov, V. V., Berezina, O., Velikodvorskaya, G. A., and Schwarz, W. H. (2006) Bacterial acetone and butanol production by industrial fermentation in the Soviet Union: use of hydrolyzed agricultural waste for biorefinery. *Appl. Microbiol. Biotechnol.*, 71(5), 587–597. doi:10.1007/s00253-006-0445-z

7

Microbial Metabolic Pathways in the Production of Valued-added Products

Gilberto V. de Melo Pereira[1], Ana M. Finco[1], Luiz A. J. Letti[1], Susan Grace Karp[1], Maria G. B. Pagnoncelli[2], Juliana de Oliveira[1], Vanete Thomaz Soccol[1], Satinder Kaur Brar[3], and Carlos Ricardo Soccol[1]

[1] *Bioprocess Engineering and Biotechnology Department, Federal University of Paraná, Curitiba, Brazil*
[2] *Bioprocess Engineering and Biotechnology Department, Federal University of Technology - Paraná (UTFPR), Dois Vizinhos, Brazil*
[3] *Institut national de la recherche scientifique, Centre - Eau Terre Environnement, Québec, Canada*

7.1 Introduction

Currently, humankind is transitioning from an oil-based economy to a sustainable era, guided by the use of natural sources of matter and energy (Krüger et al., 2017). In this context, and considering socio-economic and environmental aspects, the use of chemicals derived from microorganisms has emerged as an important alternative for industry. The National Research Council of the United States has predicted that, by 2020, half of all organic chemicals and materials will be produced by bioconversion (Eram and Ken, 2013).

Several industrial microbial platforms, including yeast, bacteria, algae and mold, have been established for the production of value-added products such as anti-bacterial and anticancer drugs, amino acids, organic acids, vitamins, industrial chemicals and biofuels, alkaloids, terpenoids, flavonoids and polyketides (Du et al., 2011). The high demand from the biotech industries has created significant attention on the molecular, biochemical and physiological aspects of these organisms. Furthermore, the use of inexpensive and renewable raw materials (e.g. agriculture waste and effluents, industrial waste and natural gas) in bioconversion is another important aspect recently explored. Nevertheless, important aspects should be clarified before using a novel carbon source, including metabolic microbial behavior and extrinsic and intrinsic optimization factors (i.e. temperature and pH). This chapter will be dedicated to major metabolic pathways, nutritional requirements and alternative carbon sources in the production of industrially important by-products.

Microbial Sensing in Fermentation, First Edition. Edited by Satinder Kaur Brar, Ratul Kumar Das, and Saurabh Jyoti Sarma.

7.2 Microbial Molecular Structure

Microbial cells possess thousands of different molecules which are responsible for forming a defined molecular structure. This complex configuration is possible due to the cell capacity for extracting, transforming and using the energy and the nutrients from the environment (Nelson and Cox, 2013). The knowledge of these fundamental properties is essential to develop and control biotechnological processes and, consequently, production of value-added metabolites.

The increase in the number of cells is known as microbial growth. During this mechanism, several chemical reactions happen to form specific molecules. Less than 30 chemicals are fundamental in the structure of all living organisms while around 50 others can be metabolized by the cell itself. This large diversity among the microbial species can represent a wide spectrum of nutritional requirement, e.g. some microbes need complex organic compounds while others can grow with few inorganic substances.

In general, the microbial cells are constituted of 70–80% water and, on dry basis, 95% of the cellular mass are composed of a few major chemical elements, such as C, O, H, N, P, S, K, Mg, Ca and Fe. These elements are responsible for the synthesis of cellular material and normal functioning of cellular structure. The macroelements (C, O, H, N, P and S) are used in large amounts by microorganisms, while other elements are present in minor (K, Mg, Ca, Na and Fe) or traces quantities (Mn, Zn, Co, Mo, Ni, Cu). The amount of an element used by the cell has no correlation with its relative importance. For example, some trace elements can be cofactors for certain enzymes and can also regulate certain metabolic pathways (Chan, 2003). Figure 7.1 illustrates the classification of chemical elements as well as relative abundance of essential and non-essential elements for a microbial cell.

Carbon is the main chemical element of the cellular material, present in all macromolecule structure (carbohydrates, lipids and proteins). Nitrogen is another essential element present in amino acids and nucleic acids. Hydrogen and oxygen are components of many compounds and cellular water, while phosphorus is an important element in the regulation of cell metabolism and cell wall formation. Sulfur, potassium and magnesium are cofactors of many enzymatic processes. Furthermore, sulfur is present in proteins; magnesium in cell wall and membranes; and calcium in the cell wall support. Finally, trace elements (e.g. Cu,Zn and Mo) are essential for many metabolic reactions (Chan, 2003; Liu, 2017; Madigan, 2012).

Furthermore, the peculiarities relative to the cell habitat influence the nutritional requirements and energy source. Some organisms can use the energy source present in the light (named phototrophic), while for others, the energy must come from the oxidative reactions of chemical compounds (named chemotrophic). From another point of view, when the carbon source comes from organic nutrients, the organism is called heterotrophic. Finally, the microorganisms that use carbon dioxide as the major or even sole carbon source are called autotrophic. Chemotrophic (e.g. bacteria and fungi) and photoautrotrophs (e.g. algae) are the main microbial groups in industrial fermentation processes.

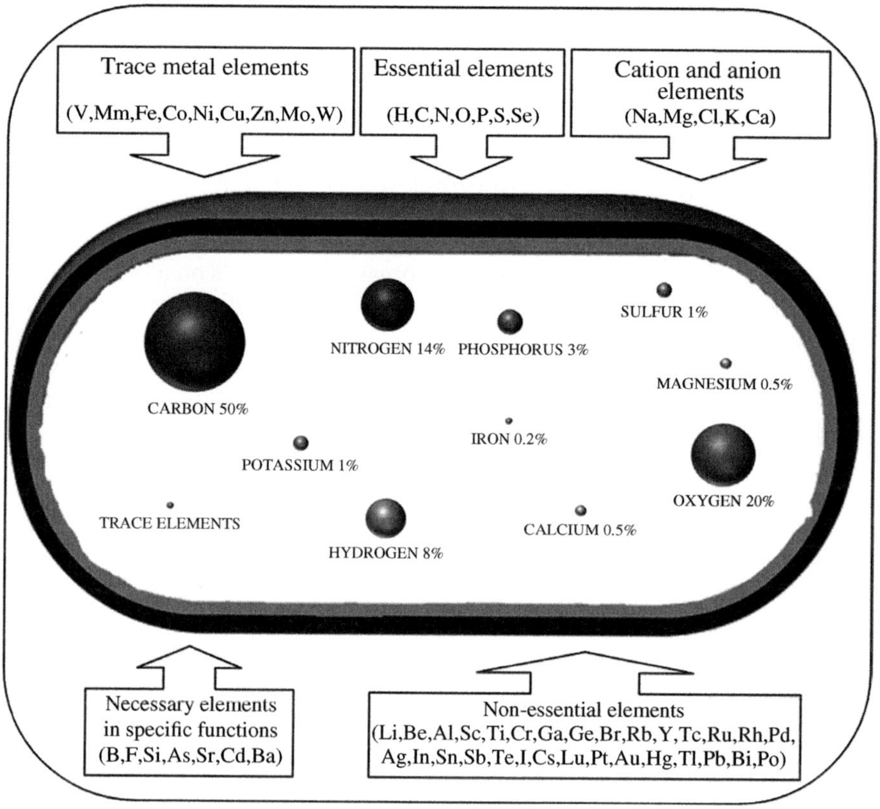

Figure 7.1 Chemical elements and their relative abundance in microbial cells. (*See insert for color representation of this figure.*)

Considering the natural compounds, the most common carbon sources in biotechnological applications are starch, molasses, whey, corn syrup and cellulose waste. Hydrogen and oxygen requirements can be supplied by media containing organic compounds. The most important nitrogen sources are ammonia or the ammonium salts, proteins, peptides and amino acids. Inorganic phosphate salts (KH_2PO_4 and K_2HPO_4) are the most common phosphate sources, besides glycerophosphates. Sulfate salts, such as $(NH_4)_2SO_4$, are the most common sulfur source.

The living organisms can extract energy or use elements as nutrients from different sources. Some classic metabolic pathways are found for different microbial species, while others are exclusive for specific groups (e.g. lipid accumulation or rare sugars synthesis). Understanding these metabolic pathways is essential to control biotechnological process through the nutritional and environmental regulation. There are a relatively low number of central pathways. In the catabolic process, the macromolecules (proteins, lipids and polysaccharides) are broken down into smaller molecules, releasing energy. Some central metabolic pathways include Embden-Meyerhof-Parnas, Pentose phosphate pathway (also called the phosphogluconate pathway), Entner-Doudoroff, Citric Acid Cycle (also known as the tricarboxylic acid cycle or Krebs cycle) and Glyoxylate Cycle (Nelson and Cox, 2013).

7.3 Biomass Production

Among the thousands of microbial compounds, products derived from biomass are one such emerging class of molecules. These include oils, proteins, peptides, lipopeptides,vitamins, among others. The production of these compounds via microbial cells have advantages over chemical synthesis, such as biodegradability, low toxicity, thermal and pH stability, production from renewable substrates and the ability to structural modification through genetic engineering or biochemical techniques. The choice of a particular microorganism depends on a complicated balance of the various nutritional, engineering, technological and economic factors (Mateles, and Tannenbaum 1968; Tannenbaum and Wang 1975; Anupama and Ravindra, 2000). Next, the main metabolic pathway for the production of proteins and lipids by microorganisms, and the use of alternative carbon sources, will be detailed.

7.3.1 Single Cell Oil

Single cell oil can be defined as the lipid obtained from microorganisms, including yeast, microalgae, bacteria and filamentous fungi, capable of accumulating more than 20% of lipids in relation to their dry weight. These lipids have emerged as an alternative to the production of fatty acids intended for the food industry or the biodiesel production. Due to their wide availability and ability to accumulate large amounts of lipids in a short time, single-celled organisms have been effective in the production of lipids, explored as a sustainable and clean alternative to fish oil (Finco et al., 2017).

The production of lipids by microorganisms is associated with important physiological functions, such as cell division, response to environmental stress and storage of energy for maintenance of viability. When growing under conditions of nutrient limitation, and with sufficient or excess carbon sources, oleaginous microorganisms silence the process of replication and use the available carbon to synthesize lipids (Huang et al., 2013; Garay et al., 2014). The accumulated lipid, serving as endogenous carbon and energy sources during starvation periods, might be a potential adaptation mechanism for coping with nutrient limitation, which is a frequent stress factor challenged by microorganisms in their natural habitats (Kalscheuer et al., 2007).

Metabolically, lipid biosynthesis is a secondary anabolic activity, conducted after essential nutrient depletion (usually nitrogen) in the medium. Due to nutrient exhaustion, the carbon flow is directed towards the accumulation of intracellular citric acid that is used as an acetyl-CoA donor in the cytoplasm (Papanikolaou and Aggelis, 2011a,b). Thus, a large intracellular supply of acetyl-CoA and NADPH is required for the synthesis of lipid. The species of oil-producing microorganisms have an enzyme, namely ATP citrate lyase, which is not found in non-oleaginous organisms. This enzyme catalyzes the reaction between citrate and coenzyme A (CoA), forming acetyl-CoA and oxaloacetate (Jin et al., 2015). In general, the limitation of nitrogen induces a cascade of events that results in the accumulation of lipids, illustrated in Figure 7.2. Initially, there is an increase in AMP deaminase activity, decreasing the AMP content in the cell. This decrease

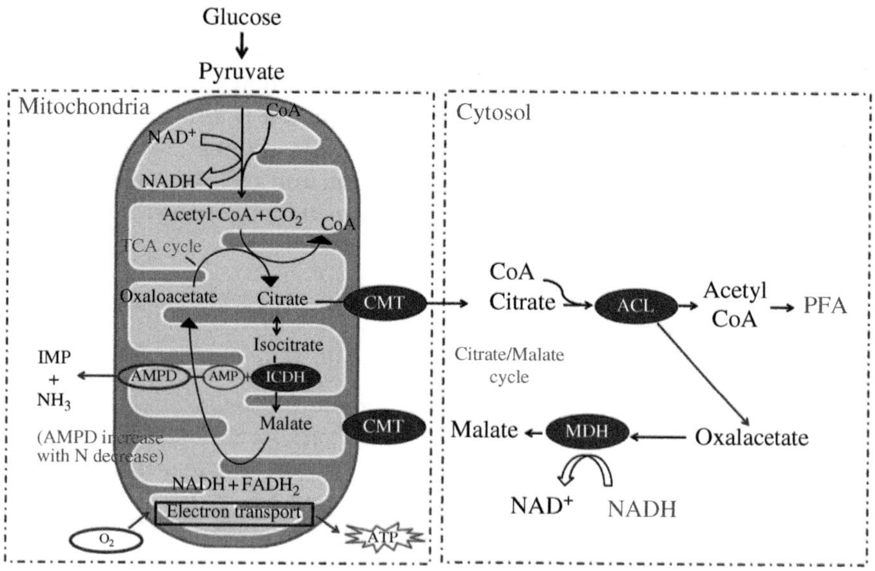

Figure 7.2 General mechanism of lipid accumulation by oil-producing microorganisms. AMPD = AMP deaminase; ICDH = isocitrate dehydrogenase; CMT = citrate–malate translocase; ACL = ATP citrate lyase; PFA = Polyunsaturated fatty acids; MDH = malate dehydrogenase. *Source:* Adapted from Lin (2015). (*See insert for color representation of this figure.*)

in the amount of AMP results in inhibition of isocitrate production, with accumulation of citrate in the mitochondria. The citrate is exported from the mitochondria to the cytosol, which is used by ATP citrate lyase in the formation of acetyl CoA and oxaloacetate. Acetyl CoA is used in the synthesis of fatty acids (the stored carbon source), while oxaloacetate is converted into malate. The oxaloacetate to malate conversion in the cytosol releases NAPH, which has protective effect against oxidative cell stress.

The cultivation process for microbial oil production is divided into two stages: (i) biomass growth; and (ii) lipid accumulation. In the first phase, the culture medium is rich in nutrients which facilitate cell reproduction; then, in the second stage, the organism undergoes a stressful condition which is the lack of nitrogen, leading to the accumulation of lipids as a store of energy (Cohen and Ratledge, 2010). Some authors report a lipid turnover stage, where part of the shorter chain lipids is consumed by the microorganism, resulting in a relative content increase in fatty acids, such as docosahexaenoic acid (Ryu et al., 2013; Chang et al., 2013).

When analyzing the production cost of microbial oils, the carbon source accounts for 60–75% of total cost (Finco et al., 2017). In this sense, it is necessary to explore alternative sources, such as the use of agro-industrial and agro-food wastes. There are four main classes of low-cost substrates so far used for the production of microbial oils, viz., (1) sources rich in monosaccharides or disaccharides (such as molasses, sugarcane juice and industrial sugars); (2) starchy feedstocks – which comprise an important fraction of agro-industrial residues, such as cereal and tuber processing residues; (3) the harder-to-consume lignocellulosic substrates, produced in large volumes in industries, such as sugarcane (bagasse) food (rice and wheat straw, rice hull); and (4) glycerol, which is a byproduct of biodiesel synthesis (Finco et al., 2017). Table 7.1 shows different sources of carbon and microbial species used for the production of several by-products including single cell oil. However, sometimes, it is difficult to achieve nitrogen limitation control in such natural medium. Thus, other aspects can be exploited for lipid accumulation, such as magnesium, iron and phosphorus nutrient limitations, and stressful pH, O_2 concentration and temperature conditions (Cohen and Ratledge, 2005).

7.3.2 Single Cell Protein

The protein extract from pure cultures of algae (*Spirulina, Chlorella, Alaria, Ulva, Durvillea*), yeasts (*Candida tropicalis, Candida krusei, Schwanniomyces occidentalis Saccharomyces cerevisiae, Pichia pastoris*), fungi (*Aspergillus niger, Fusarium graminearum, Penicillium cyclopium, Trichoderma album*) or bacteria (*Rhodopseudomonas gelatinosus, Cellulomonas* spp., *Brevibacterium* spp., *Streptomyces* spp.) can be used as protein supplement in human foods or animal feeds (Anupama and Ravindra, 2000). The single cell protein usually has a high nutritive value due to higher protein, vitamin, essential amino acids and lipid content. However, there is a doubt that it can replace the conventional protein sources due to their high nucleic acid content and slower digestibility. They can also be considered as foreign material by the body, which may, subsequently, result in allergic reactions (Nasseri et al., 2011).

The microbial protein content is not only dependent on species variation, but can also be largely influenced by cultural conditions. Unlike the accumulation of oils, the maximum yield of protein is usually reported in the optimal conditions of microbial growth. Thus, if microorganisms are grown under nitrogen-limited conditions in the presence of excess carbon substrate, the protein content is diminished. Subject to these variations among species and cultural conditions, yeasts will generally have crude protein content of 50–55%, bacteria of 50–80%, fungi of 15–45% and algae of 20–80% (Matelbs and Tannenbaum, 1968; Anupama and Ravindra, 2000). Batch fermentations are clearly inadequate for protein production, beacause the conditions in the reaction medium change with time (Oura, 1983), while fed-batch is recommended, since they involve the control of the carbon source supply through feeding rates (Nasseri et al., 2011).

Many feedstocks have been used for the production of single cell protein (Table 7.1). They can be classified as whey starch crops (grains and tubers), sugar crops (sucrose-containing crops) or sugar-containing residues and lignocellulosic materials containing cellulose and hemicelluloses. Other potential substrates for single cell protein production include bagasse, citrus wastes, sulphite waste liquor, molasses, animal manure, whey, starch, sewage, among others.

The use of C1 compounds, particularly methanol and methane, is of great economic interest for the production of single cell protein, because these are common products of industrial operations. Microorganisms that use methane and methanol as substrates are classified as methylotrophic microorganisms. Yeasts of the genus *Hansenula, Pichia, Candida* and *Torulopsis* have great methylotrophic potential (Faber et al., 1995). They all share a specific methanol

Table 7.1 Synthetic and alternative carbon sources used in the formation of different microbial-derived products.

Carbon source	Organism used	Product	Reference
Methanol	*Streptomyces* spp.	Single cell protein	Singh (1998)
Agricultural wastes	*Cellulomonas* spp.	Single cell protein	Callihan and Clemmer (1979)
Manure, Animal wastes	*Pseudomonas fluorescens*	Single cell protein	Shuler (1979)
Pineapple waste	*Saccharomyces cerevisiae*	Single cell protein	Dhanasekaran (2011)
Glycerol	*Cryptococcus curvatus*	Single cell oil	Meesters (1996)
Xylose	*Mortierella isabellina*	Single cell oil	Fakas (2009)
Glucose	*Mortierella alpina*	Single cell oil	Nie (2013)
Sugar beet molasses and sucrose	*Rhodococcus opacus*	Single cell oil	Voss and Steinbüchel (2001)
Cane molasses	*Schizochytrium sp.*	Single cell oil	Ren (2013)
Corn straw	*Trichoderma harzianum*	Single cell oil	Wang (2012)
Dulcitol	*Arthrobacter globiformis*	Rare sugar (D-tagatose)	Jayaraman and Gautam (2016)
D-galactose	*Geobacillus stearothermophilus*	Rare sugar (D-tagatose)	Kim et al. (2002)
Galactitol	*Pseudomonas sp.*	Rare sugar (D-Sorbose)	Khan et al.(1992)
L-mannitol	*Pseudomonas sp.*	Rare sugar (L-Fructose)	Dhawale et al. (1986)
Xylose	*Candida tropicalis*	Rare sugar (xylitol)	Tamburini et al.(2015)
Corn cob hydrolysate	*Candida magnoliae*	Rare sugar (xylitol)	Tada et al. (2012)
Corn straw	*Rhizopus oryzae*	Fumaric acid	Xu (2010)
Glucose	*Saccharomyces cerevisiae*	Fumaric acid	Xu (2013)
Glucose	*Escherichia coli*	Fumaric acid	Song (2013)

(*Continued*)

Table 7.1 (Continued)

Carbon source	Organism used	Product	Reference
Glucose	*Rhizopus oryzae*	Fumaric acid	Gu (2014)
Olive-mill wastewater	*Yarrowia lipolytica*	Citric acid	Papanikolaou (2008)
Glicerol	*Yarrowia lipolytica*	Citric acid	Rywińska and Rymowicz (2010)
Parkia biglobosa	*Aspergillus niger*	Citric acid	Auta et al. (2014)
Liquefaction slurry	*Aspergillus niger*	Citric acid	Wang (2017)
Corncob hydrolysate	*Actinobacillus succinogenes*	Succinic acid	Zhao and Shah (2016)
Crude Glycerol	*Yarrowia lipolytica*	Succinic acid	Gao (2016)
Crop stalk wastes	*Actinobacillus succinogenes*	Succinic acid	Yang et al.(2016)
Corn fiber	*Actinobacillus succinogenes*	Succinic acid	Chen et al. (2011)
Glucose	*Saccharomyces cerevisiae*	Methylxanthines (Caffeine, theophylline and 3-methylxanthine)	McKeague (2016)
Rye	*Claviceps* sp.	Ergot alkaloids	Hulvová (2013)
Sucrose, glucose, fructose, lactose, mannitol and soluble starch	*Curvularia* sp.	Curvulamine (alkaloid)	Yang et al. (2016)
Potato Dextrose Liquid (PDB)	*Fusarium* sp.	Peimisine and peiminine (alkaloids)	Pan et al. (2014)
Glucose	*Saccharomyces cerevisiae*	R-phenylacetylcarbinol (precursor of ephedrine)	Bruder and Boles (2017)
Post fermented dark tea	*Fungal consortium*	Metabolites of atechins	Zhu et al.(2015)
Rice bran	*Lactobacillus acidophilus* and *L. plantarum*	Phenolic compounds and flavonoids	Zhang (2017)

Cocoa beans	*Microbial consortium*	Flavonols, procyanidins, anthocyanines	Mayorga-Gross et al.(2016)
Masticated mango and banana	*Pigs faecal inoculum*	Catablolites of polyphenols	Low et al. (2016)
Wheat grains	*Aspergillus awamorinakazawa*	Phenolic compounds, flavonoids	Sandhu et al.(2016)
Apples	*Saccharomyces cerevisiae, S. bayanus, Torulospora delbrueckii* and *Oenococcus oeni*	Phenolic compounds and organic acids	Laaksonen et al. (2017)
Tea extracts	*Lactobacillus plantarum, L. acidophilus* and *L. rhamnosus*	Polyphenols	Zhao and Shah (2016)
Spent coffee	*Human gut microflora*	Polyphenols	Campos-Vega et al. (2015)
Black tea	*Saccharomyces cerevisiae*	Phenolic compounds (catechin, epicatechin and galic acid)	John et al. (2014)
Sourdoughs of wheat	*Lactobacillus plantarum*	Polyphenols and phenolic compounds (gallic acid, gallocatechins, flavonoids and organic acids)	Ferri et al.(2016)
Gluten free pasta samples	*Inoculum prepared with pig fresh faeces*	Low molecular weight phenolic acids	Rocchetti et al.(2017)
Wheat and soybean bran	*Xylaria nigripes*	Phenols and flavonoids	Divate et al.(2017)
Red beans	*Bacillus subtilis and Lactobacillus delbrueckii*	Phenolic compounds, anthocyanin, flavonoids and vitamins	Jhan et al. (2015)
Plum pomaces	*Aspergillus niger and Rhizopus oligosporus*	Phenolic compounds and flavonoids	Dulf (2016)
Kodo millet	*Penicillium roqueforti*	Phenolic componuds and flavonoids (mainly sinapicacid, caffeic acid and naringenin)	Dwivedi et al.(2015)

(*Continued*)

Table 7.1 (Continued)

Carbon source	Organism used	Product	Reference
Pomegranate juice, pulp and peel extract	*Colonic fermentation with human faces*	Catabolites of phenolic compounds	Mosele et al. (2015)
Brewing waste	*Lactobacillus plantarum*	Polyphenolic compounds	Gupta et al. (2013)
Flour of common buckwheat	*Lactobacillus heilongjiangensis and Pediococcus parvulus*	Quercetin	Luksic et al.(2016)
Cherry juice and broccoli puree	*Lactobacillus plantarum, L. spicheri, L.fermentum* and *L. reuteri*	Organic acids, phenolic acids and flavonoids	Filannino (2015)
Waste cooking oil	*Blakeslea trispora*	Carotene	Nanou and Roukas (2016)
Wheat straw	*Saccharomyces cerevisiae and Rhodotorula glutinis*	Carotene enriched biomass	Petrik et al.(2013)
Tomato byproducts	*Fusarium solani*	Lycopene	Azabou et al. (2016)
YM broth	*Rodothorula glutinis*	Carotenoids	Hernández-Almanza et al. (2014)
Glucose	*Blakeslea trispora*	Carotenes	Nanou et al. (2012)
Crude glycerol	*Thermotoga neapolitana*	Hydrogen	Ngo (2011)
Crude glycerol	*E. aerogenes*	Hydrogen	Sakai and Yagishita (2007)
Sugarcane bagasse	*Caldicellulosiruptor saccharolyticus*	Hydrogen	Ivanova (2009)
Food waste	*Anaerobic sludge*	Hydrogen	Shin (2004)
Wheat straw hydrolysate	*Clostridium beijerinckii*	Butanol	Qureshi (2007)
Corn stover and switchgrass hydrolysate	*C. beijerinckii*	Butanol	Qureshi (2010)
Wheat straw	*Saccharomyces cerevisiae*	Ethanol	Chent et al. (2015)

Wheat straw	*Kluyveromyces marxianus*	Ethanol	Tomás-Pejó (2009)
Rice straw	*Saccharomyces cerevisiae* and *Pichia stipitis*	Ethanol	Park (2010)
Rice straw	*Saccharomyces cerevisiae* and *Zymomonas mobilis*	Ethanol	Das (2013)
Rice straw	*Clostridium acetobutylicum*	Butanol	Moradi (2013)
Barley straw	*Clostridium beijerinckii*	Butanol	Qureshi (2010)
Wheat straw	*Mixed culture*	Hydrogen	Cao (2013)
Barley straw	*Caldicellulosiruptor saccharolytic*	Hydrogen	Panagiotopoulos (2012)
Cassava starch	*Saccharomyces cerevisiae*	Ethanol	Choi (2010)
Wood	*Saccharomyces cerevisiae*	Ethanol	Gupta et al. (2013)
Cane molasses	*Saccharomyces cerevisiae*	Ethanol	Kirdponpattara and Phisalaphong (2013)

utilization pathway involving several unique enzymes. The initial reactions take place in specialized microbodies, the peroxisomes, followed by subsequent metabolic steps in the cytoplasm (Macauley-Patrick et al., 2005). Peroxisomes have the three key enzymes for methanol metabolism: alcohol oxidase, catalase and dihydroxyacetone synthase. The subsequent reactions of methanol assimilation and dissimilation are localized in the cytosol (see details in Cereghino and Cregg, 2000; Jahic, 2003; Gellissen, 2000). The low solubility of C1 compounds in the fermentation medium is, however, one of the obstacles for single cell production, which requires highly efficient technology for the fermentation system.

Microbial consortium can also be used for single cell protein production, through the fermentation of kefir grains. Kefir grain consists of a polysaccharide composed of a complex microbial association among bacteria and yeasts, which works as a starter culture for milk fermentation. The main polysaccharide of kefir is known as kefiran — a water-soluble glucogalactan exopolysaccharide (EPS) produced by the bacterium *Lactobacillus kefiranofaciens* present in kefir grains (Zajšek et al., 2011; Ghasemlou et al., 2011; Blandón et al., 2016). Cultivation of kefir microorganisms on milk, as well as on various substrates (mixtures of molasses, raisin extracts and glucose-containing raw materials), has been studied (Garrote et al., 1998; Athanasiadis et al., 2001; Seydim et al., 2000). Single cell protein was produced with high productivity by aerobic fermentation of cheese whey by kefir microorganisms (Paraskevopoulou et al., 2003). Nevertheless, important aspects to enable the production of single cell protein by kefir fermentation must be clarified for commercial applications, such as medium composition and structural characteristics of the protein obtained.

7.4 Enzymes

It is known that the production of enzymes can be influenced by many environmental conditions, such as carbon and nitrogen sources, C:N ratio, micronutrients, pH and temperature. Some of these factors are directly related to cell multiplication, and some can act specifically on the expression of particular enzymes. Adaptation to different environmental conditions, in many cases, requires the modulation of gene expression, leading to changes in metabolic pathways, or to the synthesis of molecules that are secreted and act outside the cell, as in the case of the production of extracellular hydrolytic enzymes such as cellulases and hemicellulases (Mello-de-Sousa et al., 2011). The regulation of genes that encode cellulases and hemicellulases include five transcription factors (TFs): positive regulators, such as XYR1, ACE2, and the HAP2/3/5 complex, and negative regulators, such as ACE1 and the carbon catabolic repressor CRE1 (Castro et al., 2014).

The carbon source may exert a considerable influence in enzyme expression and yield. For example, the presence of simple and readily metabolizable carbon sources, such as glucose induces a mechanism known as carbon catabolic repression. This mechanism, which inhibits the synthesis of enzymes involved in the hydrolysis of carbon sources other than the preferred one, is mediated by the

CreA/CRE1 transcription repressor which down-regulates several glycosyl hydrolase genes in *Trichoderma reesei* and *Aspergillus* species (Mello-de-Sousa et al., 2011).

Castro et al. (2014) demonstrated the effect of the carbon source on the expression pattern of cellulolytic and xylanolytic genes regulated by the transcription factors XYR1 and CRE1 in *Trichoderma reesei*. The authors used three different carbon sources (glucose, cellulose and sophorose) to analyze 22 genes of cellulases (CMCase and β-glucosidase) and xylanases. In the presence of glucose, more than 80% of the evaluated genes showed increased expression levels in the *cre1* knock-out mutant (*Δcre1*) compared to the parental *T. reesei* QM9414 strain. In the *xyr1* knock-out mutant (*Δxyr1*), the expression of cellulase and xylanase genes was reduced compared to the parental strain, with cellulose or sophorose as the carbon sources. In the presence of glucose, approximately 60% of the analyzed genes had increased expression in the *Δxyr1* mutant compared to the parental strain.

The pH is one of the most important factors for fungal growth and enzyme production. This factor affects not only enzyme synthesis but also stability, since it can lead to enzyme denaturation. At the same time, fungi exhibit a versatile metabolism, capable of adapting to environmental changes in pH. The regulation of ambient pH by the terminal key zinc finger transcription factor PacC has been studied in *Aspergillus nidulans* and in other filamentous fungi. This regulatory system is mainly mediated by six genes besides *pacC*, named *palA*, *palB*, *palC*, *palF*, *palH* and *palI*. The products of the *pal* genes transmit the pH signal to modify the PacC polypeptide into an active form to regulate downstream genes. At alkaline pH, the PacC polypeptide results in a shorter, functional form that activates genes expressed at alkaline condition and represses genes expressed under acidic conditions (He, 2014). The self-regulated PacC system participates not only in the synthesis control of permeases and metabolites, but also of extracellular enzymes, such as hemicellulases (Mello-de-Sousa et al., 2011).

The productivity of cellulase by *Trichoderma reesei* is sensitive to the pH of the cultivation medium, indicating that this is an important factor to be investigated in the control of cellulase production by this fungus. He et al. (2014) identified a pH response transcription regulator, TrPac1, a PacC homolog, involved in cellulase gene expression in *Trichoderma reesei*. The deletion of Trpac1 led to a strong increase in cellulase production and cellulase gene expression at neutral pH, and to an increase in the abundances of the transcription factors XYR1 and ACE2.

A study developed by Mello-de-Sousa et al. (2011) established the expression profiles of *Humicola grisea* var. *thermoidea* hydrolase encoding genes (one of xylanase and eight of cellulases) and of two transcription factors, in different conditions of pH and carbon source (glucose and steam exploded sugarcane bagasse). Gene expression was evaluated by real time reverse-transcriptase polymerase chain reaction (RT-PCR). As expected, all hydrolase genes presented higher mRNA levels when sugarcane bagasse was used as the carbon source. At alkaline pH and with sugarcane bagasse as the carbon source, six cellulase encoding genes (*cbh1.1*, *cbh1.2*, *egl1*, *egl2*, *egl3* and *bgl4*) and the xylanase encoding gene (*xyn1*) showed enhanced expression, while the cellulase gene *egl4* was more substantially expressed at acidic pH. When glucose was used as the carbon source, the pH had no influence on mRNA accumulation for *cbh1.1*, *cbh1.2*, *egl1*, *egl2*, *egl3*, *bgl4* and *xyn1*. As for the expression of the transcription factors PacC, related to pH, and CreA, related to carbon catabolic repression, no significant difference was observed for PacC, while *creA* mRNA levels accumulated in the presence of glucose, as expected, especially at acid pH. When the pH was shifted from 5.0 to 8.0, *pacC* transcripts reached the highest levels using sugarcane bagasse as the carbon source, but not in the presence of glucose. This indicates that the carbon catabolic repression mechanism may influence the regulation of *pacC*. The influence of PacC and CreA over the expression of the hydrolase genes was also investigated through in vitro pull-down DNA assays and electrophoretic mobility shift assays analyses, using the DNA-binding domains from PacC and CreA proteins. The influence of the pH-mediated regulatory system on *H. grisea* glycosyl hydrolase gene regulation, which is a novel fact, was corroborated by the observed interaction of the GST::PacC fusion protein with the 5' UP regions.

Another typical example of the influence of the culture medium composition in enzyme production regulation at the level of transcription is the effect of copper on the expression of laccases, specifically on the isoforms phenoloxidase (POX) C and POX A1b, produced by the white-rot basidiomycete *Pleurotus ostreatus*. The positive effect of the addition of copper sulphate on the total laccase activity and specific production of POX C and POX A1b was demonstrated. Also, Northern blot analysis demonstrated the effect of copper on the induction of *poxc* and *poxa1b* gene transcription. This is due to the presence of metal responsive elements (MRE) in the *poxc* and *poxa1b* promoter regions, similar to the core MRE consensus sequence identified in the promoters of metallothionein genes in higher eukaryotes. By analyzing the nucleotide sequences of these regions in *P. ostreatus*, extending around 400 nucleotides upstream of the ATG, multiple putative MRE sites were identified. The interaction of specific proteins and *poxc* and *poxa1b* promoter regions was shown to be dependent on the presence of copper (Faraco et al., 2002).

7.5 Biofuels

The world demand for energy tends to increase by a range of 17–50% by 2040 (Correa et al., 2017). Fossil fuels are the major energy source used in the world today, but over-consumption can lead to serious environmental problems. In this context, biofuels, such as biodiesel, biohydrogen, bioethanol, biobutanol, biogas and biomethanol, are sustainable alternatives to meet the world energy demand. Biodiesel is produced by the transesterification of fatty acids such as vegetable oil, animal fats and microbial lipids (Sarma et al., 2014). The other biofuels are mostly generated from fermentation processes that use microorganisms capable of consuming carbohydrate-rich substrates; these can be classified according to the type of substrate used for their production, i.e. first generation biofuels, derived from food crops as raw materials, second generation biofuels, which come from non-food substrates, such as lignocellulosic wastes from the agricultural and forestry sectors, and third generation biofuels, produced from microalgae and other engineered sources (Rodionova et al., 2016).

The production of second generation biofuels from lignocellulosic biomass includes biomass pretreatment, cellulose hydrolysis, fermentation of hexoses, separation and effluent treatment (Ojeda and Kafarov, 2009). Intensive efforts have been made in recent years to develop efficient technologies for the pre-treatment of lignocellulosic substrates (Naik et al., 2010; Soccol et al., 2010). Several microorganisms are capable of degrading lignocellulosic material making them available for the biofuel production process. In a study conducted by Thakur (2013), the pre-treatment of wheat straw by *Pleurotus ostreatus* presented an efficient lignin degradation of 40%, making the production of second generation bioethanol by *Saccharomyces cerevisiae* possible (3.38 g / L ethanol yield). Many other methods have been used for pretreating lignocellulosic materials, such as steam explosion (Ramos et al., 1992; Glasser and Wright, 1998; Ramos et al., 2000; Balat et al., 2008; Hendriks and Zeeman, 2009; Hernández-Salas et al., 2009), alkali washing (Balat et al.,2008; Hendriks and Zeeman 2009; Hernández-Salas et al., 2009)), lime, alkaline hydrogen peroxide, dilute acid hydrolysis (Balat et al., 2008; Hernández-Salas et al., 2009), ammonia fiber explosion (Hendriks and Zeeman, 2009; Balat et al., 2008), liquid hot water and wet oxidation (Martín et al., 2007; Hendriks and Zeeman, 2009), among others. Each one of these methods has advantages and disadvantages and none seems to be optimal for all practical applications involving different types of lignocellulosic materials.

Pyruvate is a central intermediate in the carbon and energy metabolism for the synthesis of various chemicals and polymers. By the decarboxylation of pyruvate, ethanol can be directly produced by some ethanologenic organisms (Figure 7.3). The most commonly used organisms for industrial-scale production are *Saccharomyces cerevisiae* and *Zymomonas mobilis*. Substantial attention and effort has been dedicated to redirecting the metabolic pathways of these and other microorganisms (e.g. *Escherichia coli*, and *Klebsiella oxytoca*) towards higher yield of ethanol production, by means of metabolic engineering (Eram and Ma, 2013).

Several other biofuels (e.g. higher alcohols and fatty acids) can be produced from pyruvate or acetyl-CoA by elongation of the carbon chain. These microbial metabolic pathways include CoA dependent reverse β-oxidation and keto acid and isoprenoid pathways. Figure 7.3 shows a summary illustration of these pathways. For more details on such metabolic pathways, readers are directed to the work developed by Liao (2016).

Pyruvate can also be used for the production of biohydrogen. There are two types of fermentation pathways with hydrogen as a product (Figure 7.3). Hydrogen is either produced from formate in enterobacterial-type fermentation or from reduced ferredoxin in Clostridia-type fermentation. Additionally, hydrogen can be derived from NADH in Clostridia-type fermentations (Hallenbeck and Ghosh, 2009).

To achieve efficient production of biofuels, deeper knowledge of microbial metabolism and its regulation should be elucidated. Some strategies may be adopted to achieve these objectives, such as carbon source optimization,

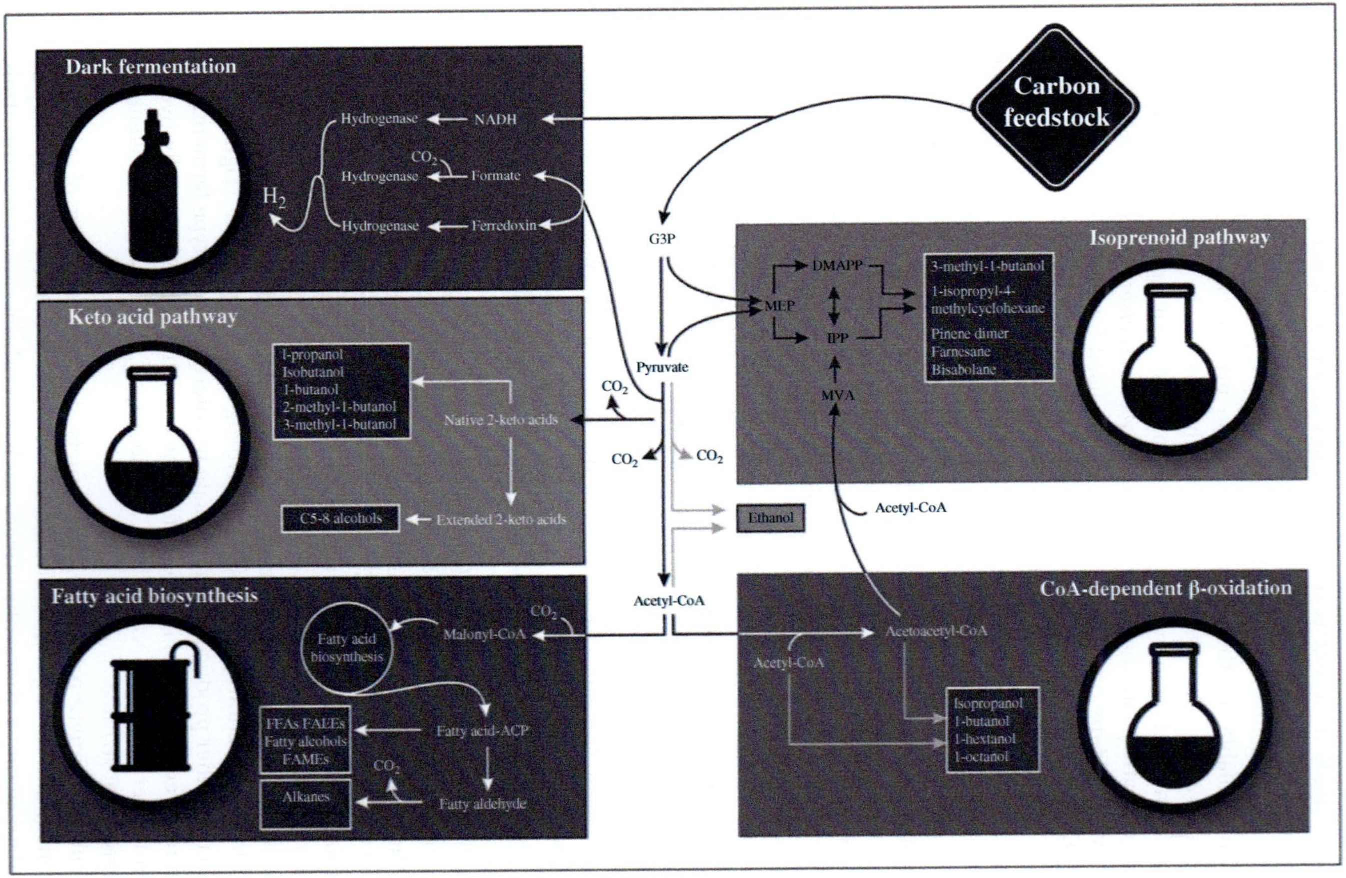

Figure 7.3 Biosynthetic pathways of biofuels (G3P: glyceraldehyde 3 phosphate; DMAPP: dimethylallyl pyrophosphate; MEP: methylerythritol 4 phosphate; IPP: Isopentenyl pyrophosphate; MVA: mevalonate; FFAs: free fatty acids; FAEEs: fatty acid ethyl esters; FAMEs: fatty acid methyl esters). (*See insert for color representation of this figure.*)

microbial improvement and the use of co-cultures. The use of micronutrients, such as iron and molybdenum as supplements in the culture medium of *Rhodobacter* (Kars et al., 2006; Uyar and Schumacher, 2009; Özgur et al., 2010) and nickel and magnesium for *Rhodopseudomonas* sp. (Liu et al., 2009) showed potential to increase hydrogen productivity. Another alternative to increase hydrogen productivity is the use of co-cultures. Asada (2006) presented a 60% higher yield using *Lactobacillus*, which converts the glucose into lactic acid, in co-cultivation with *Rhodobacter sphaeroides*, which is a photosynthetic bacterium and converts the lactate into hydrogen. According to Pachapur et al. (2015), the synergy of the co-culture system may be a strategy to improve hydrogen production without nitrogen sparging and reducing agents, which increases the possibility of field application and minimizes the production of metabolites. In addition, to improve the accumulation of lipids in microalgae, the use of different wavelengths of light has been used. Teo (2014) studied the effect of different wavelengths of light in this process. The blue wavelength showed the highest growth rate and oil production, and presented a quantity of palmitic and stearic acids of 81–83% of the total methyl esters.

The use of alternative raw materials, such as sucrose-containing feedstocks (e.g. sugar beet, sweet sorghum and sugar cane), starchy materials (e.g. wheat, corn, and barley) and lignocellulosic biomass (e.g. wood, straw, and grasses), can be an important strategy to enable the biofuels production (Balat et al., 2008). The availability of feedstocks for biofuels production can vary considerably from season to season depending on the geographic location, which could also pose difficulty in their utilization (Soccol et al., 2010). **Table 7.1** reports alternative substrates used for production of some biofuels.

7.6 Alkaloids, Terpenoids, Polyketides and Flavonoids

Secondary metabolites comprise a wide range of chemical structures, and are produced by both eukaryotes and prokaryotes. Among them, alkaloids, terpenoids, polyketides and flavonoids stand out. In general, such compounds are produced at low concentrations in native cellular environments, and do not act directly on primary cellular functions but with rather specific functions, such as biotic and abiotic stress regulation, repulsion of predators, attraction of pollinators (in plants), modulation of allopathic interactions, synergistic and antagonistic effects (Lussier et al., 2012; Zhou and Chen, 2014; Ulbrich et al., 2015; Lipko and Swiezewska, 2016).

In recent decades, humankind has been exploring the potential of these complex bioactive compounds. Applications in the agro-food industry (flavor, dietary supplement), pharmaceuticals (antioxidant, anticholesterolemic, antiobesity, antidiabetis, antiviral and antibacterial effects), chemicals (fragrance, colorants) and medicine (prevention of cardiovascular diseases, immunosuppressant activity for organ transplants and prevention/treatment of cancer) are notable (Serra et al., 2012; Wu et al., 2014; Lipko and Swiezewska, 2016).

The chemical synthesis of these bioactive compounds is, in general, impractical due to the structural complexity and the extreme reaction conditions. Many are directly extracted from plant tissues, but there are some drawbacks in this method: low concentration of the bioactive compounds inside plant cell; expensive and time-consuming processes, difficult separation/purification; and environmental risks (for example, threat to biodiversity and possible extinction of rare plant species).

A promising alternative for the production of alkaloids, terpenoids, polyketides and flavonoids has been the use of microorganisms (Wu et al., 2014; Zhou and Chen, 2014; Bond et al., 2016). In this context, the elucidation of specific metabolic pathways for each bioactive compound is crucial, as well as the characterization of enzymes involved. As an example, it can be cited that the knowledge acquired regarding polyketide production pathways has allowed the synthesis of several bioactives of commercial interest (such as naringenin, pinocembrin, resveratrol, curcuminoids and intermediates of cannabicin) in heterologous hosts (especially in *Escherichia coli*, *Saccharomyces cerevisiae* and *Streptomyces* sp.). In addition, it has led to the production of novel "unnatural" compounds because of the knowledge of such pathways combined with the use of modern molecular biology tools, such as strain engineering and genome mining (Lussier et al., 2012; Sarkar et al., 2012; Helfrish, 2014; Bond et al., 2016).

Another example is the vast knowledge gathered in the last few decades concerning biosynthesis of terpenoids. These compounds comprise a vast group of natural molecules, and they are derived from isoprene (5-carbon molecule). There are two known metabolic pathways for producing terpenoids, referred to as mevalonate pathway (MVA) and MEP (from 2-methyl-(D)-erythritol-4-phospate, also called the non-mevalonate pathway. A detailed discussion of this topic could be found in Yadav et al. (2012).

The use of recent analytical tools, such as bioinformatics, multivariate analysis strategies (such as MMME–Multivariate Modular Metabolic Engineering) and other stepwise modular engineering approaches promise further advances in the coming decades. Some of the main challenges pointed out by researchers are process scale up, cost reduction, and the expression of terpenoids and other secondary metabolites in hosts other than the classic ones (usually *Escherichia coli* and *Saccharomyces cerevisiae*) (Marienhagen and Bott, 2013; Wu et al., 2016a; Wu et al., 2016b; Lipko and Swiezewska, 2016).

Another interesting case related to microbial metabolism is flavonoids (comprising polyphenols). Polyphenols are present in several vegetables and fruits and their absorption in the human body starts in the ileum. The absorbed fraction can be methylated, sulphated and glucoronidated by the enzymatic complex of small intestine and/or liver, and can be subsequently excreted in the bile, finally reaching the colon. But the largest fraction of polyphenols passes straight through the ileum, and is metabolized by the gut microflora, being transformed into molecules of low molecular weight (through biochemical reactions of esterification, glucosidation, demethylation, dehydroxilation and decarboxylation), such as organic acids and phenolic acids, which possesses several bioactive effects, many of them beneficial for human health. The effect of these small molecules has been intensively studied in the last years, as well as the search for detailed understanding of the metabolism of the most important flavonoids by the gut microbiota (Serra et al., 2012; Galicia-Jiménez et al., 2014; Campos-Vega et al., 2015; Mosele, 2015; Ulbrich et al., 2015; Juániz et al., 2016; Tao et al., 2016; Rocchetti et al., 2017).

The production of some natural compounds has seen very recent advances, as, for example, concerning the production of benzylisoquinoline alkaloids (e.g. papaverine, morphine, noscapine, sanguinarine and barberine). Until the last few years, only papaverine has been produced from heterologous platforms, while all other benzylisoquinoline alkaloids are obtained from plant tissues. The most notable example is morphine, a potent analgesic, currently still extracted (legally) from opium. Many reasons explain the difficulty to produce morphine in bacteria and yeast cells (toxicity, lack of post transcriptional regulation). However, the main bottleneck has recently been solved: finally, all the enzymatic steps required for morphine synthesis were elucidated. After these advances, some laboratorial results using both *S. cerevisiae* and *E. coli* are encouraging, and it is expected morphine production in heterologous platforms can be feasible in the years to come (Winzer et al., 2015; Schläger and Dräger, 2016).

Much effort has been made to unravel specific metabolic pathways for the production of value-added secondary compounds. In this context, many specific precursors are added to the culture media (e. g. addition of tyrosine for production of certain polyketides), and the carbon source is usually an easily assailable sugar, as glucose. However, many researchers have also focused attention on the reuse of co- or by-products from agroindustry for production of bioactive compounds. The Table 7.1 exhibits several examples of utilization of synthetic and alternative carbon sources for biosynthesis of alkaloids, flavonoids, terpenoids and other commercially important products, and the respective microorganisms employed during the fermentation step.

7.7 Organic Acids

Organic acids serve as precursors for a variety of bulk chemicals and commercially important polymers (Werpy and Petersen, 2004; Sauer et al., 2008). The most industrially relevant organic acids include glycolic, pyruvic, lactic, succinic, citric, fumaric and malic acids. Biotechnological routes, i.e. through fermentation and/or biotransformation, of organic acid production are currently being sought, so that fossil resources could be replaced with renewable resources (Becker, 2015).

Most of the organic acids are natural products of microorganisms or, at least, natural intermediates in major metabolic pathways (Sauer et al., 2008). Fumaric, L-malic and citric acids are intermediates of the oxidative tricarboxylic acid (TCA) cycle. They are synthesized and mainly by filamentous fungi, such as *Aspergillus spp.* and *Rhizopus sp.*, under specific stress conditions (Goldberg, 2006). The production of such organic acids is mainly influenced by nitrogen limitation. After full depletion of nitrogen source, acids are accumulated and secreted into the medium (Goldberg, 2006). Ding et al. (2011) demonstrated, for example, that carbon–nitrogen ratio influences on the activity of cytosolic fumarase.

Under nitrogen limitation stress, the cytosolic fumarase presents high activity and, consequently, yield of fumaric acid. The nitrogen limitation was also seen to influence the production of citric acid by *Aspergillus niger* (Auta et al., 2014).

Succinic acid is an important organic acid which can be produced using microorganisms, such as *Actinobacillus succinogenes, Anaerobiospirillum succiniciproducens, Mannheimia succiniciproducens* and *Bacillus fragilis* (Beauprez et al., 2010), as well as recombinant microorganisms, such as *Escherichia coli, Corynebacterium glutamicum* and *Saccharomyces cerevisiae* (Jansen and Van Gulik, 2014). Some yeasts and fungi are also related as succinic acid producing microorganisms (Song and Lee, 2006; Ahn et al., 2016). Succinic acid is an intermediate compound of the tricarboxylic acid (TCA) cycle. However, it acts as fermentation end-product, in anaerobic conditions, when glucose or glycerol is used as a carbon source with the fixation of carbon dioxide. Moreover, glyoxylate pathway can form succinic acid under aerobic conditions upon adaptation to growth on acetate (Cheng et al., 2012). One bottleneck for succinic acid production by fermentative route is NADH limitation, since it required 2 mol of NADH per mol of succinic acid produced (Cheng et al., 2012; Ahn et al., 2016). Another influence on the succinic acid production is the availability of carbon dioxide source (gas or carbonate salt forms), which acts as a pH titrant and carbon dioxide source at the same time (Cheng et al., 2012; Jansen and Van Gulik, 2014).

7.8 Rare Sugars

Rare sugars, referred to as monosaccharides and their derivatives that rarely exist in nature, have various applications in the food industry (sweeteners and seasonings), agricultural chemicals (control of plant growth and insect control agent) and pharmaceuticals (conservation solutions) (Dhawale et al., 1986; Poonperm et al., 2007; Li, 2013). In addition, clinical research has shown functional proprieties associated with rare sugars, including prebiotic (Jayaraman and Gautam, 2016), antitumor (Mitani et al., 2009; Sui et al. 2005), anti-oxidative (Ishihara et al., 2011) and immunosuppressant (Hossain et al., 2000).

Rare sugars are usually produced by chemical process. More recently, the metabolism of some yeast and bacteria species has been explored as an alternative way for production of rare sugars, including D-Psicose, D-Tagatose, L-Tagatose, D-Sorbose, L-Fructose, D-fructose, D-Allose, L-Glucose, L-Talose, L-Galactose, D-ribose, L-arabinose (Itoh et al., 1994, 1995, 1996; Bhuiyan et al., 1999; Kim et al., 2003; Leang et al., 2004; Jayaraman and Gautam, 2016). Biocatalysis often offers advantages over chemical synthesis, because enzyme-catalyzed reactions are often highly enantioselective and regio-selective, in addition to being performed under mild conditions and are environmentally friendly (Li, 2013).

The biological manufacture of rare sugars has been examined. D-tagatose, a rare sugar approved by the Food and Drug Administration as a food additive, was produced by *Arthrobacter globiformis* (3.82 g/L) from dulcitol (Jayaraman and Gautam, 2016) and by *Geobacillus stearothermophilus* (230 g/L) and *Thermotoga neapolitana* (68 g/L) from D-galactose (Kim et al., 2002, 2003). Some studies have reported the production D-tagatose by *Arothrobacter, Mycobacterium, Enterobacter* and *Klebsiella* in different carbon sources, including glycerol, erythritol, and L-sorbose that could convert galactitol into tagatose (Izumori and Tsuzaki, 1988; Izumori et al., 1990; Shimonishi et al., 1995; Izumori 2002). D-Sorbose was also reported to be synthesized by *Pseudomonas* sp. from galactitol with over 70% yield (Khan et al., 1922). The possible microbial transformation route from galactitol to D-sorbose includes the substrate galactitol dehydrogenated at C-2 to afford D-tagatose, followed by C-3 epimerization (Khan et al., 1922).

Some rare sugars can be produced via bacterial oxidation metabolism. D-tagatose can be produced from dulcitol via the oxidation reaction by the acetic acid bacteria, such as *Arthrobacter globiformis* and *Gluconobacter oxydans* (Jayaraman and Gautam, 2016). L-Tagatose can be synthesized via oxidation of galactitol by *Klebsiella pneumoniae* with over 70% yield (Shimonishi et al., 1995). Huwig et al. (1997) also reported an efficient oxidation of galactitol to L-tagatose through the use of galactitol dehydrogenase (GDH, EC 1.1.1.16) from *Rhodobacter sphaeroides.* This oxidation only requires catalytic amounts of NAD+, which is regenerated in situ through the reduction of pyruvate by L-lactate dehydrogenase (EC 1.1.1.27) (Huwig et al., 1997). L-Fructose was synthesized with an expressive amount by *Pseudomonas* sp. via bacterial oxidation of L-mannitol (Dhawale et al., 1986).

Xylitol is a rare sugar alcohol that has many commercial applications in the food and pharmaceutical industries as an alternative sweetener (Mohamad et al., 2015). Xylitol is manufactured through a chemical process that has some disadvantages, including a high energy requirement, extensive purification steps and a high cost of product (Rafiqul and Sakinh, 2013).

Xylitol can be biologically produced by yeasts, bacteria, and fungi. Among these, xylose-fermenting yeasts, such as *Candida boidinii, C. guillermondii, C. parapsilosis,* and *C. tropicalis,* have great potential due to the highest yield of xylitol (Ikeuchi et al., 1999). The metabolism of D-xylose in yeast occurs through a reduction process, where D-xylose is reduced to xylitol by a xylose reductase (Winkelhausen and Kusmanova, 1998). In order to increase xylitol production, some specific parameters should be analyzed, including parameters-substrate (xylose) concentration, pH, aeration rate and temperature. Tamburini et al. (2015) evaluated the production of xylitol by *Candida tropicalis* under different conditions and using xylose as a carbon source. The maximum xylitol yield (83.66%) was reached at 60–80 g/L initial xylose concentration, pH 5.5 and at 37 °C. In another study, *Candida magnoliae* produced 18 $g\,L^{-1}$ xylitol from 25 $g\,L^{-1}$ xylose of corn cob hydrolysate (Tada et al., 2012). A mutant strain of *Candida tropicalis* was also able to produce xylitol in 80–90% yield from xylose as the sole carbon source, with a productivity of more than 3 $g\,L^{-1}\,h^{-1}$ (Jeon et al., 2011).

7.9 Conclusions

Intense research in biotechnology and an increase in human responsibility for environmental protection have contributed to the development of new microbial products. An important factor restricting the industrialization of some microbial molecules is the high fermentation costs. Numerous strategies can be applied to improve the synthesis of value-added microbial products. On a special note, the carbon source can be investigated to increase productivity and decrease production costs. Synthetic carbon source can be used to direct specific metabolic production, such as rare sugars, while low-cost substrates can be applied to produce generic metabolites including biomass and biofuels. The study of additional nutritional requirements, as well as classical metabolic pathways, is also of extreme relevance in the use of both synthetic and alternative new sources of carbon.

References

Ahn, J. H., Jang, Y. S., and Lee, S. Y. (2016) Production of succinic acid by metabolically engineered microorganisms. *Current opinion in biotechnology*, 42: 54–66.

Anupama, and Ravindra, P. (2000) Value-Added Food: Single Cell Protein. *Biotechnology Advances*, 18(6): 459–79.

Athanasiadis, I., Boskou, D., Kanellaki, M., et al. (2001) Effect of Carbohydrate Substrate on Fermentation by Kefir Yeast Supported on Delignified Cellulosic Materials. *Journal of Agricultural and Food Chemistry*, 49(2): 658–63.

Auta, H. S., Abidoye, K. T., Tahir, H., et al. (2014) Citric Acid Production by Aspergillus niger Cultivated on Parkia biglobosa Fruit Pulp. *International Scholarly Research Notices*, 2014.

Azabou, S., Abid, Y., Sebii, H. et al. (2016) Potential of the solid-state fermentation of tomato by products by *Fusarium solani pisi* for enzymatic extraction of lycopene. *LWT - Food Science and Technology*, 68:280–287.

Balat, M., Balat, H. and Öz, C. (2008) Progress in bioethanol processing. *Progress in Energy and Combustion Science*, 34(5): 551–573.

Becker, J., Lange, A., Fabarius, J., Wittmann, C. (2015) Top value platform chemicals: bio-based production of organic acids. *Current opinion in biotechnology*, 36: 168–175.

Beauprez, J. J., De Mey, M., and Soetaert, W. K. (2010) Microbial succinic acid production: natural versus metabolic engineered producers. *Process Biochemistry*, 45(7): 1103–1114.

Bhuiyan, H., Takada, G., and Izumori, K.E.N. (1999) Preparation of L-Talose and D-Gulose from L-Tagatose and D-Sorbose, Respectively, Using Immobilized L-Rhamnose Isomerase. *Journal Of Bioscience And Bioengineering*, 88(5): 567–70.

Blandón, L.M., German, A., Islan, G.R. et al. (2016) Kefiran-Alginate Gel Microspheres for Oral Delivery of Ciprofloxacin. *Colloids and Surfaces B: Biointerfaces*, 145. Elsevier B.V.: 706–15.

Bond, C., Tang, Y. and Li, L. (2016) Saccharomyces cerevisiae as a tool for mining, studying and engineering fungal polyketide synthases. *Fungal Genetics and Biology*, 89:52–61.

Bruder, S. and Boles E. (2017) Improvement of the yeast based (R)-phenylacetylcarbinol production process via reduction of by-product formation. *Biochemical Engineering Journal*, 120: 103–112.

Campos-Vega, R., Vázquez-Sánchez, K., López-Barrera, D. et al. (2015) Simulated gastrointestinal digestion and in vitro colonic fermentation of spent coffee (Coffea arabica L.): Bioaccessibility and intestinal permeability. *Food Research International*, 77:156–161.

Castro, L. Dos Santos, A.C., Pedersoli, W.R., et al. (2014) Expression Pattern of Cellulolytic and Xylanolytic Genes Regulated by Transcriptional Factors XYR1 and CRE1 Are Affected by Carbon Source in Trichoderma Reesei. *Gene Expression Patterns*, 14(2). Elsevier B.V.: 88–95.

Cereghino, J.L. and Cregg, J.M. (2000) Heterologous Protein Expression in the Methylotrophic Yeast Pichia Pastoris. *FEMS Microbiology Reviews*, 13(3): 45–66.

Chan, E.C.S. (2003) Microbial nutrition and basic metabolism. Handb. *Water Wastewater Microbiol.*, 3–33.

Chang, G., Gao, N., Tian, G. et al. (2013) Improvement of Docosahexaenoic Acid Production on Glycerol by Schizochytrium Sp. S31 with Constantly High Oxygen Transfer Coefficient. *Bioresource Technology*, 142. Elsevier Ltd: 400–406.

Chen, Z., Shi, J., Yang, X. et al. (2015) Chemical and Physical Characteristics and Antioxidant Activities of the Exopolysaccharide Produced by Tibetan Kefir Grains during Milk Fermentation. *International Dairy Journal*, 43. Elsevier Ltd: 15–21.

Cheng, K. K., Zhao, X. B., Zeng, J. et al. (2012) Biotechnological production of succinic acid: current state and perspectives. *Biofuels, Bioproducts and Biorefining*, 6(3): 302–318

Cohen, Z. and Ratledge, C. (2005) *Single Cell Oils*. Champaign, Illinois: AOCS Press.

Correa, D.F., Beyer, H. L., Possingham H. P. et al.(2017) Biodiversity impacts of bioenergy production: Microalgae vs. first generation biofuels. *Renewable and Sustainable Energy Reviews*, 74: 1131–1146.

Dhawale, M.R., Szarek, W.A., Hay, G.W. et al. (1986) Preparation of L-Fructose and D-Sorbose by the Bacterial Oxidation of L-Mannitol and L-Glucitol, Respectively. *Carbohydrate Research*, 155(C): 262–65.

Ding, Y., Li, S., Dou, C. et al. (2011) Production of fumaric acid by Rhizopus oryzae: role of carbon–nitrogen ratio. *Applied Biochemistry and Biotechnology*, 164(8): 1461–1467.

Divate, R., Wang, C-C., Chou, S-T. et al. (2017) Using wheat bran and soybean meal as solid state fermentation substances for the production of *Xylaria nigripes* with bioactivities. *Journal of the Taiwan Institute of Chemical Engineers*, 70:127–133.

Dulf, F.V., Vodnar, D.C. and Socaciu, C. (2016) Effects of solid-state fermentation with two filamentous fungi on the total phenolic contents, flavonoids, antioxidant activities and lipid fractions of plum fruit (*Prunus domestica L.*) by-products. *Food Chemistry*, 209:27–36.

Dwivedi, M., Vasantha, K.Y., Sreerama, Y.N. et al. (2015) *Nilamadana*, new fungal fermented cereal based food. *Journal of Functional Foods*, 15:217–224.

Du, J., Shao, Z. and Zhao, H.(2011) Engineering microbial factories for synthesis of value-added products. *Journal of Industrial Microbiology and Biotechnology*, 38:873–90.

Eram, M.S. and Ma, K. (2013) Decarboxylation of pyruvate to acetaldehyde for ethanol production by hyperthermophiles. *Biomolecules*, 3(3): 578–596.

Faber, K.N., Harder, W., Ab, G., et al. (1995) Methylotrophic Yeasts as Factories for the Production of Foreign Proteins. *Yeast*, 11(14): 1331–44.

Faraco, V., Giardina, P., Palmieri, G., et al. (2002) Metal-Activated Laccase Promoters. *Progress in Biotechnology*, 21(C): 105–11.

Ferri, M., Serrazanetti, D.I., Tassoni, A. et al. (2016) Improving the functional and sensorial profile of cereal-based fermented foods by selecting *Lactobacillus plantarum* strains via a metabolomics approach. *Food Research International*, 89:1095–1105.

Filannino, P., Bai, Y., Di Cagno, R. et al. (2015) Metabolism of phenolic compounds by *Lactobacillus* spp. during fermentation of cherry juice and broccoli puree. *Food Microbiology*, 46:272–279.

Finco, A. M. de Oliveira, Mamani, L.D.G., de Carvalho, J.C. et al. (2017) Technological Trends and Market Perspectives for Production of Microbial Oils Rich in Omega-3. *Critical Reviews in Biotechnology*, 37: 656–671.

Galicia-Jiménez, M.M., Rojas-Herrera, R., Sandoval-Castro, C. et al. (2014) Chemotactic responses of the rumen bacterial community towards the daidzein flavonoid. *Livestock Science*, 167:121–125.

Garay, L.A., Boundy-Mills, K.L., and German, J.B. (2014) Accumulation of High-Value Lipids in Single-Cell Microorganisms: A Mechanistic Approach and Future Perspectives. *Journal of Agricultural and Food Chemistry*, 62: 2709–27.

Garrote, G.L., Abraham, A.G., and De Antoni, G.L. (1998) Characteristics of Kefir Prepared with Different Grain : milk Ratios. *Journal of Dairy Research*, 65(1): 149–54.

Gellissen, G. (2000) Heterologous Protein Production in Methylotrophic Yeasts. *Applied Microbiology and Biotechnology*, 54(6): 741–50.

Ghasemlou, M., Khodaiyan, F., and Oromiehie, A. (2011) Rheological and Structural Characterisation of Film-Forming Solutions and Biodegradable Edible Film Made from Kefiran as Affected by Various Plasticizer Types. *International Journal of Biological Macromolecules*, 49(4): 814–21.

Glasser, W.G. and Wright, R.S. (1998) Steam-assisted biomass fractionation. II. Fractionation behavior of various biomass resources. *Biomass and Bioenergy*, 14(3): 219–235.

Goldberg, I., Rokem, J. S., Pines, O. (2006) Organic acids: old metabolites, new themes. *Journal of Chemical Technology and Biotechnology*, 81(10): 1601–1611.

Gupta, S., Jaiswal, A.K., and Abu-Ghannam, N. (2013) Optimization of fermentation conditions for the utilization of brewing waste to develop a nutraceutical rich liquid product. *Industrial Crops and Products*, 44:272– 282.

He, R., Ma, L., Li, C. et al. (2014) Trpac1, a pH Response Transcription Regulator, Is Involved in Cellulase Gene Expression in Trichoderma Reesei. *Enzyme and Microbial Technology*, 67: 17–26.

Helfrich, E.J.N., Reiter, S., and Piel, J. (2014) Recent advances in genome-based polyketide discovery. *Current Opinion in Biotechnology* 29:107–115.

Hendriks, A.T.W.M. and Zeeman, G.(2009) Pretreatments to enhance the digestibility of lignocellulosic biomass. *Bioresource Technology*, 100(1): 10–18.

Hernández-Almanza, A., Montañez-Sáenz, J., Martínez-Ávila, C. et al. (2014) Carotenoid production by *Rhodotorula glutinis* YB-252 in solid-state fermentation. *Food Bioscience*, 7:31–36.

Hernández-Salas, J.M., Villa-Ramírez, M. S., Veloz-Rendón, J. S. et al. (2009) Comparative hydrolysis and fermentation of sugarcane and agave bagasse. *Bioresource Technology*, 100(3): 1238–1245.

Hossain, M.A., Wakabayashi, H., Goda, F. et al. (2000) Effect of the Immunosuppressants FK506 and D-Allose on Allogenic Orthotopic Liver Transplantation in Rats. *Transplantation Proceedings*, 32(7): 2021–23.

Huang, C., Chen, X.-F., Xiong, L. et al. (2013) "Single Cell Oil Production from Low-Cost Substrates: The Possibility and Potential of Its Industrialization." *Biotechnology Advances*, 31(2): 129–39.

Huwig, A., Emmel, S., Jäkel, G., et al. (1997) Enzymatic Synthesis of L-Tagatose from Galactitol with Galactitol Dehydrogenase from Rhodobacter Sphaeroides D. *Carbohydrate Research*, 305(3–4): 337–39.

Ikeuchi, T., Azuma, M., Kato, J., et al.(1999) Screening of Microorganisms for Xylitol Production and Fermentation Behavior in High Concentrations of Xylose. *Biomass and Bioenergy*, 16(5): 333–39.

Ishihara, Y., Katayama, K., Sakabe, M. et al. (2011) Antioxidant Properties of Rare Sugar D-Allose: Effects on Mitochondrial Reactive Oxygen Species Production in Neuro2A Cells. *Journal of Bioscience and Bioengineering*, 112(6). The Society for Biotechnology, Japan: 638–42.

Itoh, H. and Izumori, K. (1996) Enzymatic Production of L-Tagatose and L-Fructose from L-Sorbose and L-Psicose, Respectively. *Journal of Fermentation and Bioengineering*, 81(4): 351–53.

Itoh, H., Okaya, H., Khan, A.R. et al. (1994) Purification and Characterization of D-Tagatose 3-Epimerase from Pseudomonas Sp. ST-24. *Bioscience, Biotechnology, and Biochemistry*, 58(12): 2168–71.

Itoh, H., Sato, T., Takeuchi, T. et al. (1995) Preparation of D-Sorbose from D-Tagatose by Immobilized D-Tagatose 3-Epimerase. *Journal of Fermentation and Bioengineering*, 79(2): 184–85.

Izumori, Ken. (2002) Bioproduction Strategies for Rare Hexose Sugars. *Naturwissenschaften*, 89 (3): 120–24.

Izumori, Ken, and Keiji Tsuzaki. (1988) Production of D-Tagatose from D-Galactitol by Mycobacterium Smegmatis. *Journal of Fermentation Technology*, 66 (2): 225–27.

Izumori, K., Miyuki Yamakita, Tetsuji Tsumura, et al. (1990) Production of D-Psicose from D-Talitol, D-Tagatose or. *Journal Of Fermentation And Bioengineering*, 70(1): 26–29.

Jansen, M. L. and van Gulik, W. M. (2014) Towards large scale fermentative production of succinic acid. *Current Opinion in Biotechnology*, 30: 190–197

Jayaraman, J. and Gautam, P. (2016) Evaluation of Production and Kinetics Parameters of Rare Sugar (D-Tagatose) Using Biocatalyst Arthrobacter Globiformis. *Management of Environmental Quality: An International Journal*, 27(1): 71–78.

Jeon, Y. J., Shin, H.-S., and Rogers, P.L. (2011) Xylitol Production from a Mutant Strain of Candida Tropicalis. *Letters in Applied Microbiology*, 53(1): 106–13.

Jhan, J-K., Chang, W-F., Wang, P-M. et al. (2015) Production of fermented red beans with multiple bioactivities using co-cultures of *Bacillus subtilis* and

Lactobacillus delbrueckii subsp. *Bulgaricus*. LWT - *Food Science and Technology*, 63:1281–1287.

John, M. K. M.; Thiruvengadam, M., Enkhtaivan, G. et al. (2014) Variation in major phenolic compounds and quality potential of CTC black tea elicited by *Saccharomyces cerevisiae* and its correlation with antioxidant potential. *Industrial Crops and Products*, 55:289–294.

Jin, M., Slininger, P.J., Dien, B.S. et al. (2015) Microbial Lipid-Based Lignocellulosic Biorefinery: Feasibility and Challenges. *Trends in Biotechnology*, 33(1): 43–54.

Juániz, I., Ludwig, I.A., Bresciani, L. et al. (2016) Catabolism of raw and cooked green pepper (*Capsicum annuum*) (poly)phenolic compounds after simulated gastrointestinal digestion and faecal fermentation. *Journal of Functional Foods*, 27:201–213.

Kalscheuer, R., Stöveken, T., Malkus, U. et al. (2007) Analysis of Storage Lipid Accumulation in Alcanivorax Borkumensis: Evidence for Alternative Triacylglycerol Biosynthesis Routes in Bacteria. *Journal of Bacteriology*, 189(3): 918–28.

Kars, G., Gündüz, U., Yücel, M., et al. (2006) Hydrogen production and transcriptional analysis of nif D, nif K and hup S genes in Rhodobacter sphaeroides O . U . 001 grown in media with different concentrations of molybdenum and iron. *International Journal of Hydrogen Energy*, 31, pp.1536–1544.

Khan, A.R., Takahata, S., Okaya, H., et al. (1992) 'D-Sorbose Fermentation' from Galactitol by Pseudomonas Sp. ST 24. *Journal Of Fermentation And Bioengineering*, 74(3): 149–52.

Krüger, A., Schäfers, C., Schröder, C., et al. (2017) Towards a sustainable biobased industry – Highlighting the impact of extremophiles. *New Biotechnology*, 40:144–153.

Kim, B.C., Lee, Y.H., Lee, H.S. et al. (2002) Cloning, Expression and Characterization of L-Arabinose Isomerase from Thermotoga Neapolitana: Bioconversion of D-Galactose to D-Tagatose Using the Enzyme. *FEMS Microbiology Letters*, 212(1): 121–26.

Kim, H.-J., Ryu, S.-A., Kim, P., et al. (2003) A Feasible Enzymatic Process for D-Tagatose Production by an Immobilized Thermostable L -Arabinose Isomerase in a Packed-Bed Bioreactor. *Biotechnol. Prog.* 19: 400–404.

Laaksonen, O., Kuldjärv, R., Paalme, T. et al. (2017) Impact of apple cultivar, ripening stage, fermentation type and yeast strain on phenolic composition of apple ciders. *Food Chemistry*, 233:29–37.

Leang, K., Maekawa, K., Menavuvu, B.T. et al. (2004) A Novel Enzymatic Approach to the Massproduction of L-Galactose from L-Sorbose. *Journal of Bioscience and Bioengineering*, 97(6): 383–88.

Li, Z., Gao, Y., Nakanishi, H. et al. (2013) Biosynthesis of Rare Hexoses Using Microorganisms and Related Enzymes. *Beilstein Journal of Organic Chemistry*, 9: 2434–45.

Lipko, A. and Swiezewska, E. (2016) Isoprenoid generating systems in plants—A handy toolbox how to assess contribution of the mevalonate and methylerythritol phosphate pathways to the biosynthetic process. *Progress in Lipid Research* 63:70–92.

Liu, B., Ren, N., Ding, J., et al. (2009) The effect of Ni 2 D, Fe 2 D and Mg 2 D concentration on photo-hydrogen production by Rhodopseudomonas faecalis RLD-53. *International Journal of Hydrogen Energy*, 34(2): 721–726.

Liu, S. (2017) An Overview of Biological Basics. *Bioprocess Engineering*. Elsevier.

Low, D.Y., Hodson, M.P., Williams, B.A. et al. (2016) Microbial biotransformation of polyphenols during in vitro colonic fermentation of masticated mango and banana. *Food Chemistry*, 207:214–222.

Luksic, L., Bonafaccia, G., Timoracka, M. et al. (2016) Rutin and quercetin transformation during preparation of buckwheat sourdough bread. *Journal of Cereal Science*, 69:71–76.

Lussier, F-X., Colatriano, D., Wiltshire, Z. et al. (2012) Engineering Microbes for Plant Polyketide Biosynthesis. *Computational and Structural Biotechnology Journal*, 3(4):e201210020.

Macauley-Patrick, S., Fazenda, M.L., McNeil, B., and Harvey, L.M. (2005) Heterologous Protein Production Using the Pichia Pastoris Expression System. *Yeast*, 22(4): 249–70.

Madigan, M. (2012) *Brock Biology of Microorganisms*, 13th edn. Int. Microbiol. 550–551.

Marienhagen, J. and Bott, M. (2013) Metabolic engineering of microorganisms for the synthesis of plant natural products. *Journal of Biotechnology*, 163:166– 178.

Martín, C., Klinke, H.B. and Thomsen, A.B. (2007) Wet oxidation as a pretreatment method for enhancing the enzymatic convertibility of sugarcane bagasse. *Enzyme and Microbial Technology*, 40(3): 426–432.

Mateles, R.J. and Tannenbaum, S.(1968) *Single cell protein*. MIT Press, Cambridge, MA.

Mayorga-Gross, A. L., Quirós-Guerrero, L.M., Fourny, G. et al. (2016) An untargeted metabolomic assessment of cocoa beans during fermentation. *Food Research International*, 89:901–909.

McKeague, M., Wang, Y-H., Cravens, A. et al. (2016) Engineering a microbial platform for de novo biosynthesis of diverse methylxanthines. *Metabolic Engineering*, 38:191–203.

Mello-de-Sousa, T.M., Silva-Pereira, I., and Poças-Fonseca, M.J. (2011) Carbon Source and pH-Dependent Transcriptional Regulation of Cellulase Genes of Humicola Grisea Var. Thermoidea Grown on Sugarcane Bagasse. *Enzyme and Microbial Technology*, 48(1): 19–26.

Mitani, T., Hiroshi H., Terushige M. et al. (2009) Growth inhibition of head and neck carcinomas by d-allose. *Head and Neck*, 36(10): 1049–55.

Mohamad, N.L., Mustapa Kamal, S.M., and Mokhtar, M.N. (2015) Xylitol Biological Production: A Review of Recent Studies. *Food Reviews International*, 31(1): 74–89.

Mosele, J.I., Macià, A., Romero, M-P. et al. (2015) Application of *in vitro* gastrointestinal digestion and colonic fermentation models to pomegranate products (juice, pulp and peel extract) to study the stability and catabolism of phenolic compounds. *Journal of Functional Foods*, 14:529–540.

Naik, S.N., Goud V.V., Rout, P.K. et al. (2010) Production of first and second generation biofuels: A comprehensive review. *Renewable and Sustainable Energy Reviews*, 14(2:578–597.

Nanou, K. and Triantafyllos R. (2016) Waste cooking oil: A new substrate for carotene production by *Blakeslea trispora* in submerged fermentation. *Bioresource Technology*, 203:198–203.

Nanou, K., Triantafyllos, R., and Papadakis, E. (2012) Improved production of carotenes from synthetic medium by *Blakeslea trispora* in a bubble column reactor. *Biochemical Engineering Journal*, 67:203– 207.

Nasseri, A.T., Rasoul-Ami, S. Morowvat, M.H., et al. (2011) Single Cell Protein: Production and Process. *American Journal of Food Technology*, 6: 103–116.

Nelson, D.L. and Cox, M.M. (2013) *Lehninger Principles of Biochemistry*, 6e. W.H.Freeman, New York.

Ojeda, K. and Kafarov, V. (2009) Exergy analysis of enzymatic hydrolysis reactors for transformation of lignocellulosic biomass to bioethanol. *Chemical Engineering Journal*, 154(1–3): 390–395.

Oura, E. (1983) Biomass from Carbohydrates. In: *Biotechnology*, Dellweg, H. (Ed.). VerlagChemie GmbH, Weinheim, Germany, pp.3–42.

Özgur, E., Mars A.E., Peksel, B. et al. (2010) Biohydrogen production from beet molasses by sequential dark and photofermentation. *Int. J. Hydrogen Energy*, 35: 511–517.

Pachapur, V.L., Sarma, S.J., Brar, S.K. et al. (2015) Evidence of metabolic shift on hydrogen, ethanol and 1,3-propanediol production from crude glycerol by nitrogen sparging under micro-aerobic conditions using co-culture of Enterobacter aerogenes and Clostridium butyricum. *International Journal of Hydrogen Energy*, 40(28): 8669–8676.

Pan, F., Hou, K., Gao, F. et al. (2014) Peimisine and peiminine production by endophytic fungus *Fusarium* sp. isolated from *Fritillaria unibracteata* var. *wabensis*. *Phytomedicine*, 21:1104–1109.

Papanikolaou, S. and Aggelis, G. (2011a) Lipids of Oleaginous Yeasts. Part I: Biochemistry of Single Cell Oil Production. *European Journal of Lipid Science and Technology*, 113(8): 1031–51.

Papanikolaou, S. and Aggelis, G. (2011b) Lipids of Oleaginous Yeasts. Part II: Technology and Potential Applications. *European Journal of Lipid Science and Technology*, 113(8): 1052–73.

Paraskevopoulou, A., Athanasiadis, I., Kanellaki, M. et al. (2003) Functional Properties of Single Cell Protein Produced by Kefir Microflora. *Food Research International*, 36(5): 431–38.

Petrik, S., Kádár, Z. and Márová, I. (2013) Utilization of hydrothermally pretreated wheat straw for production of bioethanol and carotene-enriched biomass. *Bioresource Technology*, 133:370–377.

Poonperm, W., Takata, G., Ando, Y. et al. (2007) Efficient Conversion of Allitol to D-Psicose by Bacillus Pallidus Y25. *Journal of Bioscience and Bioengineering*, 103(3): 282–85.

Rafiqul, I. S. M., and Mimi Sakinah, A. M. (2013) Processes for the Production of Xylitol — A Review. *Food Reviews International*, 29 (2): 127–56.

Ramos, L.P., Carpes, S.T., Silva, F.T. et al. (2000) Comparison of the susceptibility of two hardwood species, Mimosa scabrella Benth and Eucalyptus viminalis labill, to steam explosion and enzymatic hydrolysis. *Brazilian Archives of Biology and Technology*, 43(2).

Ramos, L.P., Breuil, C., Kushner, D.J. et al. (1992) Steam Pretreatment Conditions for Effective Enzymatic-Hydrolysis and Recovery Yields of Eucalyptus-Viminalis Wood Chips. *Holzforschung*, 46(2): 149–154.

Rocchetti, G., Lucini, L., Chiodelli, G. et al. (2017) Phenolic profile and fermentation patterns of different commercial gluten-free pasta during in vitro large intestine fermentation. *Food Research International*, 97:78–86.

Rodionova, M. V., Poudyal, R.S., Tiwari, I. et al. (2016) Biofuel production: Challenges and opportunities. *International Journal of Hydrogen Energy*, 42(12): 8450–8461.

Ryu, B.G., Kim, K., Kim, J. et al. (2013) Use of Organic Waste from the Brewery Industry for High-Density Cultivation of the Docosahexaenoic Acid-Rich Microalga, Aurantiochytrium Sp. KRS101. *Bioresource Technology*, 129: 351–59.

Sandhu, K.S., Punia, S. and Kaur, M. (2016) Effect of duration of solid state fermentation by *Aspergillus awamorinakazawa* on antioxidant properties of wheat cultivars. *LWT - Food Science and Technology*, 71:323–328.

Sarkar, A., Funk, A.N., Scherlach, K. et al. (2012) Differential expression of silent polyketide biosynthesis gene clusters in chemostat cultures of *Aspergillus nidulans*. *Journal of Biotechnology*, 160:64–71.

Sarma, S.J., Brar, S.K., Bihan, Y. L. et al. (2014) Mitigation of the inhibitory effect of soap by magnesium salt treatment of crude glycerol - A novel approach for enhanced biohydrogen production from the biodiesel industry waste. *Bioresource Technology*, 151: 49–53.

Sauer, M., Porro, D., Mattanovich, D., et al. (2008) Microbial production of organic acids: expanding the markets. *Trends in Biotechnology*, 26(2): 100–108.

Schläger, S. and Dräger, B. (2016) Exploiting plant alkaloids. *Current Opinion in Biotechnology*, 37:155–164.

Serra, A., Macia, A., Romero, M-P. et al. (2012) Metabolic pathways of the colonic metabolism of flavonoids (flavonols, flavones and flavanones) and phenolic acids. *Food Chemistry*, 130:383–393.

Seydim, Z.G., Seydim, A.C., Greene, A.K., et al. (2000) Determination of Organic Acids and Volatile Flavor Substances in Kefir during Fermentation. *Journal of Food Composition and Analysis*, 13: 35–43.

Shimonishi, T., Okumura, Y., and Izumori, K. (1995) Production of L-Tagatose from Galactitol by Klebsiella Pneumoniae Strain 40b. *Journal of Fermentation and Bioengineering*, 79(6): 620–22.

Soccol, C.R., Vandenberghe, L.P.S., Medeiros A.B.P. et al. (2010) Bioethanol from lignocelluloses: Status and perspectives in Brazil. *Bioresource Technology*, 101(13): 4820–4825.

Song, H., and Lee, S. Y. (2006) Production of succinic acid by bacterial fermentation. *Enzyme and Microbial Technology*, 39(3): 352–361.

Sui, L., Dong, Y., Watanabe, Y. et al. (2005) The Inhibitory Effect and Possible Mechanisms of D-Allose on Cancer Cell Proliferation. *International Journal of Oncology*, 27(4): 907–12.

Tada, K., Kanno, T., and Horiuchi, J.I. (2012) Enhanced Production of Bioxylitol from Corn Cobs by Candida Magnoliae. *Industrial and Engineering Chemistry Research*, 51(30): 10008–14.

Tamburini, E., Costa, S., Marchetti, M.G. et al. (2015) Optimized Production of Xylitol from Xylose Using a Hyper-Acidophilic Candida Tropicalis. *Biomolecules*, 5(3): 1979–89.

Tannenbaum S.R. and Wang D.I.C. (1975) *Single cell protein*, vol. II. MIT Press, Cambridge, MA.

Tao, J-H., Duan, J-A., Jiang, S. et al. (2016) Biotransformation and metabolic profile of buddleoside with human intestinal microflora by ultrahigh-performance liquid chromatography coupled to hybrid linear ion trap/orbitrap mass spectrometer. *Journal of Chromatography B*, 1025:7–15.

Teo, C.L. et al.(2014) Enhancing growth and lipid production of marine microalgae for biodiesel production via the use of different LED wavelengths. *Bioresource Technology*, 162: 38–44.

Ulbrich, K., Reichardt, N., Braune, A. et al. (2015) The microbial degradation of onion flavonol glucosides and their roasting products by the human gut bacteria *Eubacterium ramulus* and *Flavonifractor plautii*. *Food Research International*, 67:349–355.

Uyar, B. and Schumacher, Æ.M. (2009) Photoproduction of hydrogen by Rhodobacter capsulatus from thermophilic fermentation effluent. *Bioprocess. Biosyst. Eng.*, 32, 603–606.

Werpy T. and Petersen G. (2004) Top value added chemicals from biomass: volume I—Results of screening for potential candidates from sugars and synthesis gas. US Department of Energy.

Winkelhausen, E. and Kuzmanova, S. (1998) Microbial Conversion of D-Xylose to Xylitol. *Journal of Fermentation and Bioengineering*, 86(1): 1–14.

Winzer, T., Kern, M., King, A.J. et al. (2015) Plant science: morphinan biosynthesis in opium poppy requires a P450-oxidoreductase fusion protein. *Science*, 349:309–312.

Wu, J., Du, G., Zhou, J. et al. (2014) Systems metabolic engineering of microorganisms to achieve large-scale production of flavonoid scaffolds. *Journal of Biotechnology*, 188:72–80.

Wu, J., Zhang, X., Dong, M. et al. (2016a) Stepwise modular pathway engineering of Escherichia coli for efficientone-step production of (2S)-pinocembrin. *Journal of Biotechnology*, 231:183–192.

Wu, J., Zhang, X., Zhou, J. et al. (2016b) Efficient biosynthesis of (2S)-pinocembrin from D-glucose by integrating engineering central metabolic pathways with a pH-shift control strategy. *Bioresource Technology*, 218:999–1007.

Yadav, V.G., Mey, M., Lim, C.G. et al. (2012) The future of metabolic engineering and synthetic biology: Towards a synthetic practice. *Metabolic Engineering*, 14:233–241.

Yang, J., Jiao, R-H., Yao, L-Y. et al. (2016) Control of fungal morphology for improved production of a novel antimicrobial alkaloid by marine-derived fungus *Curvularia* sp. IFB-Z10 under submerged fermentation. *Process Biochemistry*, 51:185–194.

Zajšek, K., Kolar M. and Goršek, A. (2011). Characterisation of the Exopolysaccharide Kefiran Produced by Lactic Acid Bacteria Entrapped within Natural Kefir Grains. *International Journal of Dairy Technology*, 64 (4): 544–48.

Zhao, D. and Shah, N.P. (2016) Lactic acid bacterial fermentation modified phenolic composition in tea extracts and enhanced their antioxidant activity and cellular uptake of phenolic compounds following *in vitro* digestion. *Journal of Functional Foods*, 20:182–194.

Zhou, J; Du, G and Chen, J. (2014) Novel fermentation processes for manufacturing plant natural products. *Current Opinion in Biotechnology*, 25:17–23.
Zhu, Y-F., Chen, J-J., Ji, X-M. et al. (2015) Changes of major tea polyphenols and production of four new B-ring fission metabolites of catechins from post-fermented Jing-Wei Fu brick tea. *Food Chemistry*, 170:110–117.

8

Communication for a Collective Response to Environmental Stress: Bacterial and Fungal Perspectives

Azadeh Kermanshahi Pour

Biorefining and Remediation Laboratory, Department of Process Engineering and Applied Science, Dalhousie University, Halifax, Canada

8.1 Introduction

For a long time, microorganisms were perceived as autonomous unicellular organisms without the capability to communicate or co-ordinate a collective behavior (Williams et al., 2007). The first evidence that changed this perception was the report on co-operative behaviors in bacteria in a cell-density associated manner in 1962(Williams et al., 2007). Observing fruiting body formation in *Myxococcus xanthus* (McVittie et al., 1962), streptomycin biosynthesis and aerial mycelium formation in *Streptomyces griseus* (Khokhlov et al., 1967), induction of genetic competence in *Streptococcus pneumoniae* (Tomasz, 1965) and later control of bioluminescence in marine *Vibrio fischeri* (Nealson et al., 1970) led to the recognition of microorganisms for their synchronized behavior. This phenomenon was later referred to as "quorum sensing" (QS) by Fuqua et al. (Fuqua et al., 1994). Ever since, a growing body of evidence has been accumulated indicating that microorganisms function beyond autonomous behavior and are capable of communication to coordinate a co-operative behavior through the complex mechanism of QS (Williams et al., 2007). After almost four decades of discovery of QS in prokaryotes, an active molecule identified as "farnesol" secreted by pathogenic polymorphic fungus *Candida albicans* that self-inhibited mycelial growth, was found to be responsible for QS. This was the first report of QS activity in eukaryotes (Hornby et al., 2001).

These cell-density associated behaviors are mediated by biosynthesis and release of extracellular signaling molecules known as quorum sensing molecules (QSM) (also known as auto-inducers or pheromones) that diffuse in and out of the cells (Williams et al., 2007). The name of these signaling molecules originated from the Greek terms "pherine" (to transfer) and "hormone" (to excite) (Karlson and Luscher, 1959). The difference of pheromones and hormones are that unlike hormones, pheromones are secreted outside the cell as a means for cell communication, allowing the microorganisms to sense their own cell density (Williams

Microbial Sensing in Fermentation, First Edition. Edited by Satinder Kaur Brar, Ratul Kumar Das, and Saurabh Jyoti Sarma.

et al., 2007). Concentration of QSM increases as a result of increase in cell density up to a certain threshold concentration in which a target sensor kinase or response regulator is activated, resulting in regulating quorum sensing-associated gene transcription (e.g. repression or expression of genes) (De Sordi and Muhlschlegel, 2009; Miller and Bassler, 2001; Williams et al., 2007). Modulation of gene expression leads to a range of synchronized behaviors including secretion of virulence factors, biofilm formation, antibiotic production, controlling growth, pigment production and morphogenesis (Casadevall and Pirofski, 1999; Miller and Bassler, 2001; Taga and Bassler, 2003; Waters and Bassler, 2005; Albuquerque and Casadevall, 2012; Williams et al., 2007). Collectively, all of the above behavioral changes facilitate adaptation of microorganisms to the surrounding environment including improving the chance of survival, access to nutrients, competence and coordinated pathogenic attack (De Sordi and Muhlschlegel, 2009; Williams et al., 2007). Discovery of QS systems in various bacterial and fungal species indicates the significance of QS for the survival of microorganisms in different environments (Hogan, 2006).

There are also situations that bacteria take advantage of diffusible signal rather than cell population to regulate a collective behavior (Williams et al., 2007). In some cases where production and degradation rate of signal molecules influence the behavioral change, the term “diffusion sensing” or “compartment sensing” is used (Redfield, 2002; Winzer et al., 2002). Criteria to differentiate between the metabolites and QSM include:

1) The production of the quorum sensing signals occurs during specific stages of growth in response to environmental and physiological conditions;
2) the QSM is secreted from the cell to the extracellular environment and is recognized by a specific cell receptor;
3) reaching to a critical threshold concentration of the QSM leads to a collective response by microorganisms; and
4) physiological changes in cell in response to the toxicity of a molecule is not an indication of the presence of QSM (Williams et al., 2007; Winzer et al., 2002).

Most QS systems identified are species-specific or group-specific (De Sordi and Muhlschlegel, 2009;Taga and Bassler, 2003) and majority of the work on QS is focused to understand the complex mechanism of QS and its effect on cell physiology of a specific species (Taga and Bassler, 2003). However, in many situations, QSMs are also responsible for more complex interspecies interactions (Taga and Bassler, 2003; De Sordi and Muhlschlegel, 2009). Although understanding the role of QS in the interaction of species in different microbial communities is complex, this is important since the natural environment consists of poly-microbial communities (Taga and Bassler, 2003, De Sordi and Muhlschlegel, 2009). Additionally, co-culture is employed in the controlled fermentation for production of chemicals of commodity value. For instance, QS determines the productivity and yield in the fermentative production of acetic acid in which bacterial co-culture is used (Iida, Ohnishi, and Horinouchi, 2008a), indicating the significance of studying of QS in poly-microbial communities.

In the context of inter-kingdom species, eukaryotes have a range of responses to *N*-acyl homoserine lactones (AHLs), common QSM of Gram-negative bacteria. For example, halogenated furanone produced by the marine macroalgae *Delisea pulchra* acts as a competitive inhibitor of the bacterium's AHL-based QS system of *Serratia liquefaciens* (Rasmussen et al., 2000). Halogenated furanones have similar structure to AHLs and are able to prevent the interaction of AHL signal of *S. liquefacients* with its associated LuxR-type protein, resulting in preventing motility in *S. liquefacients* and inhibiting bacterial colonization on the macroalga (Givskov et al., 1996; Rasmussen et al., 2000).

Anti-quorum sensing is also a phenomenon of interest that was reported in both bacteria and eukaryotes, which is speculated to combat bacteria that rely on QS (Taga and Bassler, 2003). For example, some bacteria and fungi are capable of disruption of AHLs by their own QS systems or secretion of lactonase enzyme to hydrolyze the lactone ring of AHLs (Taga and Bassler, 2003; Dudler and Eberl, 2006; Cugini et al., 2007).

Significant research is directed to understand QS at the molecular level and its biological significance due to its potential implications in the area of medicinal chemistry and fermentation technologies. Amongst the most important QS-associated phenotypes are polymorphism and biofilm formation on infected surfaces (Albuquerque and Casadevall, 2012; Blankenship and Mitchell, 2006; De Sordi and Muhlschlegel, 2009; Ramage et al., 2002). QSMs that are toxic to the host cells and can modify the host immune system are considered virulence factors (Winzer and Williams, 2001). During the infection of a host, pathogenic bacteria coordinate secretion of virulence factors, which results in enhancing their survival in response to immune response of the host cells (de Kievit and Iglewski, 2000; Williams et al., 2007). Regulation of the virulence gene expression can occur in both bacteria (Casadevall and Pirofski, 1999) and fungi (San-Blas et al., 2000). On the other hand, biofilm is generally more resistant to the environmental changes and therefore, pathogenic biofilm is more resistant to antibiotics and host immune factors compared with planktonic cells (Blankenship and Mitchell, 2006, Nickerson et al., 2006). Identification of QSM and understanding the genes that are associated with regulating biofilm organization provide insight into developing medicinal approaches to overcome pathogenic biofilm formation and antifungal and antibiotic resistance of biofilm (Cao et al., 2005; De Sordi and Muhlschlegel, 2009). Chemical agents and mechanisms that interfere with QS pathways (e.g. biofilm development of pathogenic bacteria and fungi) can be the target for antifungal and antibacterial drug development (Rice et al., 2005). For example, synthetic furanone binds to AHL, the QSM in *P. aeruginosa*, resulting in clearance of *P. aeruginosa* infections in mice (Wu et al., 2004). Farnesol, QSM in *C. albicans* inhibits biofilm formation through a molecular mechanism that is unknown and therefore, QS has an important role in *C. albicans* antifungal drug resistance (Cao et al., 2005).

The role of QS in fermentation optimization has also been recognized (Iida et al., 2008a). QS in acetic acid bacteria is associated with repressing oxidative fermentation of several bioproducts (e.g. acetic acid and gluconic acid) as a response to their toxic effects on the acetic acid bacteria, leading to reduced growth rate and product yield (Iida et al., 2008b). Therefore, manipulation of the quorum-sensing system is applicable in the area of industrial fermentation and informs strategy development for strain improvement and fermentation optimization (Iida et al., 2008b, Lee et al., 2008).

In addition to QS that is triggered in response to the population density, other environmental signals (e.g. temperature, pH, osmolarity, oxidative stress and nutrient deprivation) are collectively sensed by the cells, leading to a response targeted for an optimum behavioral change for survival (Withers et al., 2001). For example, the integrated impact of QS and environmental cues (e.g. pH) on bio-product formation (e.g. acetoin) is demonstrated in *Serratia plymuthica* RVH1 (Moonset al., 2011). Also, in *Saccharomyces cerevisiae*, morphogenetic transitions are regulated by combination of QS and low nitrogen concentration (Chen and Fink, 2006).

In this chapter, QS, environmental cues and their associated phenotypes, particularly in the context of fermentation are reviewed. The similarities and differences of QS systems in fungi and bacteria are further discussed.

8.2 Quorum Sensing in Bacteria and the Related Phenotypes

QS controls diverse processes in bacteria including pathogenesis, biofilm development, food spoilage, metabolite formation and resistance to antimicrobial compounds (Reading and Sperandio, 2006; Van Houdt et al., 2005;Wevers et al., 2009). Three major quorum-sensing circuits are identified to describe the mechanism of QS: one in Gram-negative, one in Gram-positive and one that has been proposed to be universal (Miller and Bassler, 2001; Reading and Sperandio, 2006; Schauder and Bassler, 2001). QS systems in bacteria and fungi and their associated QSM and phenotypes have been reviewed (Taga and Bassler, 2003; Rutherford and Bassler, 2012; Papenfort and Bassler, 2016; Albuquerque and Casadevall, 2012).

Gram-negative bacteria primarily use the LuxR/I-type systems with signaling molecules being *N*-acylhomoserine lactones (AHLs) (Figure 8.1) (Parsek and Greenberg, 2000; Pearson et al., 1994; Pearson et al., 1995). Members of Enterobacteriaceae (e.g. *Escherichia*, *Salmonella*) and acidophilic extremophile (e.g. *Acidithiobacillus ferrooxidans*) are capable of producing multiple AHLs (Farah et al., 2005; Williams et al., 2007). AHL-mediated QS may result in variety of phenotypes ranging from bioluminescence to biofilm formation and production of extracellular metabolites such as protease, acetoin and butanediol (Lynch et al., 2002; Swift et al., 1999; Van Houdt et al., 2006 a,b; Van Houdt et al., 2007, Moons et al., 2011).

AHLs have a conserved homoserine lactone (HSL) ring unsubstituted at β and γ-positions and at the α-position, is connected through an amide bond to acyl side chains that may contain one or more double bond (Williams et al., 2007). The acyl chain varies in length, saturation level and oxidation state. Even number of carbon (C_4–C_{18}) is most common. AHLs with C_5 and C_7 acyl chains have also been identified (Williams et al., 2007). The structure of some of the common AHLs that are employed by QS system in bacteria is shown in Figure 8.1 and the QS system and phenotypes that these chemicals regulate is elaborated in Table 8.1.

The first AHL dependent LuxR/I QS-dependent phenotype, discovered was bioluminescence in *Vibrio fischeri*. AHL-dependent QS was later found to be

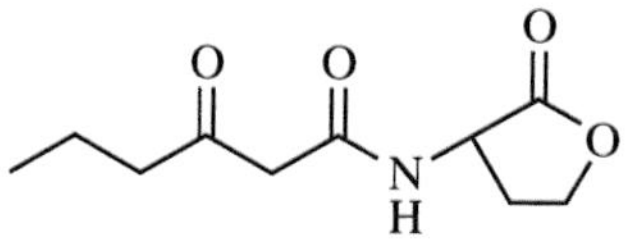

3-oxo-C_6HSL

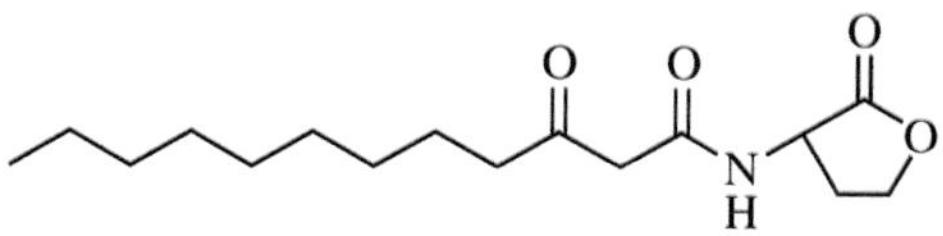

3-oxo-C_{12}-HSL

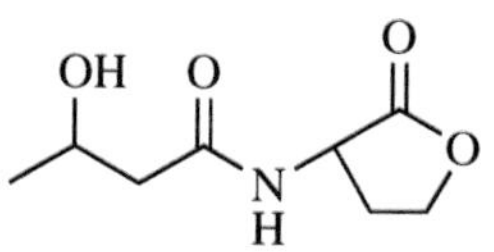

3-hydroxy-C_4-HSL

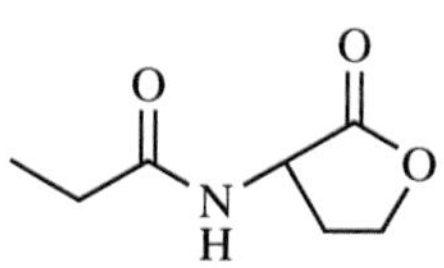

C_4-HSL

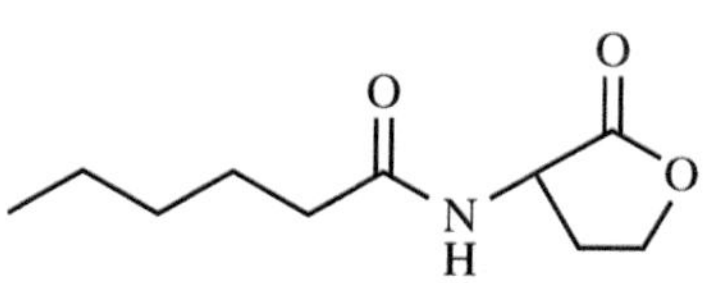

C_4-HSL

Figure 8.1 Chemical structures of common Quorum Sensing Molecules (QSM) produced by bacteria, involved in modulating phenotypes shown in Table 8.1.

common in Gram-negative bacteria (Nealson et al., 1970) and currently *Vibrio fischeri* stands as a model for AHL-type QS (Van Houdt et al., 2007). AHL synthesized by the LuxI-type proteins (AHL synthases) diffuse across the cell membrane and accumulates in the microbial broth. Charged acyl-acyl carrier protein (acyl-ACP) and S-adenosylmethionine are used by LuxI protein family as the sources of the acyl side chain and the HSL ring moiety, respectively (Jiang et al., 1998; More et al., 1996; Parsek et al., 1999). In high cell density, when AHL reaches threshold concentration, it diffuses across the bacterial envelope and binds with transcriptional regulator LuxR and activates LuxR protein (*lux* operon) (Van Houdt et al., 2007). AHL-activated LuxR protein functions as a transcriptional activator triggering a response, which results in a bioluminescence provided that the bacterial population exceeds a certain cell density threshold (Eberhard et al., 1981; Fuqua et al.,1996).

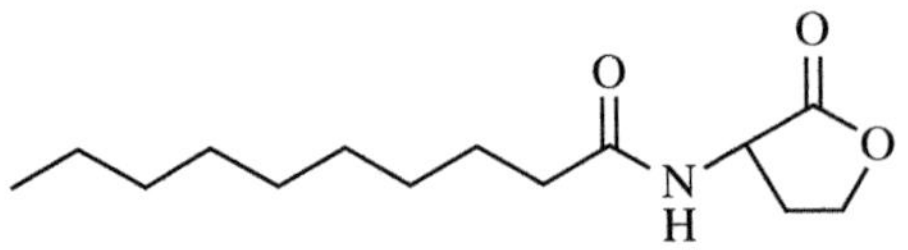

C_{10}-HSL

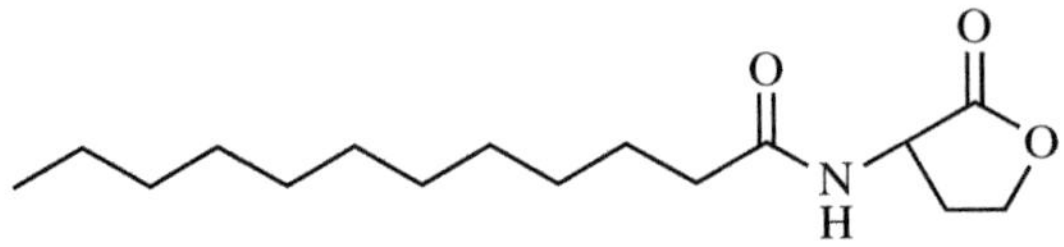

C_{12}-HSL

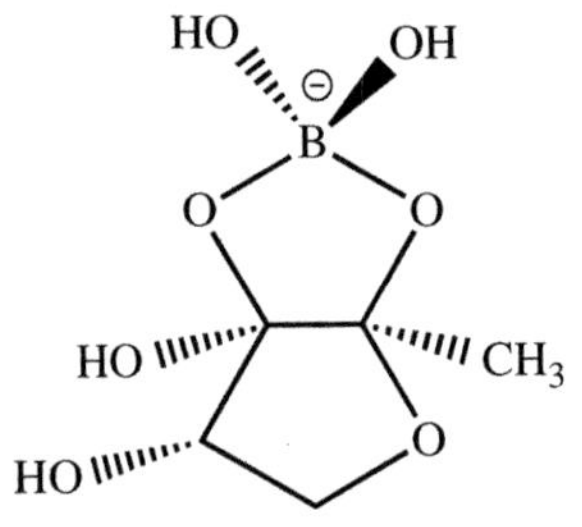

AI-2

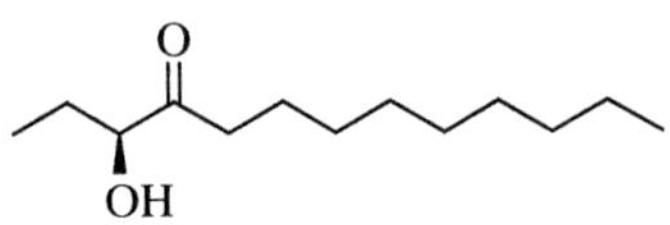

CAI-1

Figure 8.1 (Continued)

After the discovery of the bioluminescence as the first QS phenotype that was found to be controlled by LuxR/I quorum-sensing system, LuxR/I has been found to be responsible for regulating the transcription of several genes involved in variety of phenotypes in Gram-negative bacteria including production of antibiotics in *Erwinia*, motility in *Yersinia pseudotuberculosis* (de Kievit and Iglewski, 2000), pathogenesis and biofilm formation in *Pseudomonas aeruginosa* (Davies et al., 1998; Matthew R. Parsek et al., 1999), butanediol fermentation in *Aeromonas hydrophila* AH-1N (Van Houdt et al., 2007), *Serratia plymuthica* RVH1 and *Serratia marcescens* MG1 (Van Houdt et al., 2007, Van Houdt et al., 2006a,b). In all of these LuxR/I homologue systems, the bacteria produce an AHL autoinducer, which binds to the LuxR homologue protein and regulates the transcription of several genes involved in the above phenotypes (Williams et al., 2007).

In the opportunistic human pathogen, *P. aeruginosa*, AHL-dependent and 4-quinolone dependent QS systems are integrated. *P. aeruginosa* produces two AHLs, *N*-(3-oxododecanoyl)-L-homoserine lactone (3-oxo-C_{12}-HSL) and N-butanoyl-L-homoserine lactone (C_4- HSL) (Pearson et al., 1994; Pearson et al., 1995). LasI (Lux I homologue) directs the synthesis of 3-oxo-C_{12}-HSL, which activates LasR (LuxR

Table 8.1 Different Quorum Sensing Systems and influential environmental cues in bacteria and their associated phenotypes.

Microorganisms	Major QSM	Regulatory Proteins	Controlling Environmental Cues	QS Phenotype	References
Vibrio fischeri	3-oxo-C_6-HSL	LuxI/LuxR AinS	-	Bioluminescence	Eberhard et al., 1981
Pseudomonas aeruginosa	3-oxo-C_{12}-HSL C_4-HSL	LasI/LasR RhlI /RhlR	-	Virulence regulation, Rhamnolipid and elastase production	de Kievit & Iglewski, 2000; Gambello & Iglewski, 1991
Vibrio harveyi	3-hydroxy-C_4-HSL AI-2	LuxL,M/LuxN LuxS/LuxPQ	-	Bioluminescence	Bassler at al., 1994; Chen et al., 2002
Vibrio cholerae	AI-2	AphA/AlsR(LysR-type protein)	pH, acetate concentration	Acetoin production, Biofilm formation, Motility	Kovacikova et al., 2005
Serratia plymuthica RVH1	C_4-HSL	*budR/budABC* LysR	pH	Butanediol fermentaion	Van Houdt et al., 2006; Moons et al., 2011
Serratia marcescens MG1	C_4-HSL C_6-HSL		pH	Butanediol fermentation	Van Houdt et al., 2006
Aeromonas hydrophila AH-N	C_4-HSL	*ahy*R/*ahy*I	Temperature	Butanediol fermentation	Van Houdt et al., 2007
Gluconacetobacter intermedius	C_{10}-HSL C_{12}-HSL	GinI/GinR	-	Butanediol fermentation	Iida et al., 2008
L. bulgaricus ATCC BAA-365	Peptide II	-	-	Autolysis	Pang et al., 2016

homologue), leading to regulating the production of virulence factors and gaining competitive survival advantage in the presence of other microbial species (Williams et al., 2007). RhlI directs the synthesis of *N*-butanoyl-L-homoserine (C_4-HSL), which activates RhlR, inducing the production of rhamnolipid and elastase (de Kievit and Iglewski, 2000; Gambello and Iglewski, 1991).

In *Aeromonas hydrophila*, AhyI (LuxI homologue) produces C_4-HSL and in combination with the LuxR homologue, AhyR, modulates phenotypes such as biofilm development and production of an extracellular protease (Lynch et al., 2002, Swift et al., 1999).

In some Gram-negative bacteria, the biosynthesis of AHLs does not only depend on LuxI homologues. For example, the LuxM family is AHL synthases, which were originally discovered in the marine bacterium *Vibrio harveyi*, (Bassler et al., 1993) andare responsible for biosynthesis of AHL. In *V. harveyi*, LuxL and Lux M are required for the synthesis of QSMs and LuxN (LuxR homologue) is required for the response to the QSM (Bassler et al., 1994). LuxM family has later been found in other Vibrio species (Hanzelka et al., 1999, Milton et al., 2001). LuxM and LuxI homologues have been shown to co-exist in some vibrios such as *V. fischeri* (AinS; homologue of LuxM and LuxI) (Hanzelka et al., 1999) and *V. anguillarum* (VanI; homologue of LuxI and VanM; homologue of Lux M) (Milton et al., 2001)). Ains drives the synthesis of *N*-octanoylhomoserine in *V. fisheri* and VanI drives the synthesis of N-hexanoylhomoserine lactone (C_6-HSL) and N-(3-hydroxyhexanoyl) homoserine lactone (3-hydroxy-C_6-HSL) in *V. anguillarum.* LuxI proteins are genetically linked to cognate transcriptional regulators, LuxR. However, LuxM genes are associated with genes coding for histidine protein kinase sensors, which interacts with AHLs in the periplasm, trigger a phosphorelay cascade resulting in transcriptional activation of the related quorum sensing dependent genes (Camara et al., 2002a; Croxatto et al., 2004).

A third potential AHL synthase (HdtS), which does not belong to either the LuxI or LuxM families has been identified in *Pseudomonas fluorescens* F113 (Laue et al., 2000). N-(3-hydroxy-7-cis-tetradecenoyl) homoserine lactone, N-decanoylhomoserine lactone (C_{10}-HSL) and N-hexanoylhomoserine lactone (C_6-HSL) are produced by *Pseudomonas fluorescens* F113 via HdtS and these AHLs may be involved in regulating biocontrol properties of this strain (Laue et al., 2000).

In Gram-positive bacteria biofilm formation, colonization, toxin and protease secretion are controlled by QS (Novick, 2003). Unlike Gram-negative bacteria in which low molecular weight QSM (e.g. AHL) is synthesized, Gram-positive bacteria employ post-translationally modified peptides for communication, known as autoinducing polypeptides (AIPs) that range from 5 to 34 amino acids in length (Williams et al., 2007). Gram-positive bacteria typically employ a two-component circuits that consist of a membrane located histidine protein kinase (HPK) and AIPs as QSM (Pang et al., 2016).

AIPs that are produced in the cytoplasm as precursor peptides are further cleaved, modified and then transported from bacteria by transporters (Taga and Bassler, 2003). Once threshold concentration is reached in the media, AIPs interact with the external domains of membrane-bound HPK that initiate phosphorylation of the corresponding response regulator protein (Taga and

Bassler, 2003). The response regulator protein binds with DNA, resulting in the alteration of target gene transcription (Taga and Bassler, 2003). Genetic competence and sporulation in *Bacillus subtilis* (Peterson et al., 2000, Solomon et al., 1995), competence for DNA uptake in *Streptococcus pneumoniae* (Haverstein and Morrison, 1999), virulence factor expression in *Staphylococcus aureus* (Novick, 2003) and *Enterococcus faecalis* (Qin et al., 2000) are regulated by AIP-QS systems.

Autoinducer AI-2 was first identified in bioluminescent marine bacterium, *Vibrio harv*eyi and is responsible for regulating luminescence (Bassler et al., 1993; Bassler et al., 1994) and later was found to be produced by diverse variety of Gram-positive and Gram-negative bacteria and thus is known as "universal" signal for inter-species communication (Miller and Bassler, 2001, Chen et al., 2002). LuxS, synthase required for synthesis of AI-2 is widely conserved amongst Gram-negative and Gram-positive bacteria (Surette et al., 1999). Autoinducer AI-2 binds LuxP and the resulting LuxP-AI-2 complex is hypothesized to interact with LuxQ, which is a two-component hybrid sensor kinase embedded in the bacterial inner membrane (Bassler et al., 1994) to transduce the autoinducer signal (Chen et al., 2002). AI-2 is a cyclic borate diester derivative of 4,5-dihydroxy-2,3-pentanedione (DPD) (Figure 8.1) and is recognized as a compound that can provide insight into the development of antibacterial drugs that target quorum sensing (Chen et al., 2002).

For interspecies communities and interkingdom signaling system, the luxS/AI-2 and AI-3/epinephrine/norepinephrine interkingdom signaling system is employed (Reading and Sperandio, 2006).

8.3 Fermentation and Quorum Sensing in Bacteria

Microbial metabolism is a series of complex biochemical reactions that are influenced by intracellular metabolic reaction and environmental cues (Mukhopadhyay et al., 2008). Intracellular metabolites synthesized via complex metabolic reactions may diffuse across the cell membrane to their surrounding environment, resulting in the change in their surrounding environment such as change in pH or metabolite concentration. Cells sense most of extracellular signals (e.g. hormones, cytokines, extracellular matrix) via transmembrane receptors that activates complex biochemical cascades of kinases, transcription factor, etc., resulting in modulating cell physiology and phenotype (Jamora and Fuchs, 2002; El-Ali et al., 2006). The integrated effect of QS and environmental cues such as temperature, pH, physiological inducer (e.g. acetate) and type of strains in modulating phenotypes (e.g. acetoin, butanediol production) is widely recognized (Magee and Kosaric, 1987). Therefore, developing an understanding of the signaling molecules and environmental cues that influence the metabolic pathway in fermentation will guide the design of fermentation strategies and/or the development of new strains to target higher product yield and productivity (Iida et al., 2008a). As important features in the context of QS and environmental cues that influence production of acetone-butanol-ethanol (ABE), butanol and acetic acid are reviewed in the following section.

Acetone-butanol-ethanol fermentation by *Clostridium acetobutylicum* is an industrially important process in which *Clostridium acetobutylicum*, an anaerobic Grampositive bacteria, is used for the production of biofuel and solvent (Lee et al., 2008). This species takes advantage of two sequential physiological phases to produce solvent [acetone-butanol-ethanol (ABE)] (Long, Jones, and Woods, 1984). The first phase, known as acidogenic phase involves production of organic acids (acetic acid and butyric acid), leading to a substantial decrease in pH. In the second phase known as the solventogenic phase, organic acids, formed in acidogenic phase are assimilated by bacteria to produce ABE solvent to overcome the stress of low pH for survival (Grimmler et al., 2011). Solventogenesis is the result of several steps including the steps that affect the enzyme expression and steps that affect the enzyme kinetics and activity (Husemann and Papoutsakis, 1988).

A combination of low pH and high undissociated butyric acid (UBA) termed as "pH-acid effect" is realized to induce solventogenesis in batch cultures (Holt et al., 1984; Monot et al., 1984; Fond et al., 1985; Terracciano and Kashket, 1986). UBA concentration and pH are widely reported as the environmental cues that influence solventogenesis in both batch and continuous culture due to their effect on the reaction steps involved in solventogenesis (Hubert et al., 1982; Husemann and Papoutsakis, 1988).

Several reports indicate that a minimum UBA concentration is required to initiate solventogenesis (Monot et al., 1984, Husemann and Papoutsakis, 1988, Liao et al., 2015). The minimum UBA concentration at the onset of solventogenesis was reported in the range of 6–13 m*M* in the pH range between 3.6 and 5 in cultures without pH control, and also in cultures with controlled pH between 3.7 and 5 (Husemann and Papoutsakis, 1988). Monot et al. reported UBA concentration in the range of 18–21 m*M* at the start of solventogenesis for cultures with controlled pH between 4.5 and 5.5 (Monot et al., 1984). Internal UBA concentration of 13 mM and total intracellular undissociated acetic and butyric acid concentration of 40–45 mM were reported at the onset of solventogenesis by Terracciano and Kashket (Terracciano and Kashket, 1986). Internal pH affects the minimum required concentration of UBA to induce solventogenesis and lower UBA was observed at the onset of solventogenesis at higher pH (e.g. pH of 6 compared to 5) (Husemann and Papoutsakis, 1988).

Husemann and Papoutsakis studied the effect of acetoacetate (AA) or propionic acid (PA) on solventogenesis in *Clostridium acetobutylicum* fermentation. The minimum UBA to induce solventogenesis was not significantly affected in batch cultures and was around 5–6 mM in both cases that AA and PA were present or were not present (Husemann and Papoutsakis, 1988).

During the transition from the acidogenic to the solventogenic phase, 245 genes have been identified to be differentially expressed (Grimmler et al., 2011). The environmental signals, such as external pH and undissociated acids trigger the production and phosphorylation of Spo0A, the master regulator that governs the expression of a set of functional genes coding for enzymes necessary in ABE fermentation (Harris et al., 2002; Terracciano and Kashket, 1986). Phosphorylated Spo0A(Spo0A ~ P) controls the expression of downstream genes, (e.g. adc, ctfA/B, and adhE), which are essential for the transition of acidogenic-to-solventogenic phases (Liao et al., 2015).

Dissociation of the organic acid that diffuse across the cell membrane to the surrounding environment results in change in the pH (Liao et al., 2015). Protons and undissociated forms of organic acid result in cellular stress and a change in gene expression (Liao et al., 2015). Metabolites including acetate, butyrate, butanol and pH were found to exhibit synergistic inhibitory effects on the cell growth of *C. acetobutylicum* (Yang and Tsao, 1994). The solvent production of *C. acetobutylicum* was attributed to the cells response to the environmental stress and strategy for survival. Genes associated with stress are shown to be activated during the phase transition of acidogenic to solventogenic phase to increase their tolerance to environmental stress (Tomas et al., 2004). pH is an environmental cue that is significantly important since failing to control the pH leads to incomplete sugar utilization and repress the transition of acidogenesis to solventogenesis (Monot et al., 1984, Maddox et al., 2000, Liao et al., 2015).

Product inhibition is a factor that negatively impacts the fermentation synergistic inhibitory effects on the cell growth of *C. acetobutylicum* and identified as a major factor that challenges the economics of acetone-butanol via fermentation (Yang and Tsao, 1994). Butanol has been reported to lower the cellular ATP levels of *C. acetobutylicum* and to disrupt various membrane-related functions (Linden et al., 1981). Inhibition affects many cellular structures and processes including membrane composition and solute transport processes (Yang and Tsao, 1994).

Collectively, environmental cues including low pH and accumulation of organic acid and butanol results in low product formation rate and dilute final product concentration, all of which lead to an energy intensive product recovery process (Marlatt and Datta, 1986). Non-glucose-limited conditions such as phosphate-limiting and iron-limiting conditions are amongst the environmental factors that are found to favor butanol production (Bahl et al., 1986; Roos al., 1985).

Butanediol and acetoin fermentation is another bioprocess that is influenced by QS phenomena. Fermentation of 2,3-butanediol as an alternative for synthetic approach that relies on depleting petroleum-based feedstock is gaining growing interest due to its application in the manufacture of liquid fuel additive, printing inks, perfumes, fumigants, plasticizers, and pharmaceuticals (Garg and Jain, 1995; Van Houdt et al., 2007; Huang and Ouyang, 2011).

Several Gram-negative bacteria, *Klebsiella, Enterobacter, Serratia, Erwinia* in the family of enterobacteriaceaeas well as *Bacillus subtilis* in the family of Bacillaceae and *Aeromonas* in the family of Aeromonadaceae are capable of switching from mixed acid to acetoin and 2,3-butanediol to overcome lethal acidification by redirecting pyruvate from glycolysis pathway into neutral end-products such as butanediol as shown in Figure 8.2 (Johansen et al., 1975; Mayer et al., 1995; Renna et al., 1993). Acetoin and 2,3-butanediol that are released to the media, can be re-utilized when other carbon sources have been consumed (Grundy et al., 1993, Ali et al., 2001). The 2,3-butanediol pathway also may participate in the regulation of the NAD/NADH ratio in bacteria (Johansen et al., 1975).

Switching to 2,3-butanediol and other neutral end products to overcome lethal acidification is regulated by QS and environmental cues such as pH and the presence of acetate (Godtfredsen, et al., 1983; Tsau et al., 1992; Van Houdt et al.,

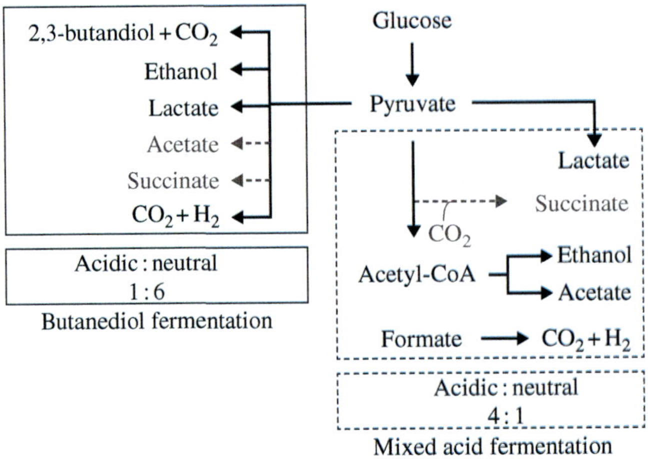

Figure 8.2 Schematic overview of butanediol and mixed acid fermentation in *Enterobacteriaceae*.

2006a,b; Moon et al., 2010). Some members of *Enterobacteriaceae* including *Escherichia, Salmonella,* and *Shigella* are not capable of diverting pyruvate from glycolysis into butanol and use mixed-acid pathway to ferment sugars into acids (e.g. acetate, lactate, succinate, and formate) resulting in acidification (Huang and Ouyang, 2011).

Butanediol is produced by first converting two molecules of pyruvate into α-acetolactate by α-acetolactate synthase (ALsS), after which α-acetolactate is converted to acetoin by α-acetolactate decarboxylase (ALsD). ALsS and ALsD are encoded by the *als*S and *als*D genes, respectively (Renna et al., 1993). Finally, acetoin can be converted reversibly to 2,3-butanediol through the action of acetoin reductase (AR) (Loken and Stormer, 1970). Often, the structural genes for 2,3-butanediol production is organized into an operon that is regulated by a neighboring and divergently transcribed LysR homologue, which is encoded by *als*R gene that regulates the expression of *als*S and *als*D (Renna et al., 1993, Henikoff et al., 1988).

Cultivation of *Bacillus subtilis* using fermentable carbon sources such as glucose results in production and release of acetoin into the media. ALsS and ALsD enzymes, involved in biosynthesis of acetoin were synthesized in significant amount at stationary phase by *B. subtilis* (Renna et al., 1993). Holtzclaw and Chapman have shown that AlsS activity is inducible during logarithmic growth by addition of high concentrations of acetate to the growth medium, indicating the integration of environmental cues into QS mechanism (Holtzclaw and Chapman, 1975). AlsR is believed to be present in the cells at all stages of growth, but in the absence of inducing ligand, it is not capable to initiate *als*S transcription (Renna et al., 1993). Most of the LysR family proteins are activated by inducing ligands. Therefore, it was hypothesized that changes in intracellular pH or concentration of acetate or one of its derivatives is an inducing ligand for the AlsR protein (Renna et al., 1993).

In *Klebsiella terrigena*, the LysR-type transcriptional activator, BudR regulates part of 2,3-butanediol pathway (Mayer et al., 1995). Structural genes homologous to *alsS* and *alsD*, involved in acetoin production were designated as *budB* and *budA*, respectively and LysR gene homologue was designated as *budR* (Mayer et al., 1995). These genes are organized in *bud ABC* operon (Blomqvist et al., 1993). The expression of the *bud ABC* operon is optimal in the presence of acetate, at low medium pH, and under anaerobic conditions (Magee and Kosaric, 1987). Similar to *Bacillus subtilis*, LysR-type activator protein in *Klebsiella terrigena*, is influenced by changes in intracellular pH and/or that responds to acetates (Renna et al., 1993, Mayer et al., 1995).

In *Vibrio cholera*, LysR-type transcriptional activator AlsR and transcriptional activator AphA are involved in production of acetoin and 2,3-butanediol (Kovacikova et al., 2005). Deletion of AphA showed to increase the production of acetoin. Dual regulation of motility, biofilm formation and acetoin production by acetate and quorum sensing through LysR-type regulator AlsR and AphA in *Vibrio cholera* was demonstrated (Kovacikova et al., 2005). The AphA activator is regulated by multiple QS systems that act in parallel. System 1 uses CAI-1 autoinducer identified as (*S*)-3-hydroxytridecan-4-one (Higgins, Pomianek et al., 2007) and System 2 uses AI-2, a furanosyl borate diester (Figure 8.1). Systems 1 and 2 both are involved in regulating AphA (Kovacikova et al., 2005; Miller et al., 2002; Zhu et al., 2002). Homologues of these QS systems were first identified in *Vibrio harveyi* in which bioluminescence is controlled by AHL, produced by the LuxM family of AHL synthases and AI-2 (furanosyl borate diester), which are sensed by LuxN and LuxQ, respectively (Chen et al., 2002).

2,3-Butaneol production in *Serratia plymuthica* RVH1 is demonstrated to be regulated by LuxI/LuxR homologous QM system, SplIR (Van Houdtet al., 2006 b). The SplIR QM system also regulates production of extracellular chitinase, protease, nuclease, and antibacterial compound (van Houdt et al., 2006a). AHL, synthesized by LuxI protein regulates these phenotypes and AHL-activated LuxR protein is the transcriptional activator (Van Houdt et al., 2006b). 2,3-butanediol production in *Serratia marcescens* MG1 is also regulated by AHL-dependent LuxI/LuxR QS system (Van Houdt et al., 2006a). Mutants that were deficient in the biosynthesis of C_4-HSL and *N*-hexanoyl-L-homoserine (C_6-HSL) resulted in enhanced acid production, leading to early growth arrest. The phenotype was reversed by the addition of synthetic AHLs (Van Houdt et al., 2006a).

Further research into investigating the parameters that control the acetoin production in *Serratia plymuthica* RVH1 showed that in addition to an AHL-based QS (Van Houdt et al., 2006b), pH of the medium also contributes into regulation of acetoin and 2,3-butanediol production, indicating the presence of a dual regulation (Moons et al., 2011).

Consistent with the earlier observation in *Bacillus subtilis*, in *Klebsiella terrigena* and *Vibrio cholerae* integrated QS and environmental cues (e.g. pH, presence of acetate) are also involved in regulation of LysR type (Renna et al., 1993, Mayer et al., 1995, Kovacikova et al., 2005). Additionally, in cultures of *Serratia plymuthica* RVH1, under acidic conditions (pH 5.5), acetoin was detected earlier and reached to a higher concentration compared to the cultures starting at a pH of 7.0 (buffered or unbuffered). All of these findings indicated that QS is not the only cue involved in regulating acetoin production, but pH as an environmental cue is also involved in controlling acetoin production (Moons et al., 2011).

Butanol production in *Aeromonas hydrophila* AH-1N (afacultative anaerobic, Gram-negative bacteria in Aeromonadaceae family) was also demonstrated to be the result of AHL-mediated QS with the major QSM being C_4-HSL (Van Houdt et al., 2007). Similar to *Serratia plymuthica* and *Klebsiella terrigena, A. hydrophila* switches to butanediol fermentation to reducemedium acidification to allow further growth through an AHL-mediated quorum sensing mechanism (Van Houdt et al., 2007).

LuxI/LuxR homologues are designated as AhyI/AhyR in *A. hydrophila* (Van Houdt et al., 2007). Growth curve of *A. hydrophila* AH-1N follows a biphasic growth in the presence of 0.5% glucose at 30 °C in which strong acidification in the first phase as a result of fermentation of glucose to acidic compounds leads to butanediol fermentation to overcome the environmental stress imposed by strong acidification and low pH encountered in the first phase. Growth of *A. hydrophila* mutants deficient in *ahy*I gene, which could not produce C_4-HSL in the presence of 5% glucose, resulted in acidification and further blocking the growth in the second phase and even cell death (Van Houdt et al., 2007). Addition of C_4-HSL to the cultivation broth of *ahyI* mutant resulted in more rapid growth rate and higher biomass concentration in the second growth phase compared to the parent strain (Rob Van Houdt et al., 2007).

However, biphasic growth was not observed at 37 °C, which was hypothesized to be the result of lower pH (5.1) observed at 37 °C compared to the pH at 30 °C, leading to early cell growth arrest, suppression of production of C_4-HSL and therefore the threshold concentration to switch to butanediol production was not reached (Van Houdt et al., 2007).

Martinez et al. showed that *A. hydrophila* 519 does not produce C_4-HSL in LB at 37 °C. However, *A. hydrophila* AH-1N produced C_4-HSL in LB at 37 °C in LB (Swift et al., 1999), which indicates that the effect of temperature on C_4-HSL production is strain dependent (Van Houdt et al., 2007). An integrated effect of environmental cues and QS indicates that these cues must be considered simultaneously for regulating 2,3-butanediol production (Moons et al., 2011).

Acetic acid bacteria are Gram-negative, obligatory aerobic bacteria, capable of oxidizing ethanol and sugars into vinegar (acetic acid) using the two membrane-bound enzymes, alcohol dehydrogenase and aldehyde dehydrogenase. These bacteria are known for their high tolerance to ethanol and acetic acid (Iida et al., 2008b). *Gluconacetobacter polyoxogenes and Gluconacetobacter intermedius* NCI1051 are amongst the Gram-negative acetic acid bacteria that employ AHL-mediated QS system with LuxI and LuxR homologues, designated as GinI and GinR to regulate gene expression. Homologous of *luxI* and *luxR* genes in *Gluconacetobacter intermedius* are designated as *ginI* and *ginR* and three types of AHLs, *N*-decanoyl-L-homoserine lactone (C_{10}HSL), *N*-dodecanoyl-L-homoserine lactone (C_{12}-HSL), and an *N*-dodecanoyl-L-homoserine lactone with a single unsaturated bond in its acyl chain (Iida et al., 2008b).

The GinI/GinR quorum-sensing system (homologue of luxI/R) in *Gluconacetobacter intermedius* NCI1051 controls the expression of *ginA* gene, which in turn represses oxidative fermentation, including acetic acid and gluconic acid fermentation (Iida et al., 2008b). *GinI* gene, which encodes AHL synthase is also activated by GinR. Most acetic acid bacteria have a QS system homologous to GinI/GinR (Iida et al., 2008b).

The **lactic acid bacteria** (LAB) are important Gram-positive bacteria in food fermentation due to their pre-biotic effects in human and have wide applications in dairy, meat and vegetable products. Autolysis of LAB during the fermentation was demonstrated to occur when a certain cell density is achieved and is considered a QS phenotype (Pang et al., 2014). Understanding the mechanism of QS dependent autolysis in LAB allows optimization of the fermentation by reducing the autolysis rate (Pang et al., 2014). Similar to many Gram-positive bacteria, LAB use a two-component system (TCS) consisting of a membrane located histidine protein kinase (HPK), which monitors signal molecules, and a response regulator (RR), which modulates the expression of specific genes (Li and Tian, 2012; Taillefumier and Wingreen, 2015, Castillo-Juarez et al., 2015). Autolysis of *Lactobacillus bulgaricus* ATCC BAA-365 was found to be a QS-associated phenotype regulated by TCS (LBUL_RS00115/LBUL_RS00110) with Peptide II, identified as QSM via bioinformatics analysis. The autolysis rate of *L. bulgaricus* ATCC BAA-365 with 2 μ M peptide II sharply increased compared with BAA-365 with no peptide, leading to significant decrease in the density of *L. bulgaricus* in the presence of peptide II (Pang et al., 2016).

8.4 Quorum Sensing in Fungi and the Related Phenotypes

Similar to bacteria, fungi also communicate through signaling molecules in response to environmental cues (Albuquerque and Casadevall, 2012). Variety of phenotypes are found to be controlled by QS in fungi including germination, formation of the mycelial network, coordination of colony development and sexual and asexual reproduction (Table 8.2). QS-dependent phenotypes in fungi and their related QSM are reviewed by Chen and Fink (Chen and Fink, 2006), Hogan, (Hogan, 2006) and Albuquerque and Casadevall (Albuquerque and Casadevall, 2012).

Many fungi can grow as a unicellular yeast form or as a filamentous form (Lengeler et al., 2000). Fungal biofilm formation involves three steps of adhesion, biofilm growth, and maturation (Cao et al., 2005; Chandra et al., 2001). The most important QS dependent phenotype reported for fungi is morphological transition or the transition between the filamentous and the solitary yeast form (Chen and Fink, 2006, Nickerson et al., 2006). Morphological transition has been found to be regulated by QS and environmental cues in many fungal species including pathogenic (*Candida albicans, Candida glabrata, Cryptococcus neoformans*) and nonpathogenic (*S. cerevisiae*) fungi and therefore the role of QS in biofilm formation was the focus of several studies (Hornby et al., 2001, Hornby et al., 2004, Chen et al., 2004, Chen and Fink, 2006).

QS in eukaryotic cells was first reported in 2001 with the discovery of farnesol as QSM that controls filamentation in the pathogenic and polymorphic fungus *C. albicans* (Hornby et al., 2001) and as such QS research in fungi is at earlier stage compared to bacterial QS (Albuquerque and Casadevall, 2012;

Table 8.2 Different Quorum Sensing Systems in fungi and their associated phenotypes.

Species	QSM	Chemical Structure	QS phenotype	Reference
C. albicans	Farnesol	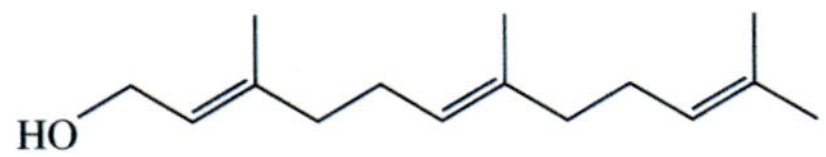	Inhibiting filamentation	Hornby et al., 2001; Mosel et al., 2005
			Inhibiting biofilm formation Regulating virulence trait	Ramage et al., 2002
	Farnesoic acid	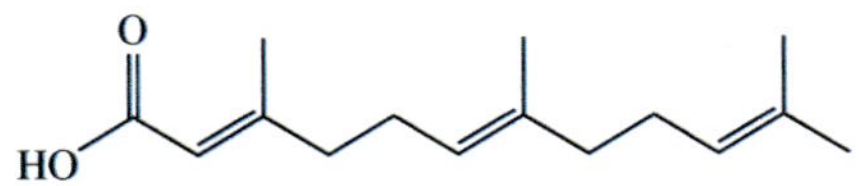	Inhibiting biofilm formation, the same effect as farnesol but less intense	San-Blas et al., 2000; Nickerson et al., 2006
	Tyrosol	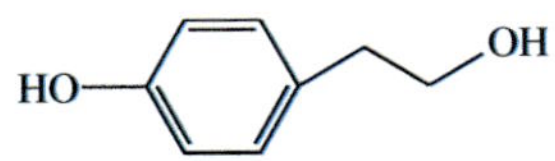	Regulating virulence trait	Chen et al., 2004
S. cerevisiae	Phenylethanol	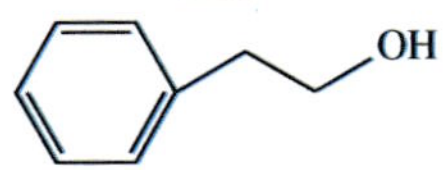	Cell-dependent polymorphism(morphological transition), inducing pseudohyphal growth	Chen and Fink 2006
	Tryptophol	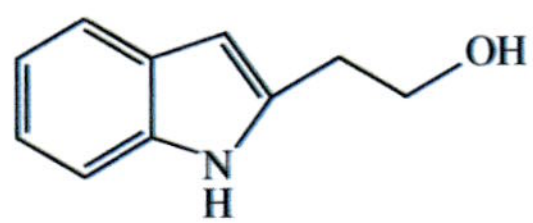	Cell-dependent polymorphism(morphological transition), inducing pseudohyphal growth	Chen and Fink 2006
	A-factor	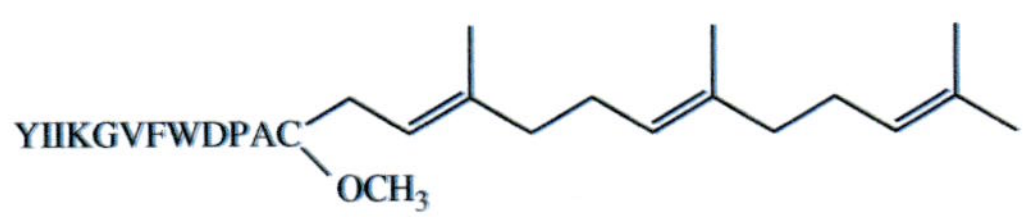	Induce matting and morphogenesis	Pan et al., 2000
Cryptococcus neformans	11 mer peptide	-	Biofilm formation on solid	Lee et al., 2007

Chen et al., 2004). Farnesol prevents the yeast-to-mycelium conversion without altering cellular growth rates (Hornby et al., 2001) and showed to suppress mycelium formation in a variety of *C. albicans* strains (Mosel et al., 2005).

In addition to its role as QSM, farnesol is also identified to play multiple roles in *C. albicans* physiology (Cao et al., 2005). Analysis of differential gene expression through cDNA microarray and independent reverse transcription-PCR analysis in *C. albicans* biofilm exposed to farnesol revealed that gene related to drug resistance (e.g. *FCR1* and *PDR16*), cell wall maintenance (e.g. *CHT2* and *CHT3*), iron transport (e.g. *FTR2*) heat shock proteins (e.g. *HSP70, HSP90, HSP104, CaMSI3*, and *SSA2*) and cell surface hydrophobicity (*CSH1*) were differentially expressed in addition to the genes associated with hyphae formation (e.g. *TUP1, CRK1*, and *PDE2*)(Cao et al., 2005). A decrease in cell surface hydrophobicity and down regulation of cell surface hydrophobicity-associated gene (*CSH1*) in the farnesol-treated *C. albicans* compared to the control group (not treated with farnesol) was also observed (Cao et al., 2005).

Discovery of the role of farnesol in *C. albicans* filamentaion further inspired more research in the area of fungal QS, leading to the discovery of farnesoic acid (Oh et al., 2001) and tyrosol in *C. albicans* (Chen et al., 2004) and 2-phenylethanol, tryptophol, and tyrosol as QSMs in *Saccharomyces cerevisiae* (Avbelj et al., 2015; Chen and Fink, 2006; Wuster and Babu, 2010).

Tyrosol in *C. albicans* showed to reduce the lag phase of growth in a low-density culture without any effect on exponential growth (Chen et al., 2004). Unlike farnesol that inhibits biofilm formation, tyrosol induces formation of germ tubes (yeast-to hypha formation) in yeast (Chen et al., 2004) and promotes the development of hyphae in the early stage of biofilm formation (Alem et al., 2006).

Aromatic alcohols (2-phenylethanol, tryptophol, and tyrosol) produced by *Saccharomyces* induce morphogenesis by expressing FLO11 through a Tpk2p-dependent mechanism, resulting in induction of pseudohyphae formation under low-nitrogen concentration (Chen and Fink, 2006).The low-nitrogen signal that evokes the morphogenetic transitions is transmittedby elements of mitogen-activated protein kinase (MAPK) and protein kinase A (PKA) pathways, resulting in the enhanced transcription of genes required for filamentation (Lengeler et al., 2000, Gancedo, 2001). Therefore, QS mechanism in *S. cerevisiae* links the morphogenesis, environmental cues and cell density (Chen and Fink, 2006).

An oligopeptide composed of 11amino acids was identified as QSM in *Cryptococcus neoformans*, pathogenic yeast, indicating the presence of a diverse mechanism of cell-to-cell communication in the Fungi (Lee et al., 2007). *C. neoformans* tup1D strain failed to form colonies on agar media at an inoculum size of lower than 10^3 cells. Inability to grow at low cell density was identified to be a QS-associated phenotype since addition of culture filtrate from a high-density tup1D strain that contained 11-mer peptide led to growth (Lee et al., 2007).

The morphological change from unicellular budding yeasts to hyphae and pseudohyphae is demonstrated to be the result of changes in transcription (Uhl et al., 2003), influenced by QS, as well as environmental conditions including pH, temperature, glucose levels, nitrogen source, starvation, carbon dioxide level, inoculum size, chelating agents and transition metals. L-proline, N-acetylglucosamine, and serum are also common chemical triggers for germ tube formation (Nickerson et al., 2006; Romano, 1966). Both nonpathogenic (*Saccharomyces cerevisiae*) and pathogenic (*C. albicans, C. glabrata,* and *C. neoformans*) fungi switch from the yeast to the filamentous form under nitrogen starvation condition (Gimeno et al., 1992; Wickes et al., 1996; Biswas and Morschhauser, 2005). A network of signal transduction pathway whose activity is coordinated by transcription regulators transduces the environmental changes into a change in morphology (Sudbery et al., 2004, Liu, 2001).

Coordination of mating and morphogenesis has been observed in several fungal species. In *S. cerevisiae,* members of the same mitogen-activated protein kinase (MAPK) signaling cascade are involved in pheromone response, agar invasion and biofilm formation with A-factor being mating pheromones that induce both matting and morphogenesis (Pan, Harashima, and Heitman, 2000; Roberts and Fink, 1994; Reynolds and Fink, 2001).

8.5 Fermentation and Quorum Sensing in Fungi

Saccharomyces cerevisiae hasbeen used for the last eight thousand years for production of alcoholic beverages (McGovern et al., 2004). Fermentation by yeast produces aromatic and aliphatic alcohols, known as fusel alcohols, which at higher concentration have been linked with inducing pseudohyphal growth in *S. cerevisiae,* resulting in a decrease in the growth rate (Dickinson, 2008).

These alcohols at high concentration result in off-flavors, while at low concentration, these alcohols along with their esters enhance the flavors of the fermentation product (Hazelwood et al., 2008). For example, 2-phenyl ethanol that is QSM and is responsible for morphological change from yeast to pseudohypha (Mas et al., 2014) is widely used as cosmetics and as food additive and also favorably affect wine aroma (Mas et al., 2014). Therefore, investigating QS in *S. cerevisiae* and its application in controlling fermentation has been the center of attention in recent years.

Amino acids; phenylalanine, tryptophan, and tyrosine are involved in the synthesis of fusel alcohols; 2-phenylethanol, tryptophol and tyrosol, respectively. The biosynthesis initiates by transmination, followed by decarboxylation and reduction is described by Ehrlich pathway as shown in Figure 8.3. Amino acids are converted to their alpha–keto-acid analogues, mediated by aromatic aminotransferases I and II, encoded by the genes *ARO8* and *ARO9,* respectively in the transamination step. Alpha–keto-acids analogue of amino-acids are converted to an aldehyde, catalyzed by the aromatic decarboxylase encoded by the gene *ARO10* and by pyruvate decarboxylases in the decarboxylation step. Aldehyde is converted into primary alcohol in the final step (Dickinson et al., 2008). Ammonia in the culture medium and end products; 2-phenylethanol and tryptophol are the environmental cues that regulate QS and these regulatory mechanisms also influence the transmination and decarboxylation steps of the Ehrlich pathway (Chen and Fink, 2006; Wuster and Babu, 2010).

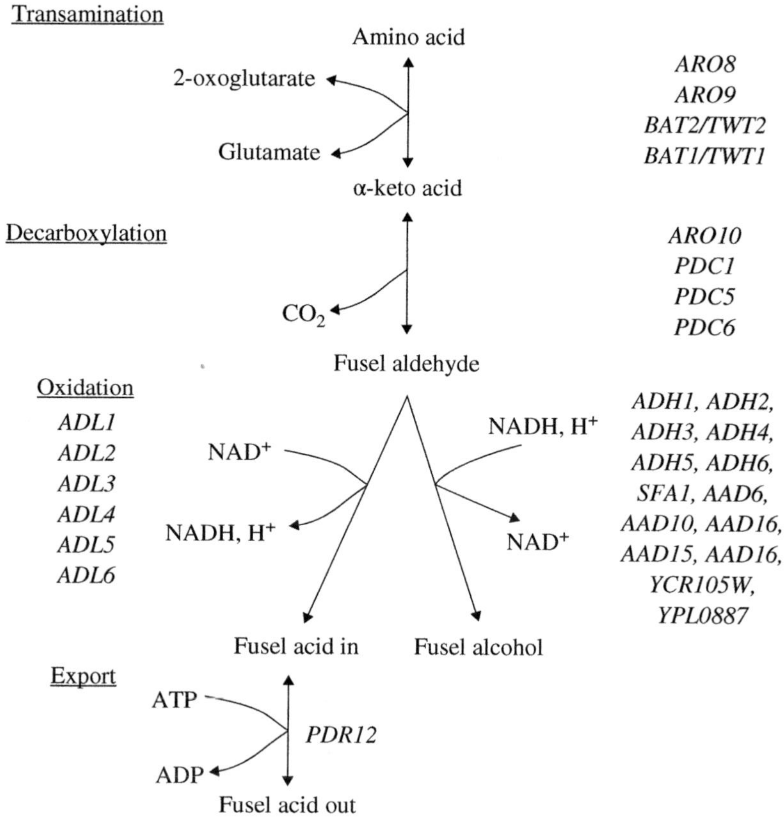

Figure 8.3 Schematic overview of Ehrlich pathway.

One of the functions of Ehrlich pathway is to generate QSM that are involved in the adaptation of yeast cells to environmental changes (Hazelwood et al., 2008; Avbelj et al., 2015). ARO9 and ARO10, two genes required for the synthesis of phenylethanol and tryptophol are up-regulated in response to the high cell density (Chen and Fink, 2006). CAT8 and MIG1 were predicted to be the key transcriptional regulators that control the differential expression of the genes affected by aromatic alcohol communication (Wuster and Babu, 2010). Tryptophol induces the expression of aminotransferases enzymes (ARO9 and ARO10 genes) via the Aro80p transcription factor (Chen and Fink, 2006). These enzymes are required for the synthesis of tryptophol, resulting in positive feedback loop and production of more aromatic alcohol per cell at higher cell density compared to lower cell density (Chen and Fink, 2006; Zupan et al., 2013).

Aromatic alcohols were found to affect the transcript abundance of hundreds of different genes in *S. cerevisiae* (Chen and Fink, 2006). Kinetics of the production of the aromatic alcohols correlates with the expression of the ARO8, ARO9, and ARO10 genes and the peak expression of the ARO genes coincides with the peak production rates of the aromatic alcohols (Avbelj et al., 2015).

Ethanol, which is a major stress that reduces the growth rate of *S. cerevisiae*, also was found to reduce the production rate of QSM; 2-phenylethanol, tryptophol and tyrosol. This phenomenon was attributed to the disrupted reactions associated with cell growth and viability (Chandler, 2004), leading to decreased fermentation productivity and ethanol yield (Aguilera et al., 2006; Ansanay-Galeote et al., 2001). Therefore, fundamental understanding of the relationship of QS and fermentation productivity can provide insight to controlling fermentation towards a higher productivity (Avbelj et al., 2015).

8.6 Quorum Sensing in Bacteria and Fungi: Similarities and Differences

The common feature of all QS systems is that they need molecular machinery to synthesize, sense, and respond to the communication molecules (Wuster and Babu, 2010). Similar phenotypes have been found to be regulated by QS in both bacteria and fungi. For example, regulation of the virulence, gene expression and morphogenesis were found to be controlled by QS in both bacteria and fungi (Casadevall and Pirofski, 1999; San-Blas et al., 2000; Chen and Fink, 2006) and the connection between mating and morphogenesis in *S. cerevisiae* is recognized to be analogue to the coordination between competence and virulence in *Streptococcus pneumoniae* and other gram-positive bacteria, which is regulated by AI-2 and oligopeptide (Henke and Bassler, 2004; Rasmussen et al., 2005). However, QS systems in bacteria and fungi vary in details such as the chemical structure of the communication molecules or genes involved in regulating the phenotypes. None of the QSM produced by bacteria was found to regulate cell-cell communication in fungi (Table 8.1 and 8.2) (De Sordi and Muhlschlegel, 2009; Tseng, 2008).

AHLs and modified peptides are common QSM in Gram-negative and Gram-positive bacteria, respectively. In fungi, however, primary alcohols such as farnesol and tyrosol in *Candida albicans*, and phenylethanol and tryptophol in *S. cerevisiae* are common QSM (Atkinson and Williams, 2009; Hogan, 2006).

Even, among the bacteria, specific QSM may be produced by a particular group of bacteria. For example, AHL-mediated quorum sensing is employed by diverse Gram-negative bacteria, however no AHL-producing Gram-positive bacteria have so far been identified (Withers et al., 2001; Camara et al., 2002b; Williams et al., 2007).

Diverse QSMs are also identified in different fungal species. Phenylethanol and tryptophol, which are QSM in *S. cerevisiae* are produced by *C. albicans* but do not induce density-dependent responses and therefore have not been recognized as QSM in *C. albicans*, indicating that these QSMs are species-specific (Chen and Fink, 2006).

The orthologues of key *S. cerevisiae* genes involved in QS was identified in other 31 fungal species, showing that the genes involved in aromatic alcohol synthesis are present in most yeast species (Wuster and Babu, 2010). The expression of transcription factor Aro80p, which are involved in the biosynthesis of aromatic alcohol are conserved within the genera *Saccharomyces* and *Candida*

(Wuster and Babu, 2010). Differentially expressed genes in response to aromatic alcohol are distributed in a clustered way across the 16 different chromosomes in *S. cerevisiae* (Wuster and Babu, 2010). The orthologues of differentially expressed genes found to be distributed in a random (non-clustered) manner in *C. albicans* (Wuster and Babu, 2010). Given that filamentation of *C. albicans* is not stimulated by tryptophol and phenylethanol (Chen and Fink, 2006), it was concluded that these genes do not act in the same pathway across different fungal species (Janga et al., 2008) and QS by aromatic alcohols is not conserved beyond the genus *Saccharomyces* (Wuster and Babu, 2010).

Collectively, the similarities of QS in fungi and bacteria are limited to the fact that similar phenotypes in fungi and bacteria have been observed to be QS-associated phenomenon. Another interesting phenomenon is the communication between bacteria and eukaryotic cells through the interaction of QSM (De Sordi and Muhlschlegel, 2009).

For example, farnesol, QSM produced by *C. Candida* exhibit antifungal and antibacterial activity against bacteria (e.g. *Aspergillus nidulans*) and other fungi (e.g. *Saccharomyces cerevisiae*) (De Sordi and Muhlschlegel, 2009; Tseng, 2008). Enhanced antibiotic susceptibility, reduced biofilm formation and lipase activity in pathogenic *Staphylococcus aureus* (Kuroda, Nagasaki, Ito, and Ohta, 2007, Jabra-Rizk, Meiller, James, and Shirtliff, 2006) and decreased viability of murine macrophases in the presence of farnesol (Abe et al., 2009) were reported.

AHLs, produced by Gram-negative bacteria are recognized by eukaryotic cells including animal cells and plants seaweed (Dudler and Eberl, 2006). Fungi such as *Penicillium* species were found to produce secondary metabolites (patulin and penicillic acid) that act as QS inhibitory (QSI) in bacteria, enhancing biofilm susceptibility to antibiotic treatment (Rasmussen et al., 2005). In a mouse pulmonary infection model, *P. Aeruginosa* was more rapidly cleared from the mice that were treated with patulin compared with the control group (Rasmussen et al., 2005).

Acknowledgment

I would like to extend my appreciation to Dr. Bahareh Asadishad for binging couple of interesting references to my attention.

References

Abe, S., Tsunashima, R., Iijima, R. et al. (2009) Suppression of anti-Candida activity of macrophages by a quorum-sensing molecule, farnesol, through induction of oxidative stress. *Microbiology and Immunology*, 53(6), 323–330.

Aguilera, F., Peinado, R. A., Millan, C. (2006) Relationship between ethanol tolerance, H+ -ATPase activity and the lipid composition of the plasma membrane in different wine yeast strains. *International Journal of Food Microbiology*, 110(1), 34–42.

Albuquerque, P. and Casadevall, A. (2012) Quorum sensing in fungi: a review. *Medical Mycology*, 50(4), 337–345.

Alem, M. A. S., Oteef, M. D. Y., Flowers, T. H., and Douglas, L. J. (2006) Production of Tyrosol by *Candida albicans* Biofilms and Its Role in Quorum Sensing and Biofilm Development. *Eukaryotic Cell*, 5(10), 1770–1779.

Ali, N. O., Bignon, J., Rapoport, G., and Debarbouille, M. (2001) Regulation of the acetoin catabolic pathway is controlled by sigma L in *Bacillus subtilis*. *Journal of Bacteriology*, 183(8), 2497–2504.

Ansanay-Galeote, V., Blondin, B., Dequin, S., and Sablayrolles, J.-M. (2001) Stress effect of ethanol on fermentation kinetics by stationary-phase cells of *Saccharomyces cerevisiae*. *Biotechnology Letters*, 23(9), 677–681.

Atkinson, S. and Williams, P. (2009) Quorum sensing and social networking in the microbial world. *Journal of the Royal Society Interface*, 6(40), 959–978.

Avbelj, M., Zupan, J., Kranjc, L., and Raspor, P. (2015) Quorum sensing kinetics in *Saccharomyces cerevisiae*: a symphony of ARO genes and aromatic alcohols. *Journal of Agricultural and Food Chemistry*. 63(38), 8544–8550.

Avbelj, M., Zupan, J., Kranjc, L., and Raspor, P. (2015) Quorum-Sensing Kinetics in *Saccharomyces cerevisiae*: A Symphony of ARO Genes and Aromatic Alcohols. *Journal of Agricultural and Food Chemistry*, 63(38), 8544–8550.

Bahl, H., Andersch, W., and Gottschalk, G. (1982) Continuous production of acetone and butanol by *Clostridium acetobutylicum* in a two-stage phosphate limited chemostat. *European journal of applied microbiology and biotechnology*, 15(4), 201–205.

Bahl, H., Gottwald, M., Kuhn, A. et al. (1986) Nutritional Factors Affecting the Ratio of Solvents Produced by *Clostridium acetobutylicum*. *Applied and Environmental Microbiology*, 52(1), 169–172.

Bassler, B. L., Wright, M., Showalter, R. E., and Silverman, M. R. (1993). Intercellular signalling in Vibrio harveyi: sequence and function of genes regulating expression of luminescence. *Molecular Microbiology*, 9(4), 773–786.

Bassler, B. L., Wright, M., and Silverman, M. R. (1994) Multiple signalling systems controlling expression of luminescence in Vibrio harveyi: sequence and function of genes encoding a second sensory pathway. *Molecular Microbiology*, 13(2), 273–286.

Biswas, K. and Morschhauser, J. (2005) The Mep2p ammonium permease controls nitrogen starvation-induced filamentous growth in *Candida albicans*. *Molecular Microbiology*, 56(3), 649–669.

Blankenship, J. R. and Mitchell, A. P. (2006) How to build a biofilm: a fungal perspective. *Current Opinion in Microbiology*, 9(6), 588–594.

Blomqvist, K., Nikkola, M., Lehtovaara, P. et al. (1993) Characterization of the genes of the 2,3-butanediol operons from Klebsiella terrigena and Enterobacter aerogenes. *Journal of Bacteriology*, 175(5), 1392–1404.

Camara, M., Hardman, A., Williams, P., and Milton, D. (2002a) Quorum sensing in *Vibrio cholerae*. *Nature Genetics*, 32(2), 217–218.

Camara, M., Williams, P., and Hardman, A. (2002b) Controlling infection by tuning in and turning down the volume of bacterial small-talk. *Lancet Infectios Disease*, 2(11), 667–676.

Cao, Y.Y., Cao, Y.B., Xu, Z. et al. (2005) cDNA Microarray Analysis of Differential Gene Expression in Candida albicans Biofilm Exposed to Farnesol. *Antimicrobial Agents and Chemotherapy*, 49(2), 584–589.

Casadevall, A. and Pirofski, L.A. (1999) Host-pathogen interactions: redefining the basic concepts of virulence and pathogenicity. *Infectionand Immunity*, 67(8), 3703–3713.

Castillo-Juarez, I., Maeda, T., Mandujano-Tinoco, E. A. et al. (2015) Role of quorum sensing in bacterial infections. *World Journal of Clinical Cases*, 3(7), 575–598.

Chandler, M.S., Stanley, G.A., Rogers, P., and Chambers, P. A. (2004) A genomic approach to defining the ethanol stress response in the yeast*Saccharomyces cerevisiae*. *Annals of Microbiology.*, 54, 427–454.

Chandra, J., Kuhn, D.M., Mukherjee, P.K. et al. (2001) Biofilm Formation by the Fungal Pathogen*Candida albicans*: Development, Architecture, and Drug Resistance. *Journal of Bacteriology*, 183(18), 5385–5394.

Chen, H. and Fink, G.R. (2006) Feedback control of morphogenesis in fungi by aromatic alcohols. *Genes and Development*, 20(9), 1150–1161.

Chen, H., Fujita, M., Feng, Q. et al. (2004) Tyrosol is a quorum-sensing molecule in *Candida albicans*. *Proceeding of National Academy of Science*, 101(14), 5048–5052.

Chen, X., Schauder, S., Potier, N. et al. (2002) Structural identification of a bacterial quorum-sensing signal containing boron. *Nature*, 415(6871), 545–549.

Croxatto, A., Pride, J., Hardman, A. et al. (2004). A distinctive dual-channel quorum-sensing system operates in *Vibrio anguillarum*. *Molecular Microbiology*, 52(6), 1677–1689.

Cugini, C., Calfee, M. W., Farrow, J. M., 3rd et al. (2007) Farnesol, a common sesquiterpene, inhibits PQS production in *Pseudomonas aeruginosa*. *Molecular Microbiology*, 65(4), 896–906.

Davies, D. G., Parsek, M. R., Pearson, J. P. et al. (1998) The involvement of cell-to-cell signals in the development of a bacterial biofilm. *Science*, 280(5361), 295–298.

de Kievit, T. R. and Iglewski, B.H. (2000) Bacterial quorum sensing in pathogenic relationships. *Infection and Immunity*, 68(9), 4839–4849.

De Sordi, L. and Muhlschlegel, F.A. (2009) Quorum sensing and fungal-bacterial interactions in Candida albicans: a communicative network regulating microbial coexistence and virulence. *FEMS Yeast Res*, 9(7), 990–999.

Dickinson, J.R. (2008) Filament formation in *Saccharomyces cerevisiae*-a review. *Folia Microbiol. (Praha)*, 53(1), 3–14.

Dickinson, J. R., Salgado, L. E. J., and Hewlins, M.J.E. (2003) The Catabolism of Amino Acids to Long Chain and Complex Alcohols in *Saccharomyces cerevisiae*. *Journal of Biological Chemistry*, 278(10), 8028–8034.

Dudler, R. and Eberl, L. (2006) Interactions between bacteria and eukaryotes via small molecules. *Current Opinion in Biotechnology*, 17(3), 268–273.

Eberhard, A., Burlingame, A.L., Eberhard, C. et al. (1981) Structural identification of autoinducer of Photobacterium fischeri luciferase. *Biochemistry*, 20(9), 2444–2449.

El-Ali, J., Sorger, P.K., and Jensen, K.F. (2006) Cells on chips. *Nature*, 442(7101), 403–411.

Farah, C., Vera, M., Morin, D. et al. (2005) Evidence for a Functional Quorum-Sensing Type AI-1 System in the Extremophilic Bacterium Acidithiobacillus ferrooxidans. *Applied and Environmental Microbiology*, 71(11), 7033–7040.

Fond, O., Matta-Ammouri, G., Petitdemange, H., and Engasser, J.M. (1985) The role of acids on the production of acetone and butanol by *Clostridium acetobutylicum*. *Applied Microbiology and Biotechnology*, 22(3), 195–200.

Fuqua, C., Winans, S. C., and Greenberg, E.P. (1996) Census and consensus in bacterial ecosystems: the LuxR-LuxI family of quorum-sensing transcriptional regulators. *Annual Review of Microbiology*, 50, 727–751.

Fuqua, W. C., Winans, S. C., and Greenberg, E.P. (1994) Quorum sensing in bacteria: the LuxR-LuxI family of cell density-responsive transcriptional regulators. *Journal of Bacteriology*, 176(2), 269–275.

Gambello, M.J., and Iglewski, B.H. (1991) Cloning and characterization of the Pseudomonas aeruginosa lasR gene, a transcriptional activator of elastase expression. *Journal of Bacteriology*, 173(9), 3000–3009.

Gancedo, J.M. (2001) Control of pseudohyphae formation in *Saccharomyces cerevisiae. FEMS Microbiology Reviews*, 25(1), 107–123.

Garg, S.K., and Jain, A. (1995) Fermentative production of 2,3-butanediol: A review. *Bioresource Technology*, 51(2), 103–109.

Gimeno, C.J., Ljungdahl, P.O., Styles, C.A., and Fink, G.R. (1992) Unipolar cell divisions in the yeast S. cerevisiae lead to filamentous growth: regulation by starvation and RAS. *Cell*, 68(6), 1077–1090.

Givskov, M., de Nys, R., Manefield, M. (1996) Eukaryotic interference with homoserine lactone-mediated prokaryotic signalling. *Journal of Bacteriology*, 178(22), 6618–6622.

Godtfredsen, S.E., Lorck, H., and Sigsgaard, P. (1983). On the occurrence of α-acetolactate decarboxylases among microorganisms. *Carlsberg Research Communications*, 48(3), 239–247.

Grimmler, C., Janssen, H., Krausse, D. et al. (2011) Genome-wide gene expression analysis of the switch between acidogenesis and solventogenesis in continuous cultures of *Clostridium acetobutylicum. Journal of Molecular Microbiology and Biotechnology*, 20(1), 1–15.

Grundy, F.J., Waters, D.A., Takova, T.Y., and Henkin, T.M. (1993). Identification of genes involved in utilization of acetate and acetoin in Bacillus subtilis. *Molecular Microbiology*, 10(2), 259–271.

Hanzelka, B.L., Parsek, M.R., Val, D.L. et al. (1999) Acylhomoserine lactone synthase activity of the Vibrio fischeri AinS protein. *Journal of Bacteriology*, 181(18), 5766–5770.

Harris, L.M., Welker, N.E., and Papoutsakis, E.T. (2002) Northern, Morphological, and Fermentation Analysis of spo0A Inactivation and Overexpression in *Clostridium acetobutylicum* ATCC 824. *Journal of Bacteriology*, 184(13), 3586–3597.

Haverstein, L.S. and Morrison, D.A. (1999). In: G.M.W. Dunney and S.C. Winans (Eds), *Cell-Cell Signaling in Bacteria* (pp. 9–26): Am. Soc. Microbiol., Washington, DC.

Hazelwood, L.A., Daran, J.M., van Maris, A.J. et al. (2008) The Ehrlich pathway for fusel alcohol production: a century of research on Saccharomyces cerevisiae metabolism. *Applied Environmental Microbiology*, 74(8), 2259–2266.

Henikoff, S., Haughn, G.W., Calvo, J.M., and Wallace, J.C. (1988) A large family of bacterial activator proteins. *Proceedings of the National Academy of Science U S A*, 85(18), 6602–6606.

Henke, J.M. and Bassler, B.L. (2004) Bacterial social engagements. *Trends Cell Biol.*, 14(11), 648–656.

Higgins, D.A., Pomianek, M.E., Kraml, C.M. et al. (2007) The major Vibrio cholerae autoinducer and its role in virulence factor production. *Nature*, 450(7171), 883–886.

Hogan, D.A. (2006) Talking to Themselves: Autoregulation and Quorum Sensing in Fungi. *Eukaryotic Cell*, 5(4), 613–619.

Holt, R.A., Stephens, G.M., and Morris, J.G. (1984) Production of Solvents by Clostridium acetobutylicum Cultures Maintained at Neutral pH. *Applied and Environmental Microbiology*, 48(6), 1166–1170.

Holtzclaw, W.D., and Chapman, L.F. (1975) Degradative acetolactate synthase of Bacillus subtilis: purification and properties. *Journal of Bacteriology*, 121(3), 917–922.

Hornby, J.M., Jacobitz-Kizzier, S.M., McNeel, D.J. et al. (2004) Inoculum Size Effect in Dimorphic Fungi: Extracellular Control of Yeast-Mycelium Dimorphism in Ceratocystis ulmi. *Applied and Environmental Microbiology*, 70(3), 1356–1359.

Hornby, J.M., Jensen, E.C., Lisec, A.D. et al. (2001) Quorum Sensing in the Dimorphic Fungus Candida albicans Is Mediated by Farnesol. *Applied and Environmental Microbiology*, 67(7), 2982–2992.

Husemann, M.H., and Papoutsakis, E.T. (1988) Solventogenesis in Clostridiumacetobutylicum fermentations related to carboxylic acid and proton concentrations. *Biotechnology and Bioengineering*, 32(7), 843–852.

Iida, A., Ohnishi, Y., and Horinouchi, S. (2008a) An OmpA family protein, a target of the GinI/GinR quorum-Sensing System in Gluconacetobacter intermedius, Controls Acetic Acid Fermentation. *Journal of Bacteriology*, 190 (14) 5009–5019.

Iida, A., Ohnishi, Y., and Horinouchi, S. (2008b) Control of Acetic Acid Fermentation by Quorum Sensing via N-Acylhomoserine Lactones in Gluconacetobacter intermedius. *Journal of Bacteriology*, 190(7), 2546–2555.

Jabra-Rizk, M.A., Meiller, T.F., James, C.E., & Shirtliff, M.E. (2006) Effect of Farnesol on Staphylococcus aureus Biofilm Formation and Antimicrobial Susceptibility. *Antimicrobial Agents and Chemotherapy*, 50(4), 1463–1469.

Jamora, C. and Fuchs, E. (2002). Intercellular adhesion, signalling and the cytoskeleton. *Nature Cell Biology*, 4(4), E101–108.

Janga, S.C., Collado-Vides, J., and Babu, M.M. (2008) Transcriptional regulation constrains the organization of genes on eukaryotic chromosomes. *Proceedings of the National Academy of Sciences*, 105(41), 15761–15766.

Ji, X.-J., Huang, H., and Ouyang, P.-K. (2011) Microbial 2,3-butanediol production: A state-of-the-art review. *Biotechnology Advances*, 29(3), 351–364.

Ji, X. J., Huang, H., and Ouyang, P.K. (2011) Microbial 2,3-butanediol production: a state-of-the-art review. *Biotechnology Advances*, 29(3), 351–364.

Jiang, Y., Camara, M., Chhabra, S. R. (1998) In vitro biosynthesis of the *Pseudomonas aeruginosa* quorum-sensing signal molecule N-butanoyl-L-homoserine lactone. *Molecular Microbiology*, 28(1), 193–203.

Johansen, L., Bryn, K., and Stormer, F.C. (1975) Physiological and biochemical role of the butanediol pathway in Aerobacter (Enterobacter) aerogenes. *Journal of Bacteriology*, 123(3), 1124–1130.

Karlson, P. and Luscher, M. (1959). Pheromones: A new term for a class of biologically active substances. *Nature*, 183(4653), 55–56.

Khokhlov, A.S., Tovarova, I.I., Borisova, L.N. (1967) The A-factor, responsible for streptomycin biosynthesis by mutant strains of Actinomyces streptomycini. *Dokl. Akad. Nauk. SSSR*, 177(1), 232–235.

Kovacikova, G., Lin, W., and Skorupski, K. (2005) Dual regulation of genes involved in acetoin biosynthesis and motility/biofilm formation by the virulence activator AphA and the acetate-responsive LysR-type regulator AlsR in Vibrio cholerae. *Molecular Microbiology*, 57(2), 420–433.

Kovacikova, G., Lin, W., and Skorupski, K. (2005) Dual regulation of genes involved in acetoin biosynthesis and motility/biofilm formation by the virulence activator AphA and the acetate-responsive LysR-type regulator AlsR in Vibrio cholerae. *Molecualr Microbiology*, 57(2), 420–433.

Kuroda, M., Nagasaki, S., Ito, R., and Ohta, T. (2007) Sesquiterpene farnesol as a competitive inhibitor of lipase activity of Staphylococcus aureus. *FEMS Microbiology Letters*, 273(1), 28–34.

Laue, B.E., Jiang, Y., Chhabra, S.R. et al. (2000) The biocontrol strain Pseudomonas fluorescens F113 produces the Rhizobium small bacteriocin, N-(3-hydroxy-7-cis-tetradecenoyl)homoserine lactone, via HdtS, a putative novel N-acylhomoserine lactone synthase. *Microbiology*, 146 (Pt 10), 2469–2480.

Lee, H., Chang, Y.C., Nardone, G., and Kwon-Chung, K.J. (2007) TUP1 disruption in Cryptococcus neoformans uncovers a peptide-mediated density-dependent growth phenomenon that mimics quorum sensing. *Molecuar Microbiology*, 64(3), 591–601.

Lee, S.Y., Park, J.H., Jang, S.H. (2008) Fermentative butanol production by Clostridia. *Biotechnology Bioengineering*, 101(2), 209–228.

Lengeler, K.B., Davidson, R.C., D'Souza, C. et al. (2000) Signal Transduction Cascades Regulating Fungal Development and Virulence. *Microbiology and Molecular Biology Reviews*, 64(4), 746–785.

Li, Y.H., & Tian, X. (2012) Quorum sensing and bacterial social interactions in biofilms. *Sensors (Basel)*, 12(3), 2519–2538.

Liao, C., Seo, S.O., Celik, V. (2015) Integrated, systems metabolic picture of acetone-butanol-ethanol fermentation by *Clostridium acetobutylicum*. *Proceedings of the National Academy of Sciences*, 112(27), 8505–8510.

Liu, H. (2001) Transcriptional control of dimorphism in *Candida albicans*. *Current Opinionin Microbiology*, 4(6), 728–735.

Loken, J.P. and Stormer, F.C. (1970) Acetolactate decarboxylase from Aerobacter aerogenes. Purification and properties. *European Journal of Biochemistry*, 14(1), 133–137.

Long, S., Jones, D.T., & Woods, D.R. (1984) The relationship between sporulation and solvent production in *clostridium acetobutylicum* P262. *Biotechnology Letters*, 6(8), 529–534.

Lynch, M.J., Swift, S., Kirke, D.F. (2002) The regulation of biofilm development by quorum sensing in *Aeromonas hydrophila*. *Environmental Microbiology*, 4(1), 18–28.

Maddox, I.S., Steiner, E., Hirsch, S. et al. (2000) The cause of "acid-crash" and "acidogenic fermentations" during the batch acetone-butanol-ethanol (ABE-) fermentation process. *Journal of Molecular Microbiologyand Biotechnology*, 2(1), 95–100.

Magee, R.J. and Kosaric, N. (1987) The Microbial Production of 2,3-Butanediol. *Advances in applied microbiology*, 32, 89–161.

Marlatt, J.A. and Datta, R. (1986) Acetone-butanol fermentation process development and economic evaluation. *Biotechnology Progress*, 2(1), 23–28.

Mas, A., Guillamon, J.M., Torija, M.J. et al. (2014) Bioactive Compounds Derived from the Yeast Metabolism of Aromatic Amino Acids during Alcoholic Fermentation. *BioMed Research International*, 2014, 7.

Mayer, D., Schlensog, V., and Böck, A. (1995) Identification of the transcriptional activator controlling the butanediol fermentation pathway in Klebsiella terrigena. *Journal of Bacteriology*, 177(18), 5261–5269.

McGovern, P.E., Zhang, J., Tang, J. et al. (2004) Fermented beverages of pre- and proto-historic China. *Proceeding of the National Academy of Science U S A*, 101(51), 17593–17598.

McVittie, A., Messik, F., and Zahler, S.A. (1962) Developmental Biology OF Myxococcus. *Journal of Bacteriology*, 84(3), 546–551.

Miller, M.B. and Bassler, B.L. (2001) Quorum sensing in bacteria. *Annual Review of Microbiology*, 55, 165–199.

Miller, M.B., Skorupski, K., Lenz, D.H. et al. (2002) Parallel quorum sensing systems converge to regulate virulence in *Vibrio cholerae*. *Cell*, 110(3), 303–314.

Milton, D.L., Chalker, V.J., Kirke, D. et al. (2001) The LuxM homologue VanM from Vibrio anguillarum directs the synthesis of N-(3-hydroxyhexanoyl)homoserine lactone and N-hexanoylhomoserine lactone. *Journal of Bacteriology*, 183(12), 3537–3547.

Monot, F., Engasser, J.-M., and Petitdemange, H. (1984) Influence of pH and undissociated butyric acid on the production of acetone and butanol in batch cultures of *Clostridium acetobutylicum*. *Applied Microbiology and Biotechnology*, 19(6), 422–426.

Moon, T.S., Dueber, J.E., Shiue, E., and Prather, K.L. (2010) Use of modular, synthetic scaffolds for improved production of glucaric acid in engineered *E. coli*. *Metabolic. Engineering*, 12, 298–305.

Moons, P., Van Houdt, R., Vivijs, B. et al. (2011) Integrated Regulation of Acetoin Fermentation by Quorum Sensing and pH in *Serratia plymuthica* RVH1. *Applied and Environmental Microbiology*, 77(10), 3422–3427.

More, M.I., Finger, L.D., Stryker, J.L. (1996) Enzymatic synthesis of a quorum-sensing autoinducer through use of defined substrates. *Science*, 272(5268), 1655–1658.

Mosel, D.D., Dumitru, R., Hornby, J.M. (2005) Farnesol Concentrations Required To Block Germ Tube Formation in Candida albicans in the Presence and Absence of Serum. *Applied and Environmental Microbiology*, 71(8), 4938–4940.

Mukhopadhyay, A., Redding, A.M., Rutherford, B.J., and Keasling, J.D. (2008) Importance of systems biology in engineering microbes for biofuel production. *Current Opinionin Biotechnology*, 19(3), 228–234.

Nealson, K.H., Platt, T., and Hastings, J. W. (1970) Cellular Control of the Synthesis and Activity of the Bacterial Luminescent System. *Journal of Bacteriology*, 104(1), 313–322.

Nickerson, K.W., Atkin, A.L., and Hornby, J.M. (2006) Quorum sensing in dimorphic fungi: farnesol and beyond. *Applied and Environmental Microbiology*, 72(6), 3805–3813.

Novick, R.P. (2003) Autoinduction and signal transduction in the regulation of staphylococcal virulence. *Molecular Microbiology*, 48(6), 1429–1449.

Oh, K.B., Miyazawa, H., Naito, T., and Matsuoka, H. (2001) Purification and characterization of an autoregulatory substance capable of regulating the morphological transition in Candida albicans. *Proceedings of the National Academy of Sciences U S A*, 98(8), 4664–4668.

Pan, X., Harashima, T., and Heitman, J. (2000) Signal transduction cascades regulating pseudohyphal differentiation of *Saccharomyces cerevisiae. Current Opinion in Microbiology*, 3(6), 567–572.

Pang, X., Liu, C., Lyu, P. (2016) Identification of Quorum Sensing Signal Molecule of *Lactobacillus delbrueckii* subsp. *bulgaricus. Journal of Agriculture and Food Chemistry*, 64(49), 9421–9427.

Pang, X.Y., Cui, W.M., Liu, L. (2014) Gene knockout and overexpression analysis revealed the role of N-acetylmuramidase in autolysis of *Lactobacillus delbrueckii subsp. bulgaricus* ljj-6. *PLoS ONE*, 9(8), e104829.

Papenfort, K. and Bassler, B.L. (2016) Quorum sensing signal-response systems in Gram-negative bacteria. *Nature Reviews Microbiology*, 14(9), 576–588.

Parsek, M.R. and Greenberg, E.P. (2000) Acyl-homoserine lactone quorum sensing in gram-negative bacteria: a signaling mechanism involved in associations with higher organisms. *Proceeding ofNational Academic Science U S A*, 97(16), 8789–8793.

Parsek, M.R., Val, D.L., Hanzelka, B.L. et al. (1999) Acyl homoserine-lactone quorum-sensing signal generation. *Proceedings of the National Academy of Sciences*, 96(8), 4360–4365.

Pearson, J.P., Gray, K.M., Passador, L. (1994) Structure of the autoinducer required for expression of *Pseudomonas aeruginosa* virulence genes. *Proceedings of the National Academy of Sciences U S A*, 91(1), 197–201.

Pearson, J.P., Passador, L., Iglewski, B.H., and Greenberg, E.P. (1995) A second N-acylhomoserine lactone signal produced by Pseudomonas aeruginosa. *Proceedings of the National Academy of Sciences*, 92(5), 1490–1494.

Peterson, S., Cline, R.T., Tettelin, H. et al. (2000) Gene Expression Analysis of the *Streptococcus pneumoniae* Competence Regulons by Use of DNA Microarrays. *Journal of Bacteriology*, 182(21), 6192–6202.

Qin, X., Singh, K.V., Weinstock, G.M., and Murray, B.E. (2000) Effects of Enterococcus faecalis fsr genes on production of gelatinase and a serine protease and virulence. *Infection and Immunity*, 68(5), 2579–2586.

Ramage, G., Saville, S.P., Wickes, B.L., and López-Ribot, J. L. (2002). Inhibition of *Candida albicans* Biofilm Formation by Farnesol, a Quorum-Sensing Molecule. *Applied and Environmental Microbiology*, 68(11), 5459–5463.

Rasmussen, T.B., Manefield, M., Andersen, J.B. et al. (2000) How *Delisea pulchra* furanones affect quorum sensing and swarming motility in *Serratia liquefaciens* MG1. *Microbiology*, 146 Pt 12, 3237–3244.

Rasmussen, T.B., Skindersoe, M.E., Bjarnsholt, T. (2005) Identity and effects of quorum-sensing inhibitors produced by *Penicillium* species. *Microbiology*, 151(Pt 5), 1325–1340.

Reading, N.C. and Sperandio, V. (2006) Quorum sensing: the many languages of bacteria. *FEMS Microbiology Letters*, 254(1), 1–11.

Redfield, R.J. (2002) Is quorum sensing a side effect of diffusion sensing? *Trends in Microbiology*, 10(8), 365–370.

Renna, M.C., Najimudin, N., Winik, L.R., and Zahler, S.A. (1993) Regulation of the Bacillus subtilis alsS, alsD, and alsR genes involved in post-exponential-phase production of acetoin. *Journal of Bacteriology*, 175(12), 3863–3875.

Reynolds, T.B. and Fink, G.R. (2001) Bakers' yeast, a model for fungal biofilm formation. *Science*, 291(5505), 878–881.

Rice, S.A., McDougald, D., Kumar, N., and Kjelleberg, S. (2005) The use of quorum-sensing blockers as therapeutic agents for the control of biofilm-associated infections. *Current Opinion in Investigational Drugs*, 6(2), 178–184.

Roberts, R.L. and Fink, G.R. (1994) Elements of a single MAP kinase cascade in *Saccharomyces cerevisiae* mediate two developmental programs in the same cell type: mating and invasive growth. *Genes and Development*, 8(24), 2974–2985.

Romano, A. (1966) Dimorphism. In G. C. A. a. A. S. Sussman (Ed.), *The Fungi, volume* 2. New York, NY: Academic Press.

Roos, J.W., McLaughlin, J.K., and Papoutsakis, E.T. (1985) The effect of pH on nitrogen supply, cell lysis, and solvent production in fermentations of *Clostridium acetobutylicum*. *Biotechnol Bioeng*, 27(5), 681–694.

Rutherford, S.T. and Bassler, B.L. (2012) Bacterial Quorum Sensing: Its Role in Virulence and Possibilities for Its Control. *Cold Spring Harbor Perspectives in Medicine*, 2(11).

San-Blas, G., Travassos, L.R., Fries, B.C. et al. (2000) Fungal morphogenesis and virulence. *Medical Mycology*, 38 Suppl 1, 79–86.

Schauder, S. and Bassler, B.L. (2001) The languages of bacteria. *Genes and Development*, 15(12), 1468–1480.

Solomon, J.M., Magnuson, R., Srivastava, A., and Grossman, A.D. (1995) Convergent sensing pathways mediate response to two extracellular competence factors in *Bacillus subtilis*. *Genes and Development*, 9(5), 547–558.

Sudbery, P., Gow, N., and Berman, J. (2004). The distinct morphogenic states of *Candida albicans*. *Trends in Microbiology*, 12(7), 317–324.

Surette, M.G., Miller, M.B., & Bassler, B.L. (1999) Quorum sensing in *Escherichia coli, Salmonella typhimurium*, and *Vibrio harveyi*: A new family of genes responsible for autoinducer production. *Proceedings of the National Academy of Sciences*, 96(4), 1639–1644.

Swift, S., Lynch, M.J., Fish, L. (1999) Quorum Sensing-Dependent Regulation and Blockade of Exoprotease Production in Aeromonas hydrophila. *Infection and Immunity*, 67(10), 5192–5199.

Taga, M.E. and Bassler, B.L. (2003) Chemical communication among bacteria. *Proceedings of the National Academyof Sciences U S A*, 100 Suppl 2, 14549–14554. doi:10.1073/pnas.1934514100

Taillefumier, T. and Wingreen, N.S. (2015) Optimal census by quorum sensing. *PLoS Computational Biology*, 11(5), e1004238.

Terracciano, J.S., & Kashket, E.R. (1986) Intracellular Conditions Required for Initiation of Solvent Production by Clostridium acetobutylicum. *Applied and Environmental Microbiology*, 52(1), 86–91.

Tomas, C.A., Beamish, J., and Papoutsakis, E.T. (2004) Transcriptional Analysis of Butanol Stress and Tolerance in Clostridium acetobutylicum. *Journal of Bacteriology*, 186(7), 2006–2018.

Tomasz, A. (1965) Control of the competent state in Pneumococcus by a hormone-like cell product: an example for a new type of regulatory mechanism in bacteria. *Nature*, 208(5006), 155–159.

Tsau, J.-L., Guffanti, A.A., and Montville, T.J. (1992) Conversion of Pyruvate to Acetoin Helps To Maintain pH Homeostasis in Lactobacillus plantarum. *Applied and Environmental Microbiology*, 58(3), 891–894.

Tseng, C.C. and Fink, G. (2008) Quorum sensing in fungi. In: S.C. Winans and B. Bassler (Eds), *Chemical Communication Among Bacteria* (pp. 443–451): ASM Press, Washington, DC.

Uhl, M.A., Biery, M., Craig, N., and Johnson, A.D. (2003) Haploinsufficiency-based large-scale forward genetic analysis of filamentous growth in the diploid human fungal pathogen C.albicans. *The Embo Journal*, 22(11)

Van Houdt, R., Aertsen, A., and Michiels, C.W. (2007) Quorum-sensing-dependent switch to butanediol fermentation prevents lethal medium acidification in Aeromonas hydrophila AH-1N. *Research in Microbiology*, 158(4), 379–385.

Van Houdt, R., Moons, P., Hueso Buj, M., and Michiels, C. W. (2006a) N-acyl-L-homoserine lactone quorum sensing controls butanediol fermentation in *Serratia plymuthica* RVH1 and *Serratia marcescens* MG1. *Journal of Bacteriology*, 188(12), 4570–4572.

Van Houdt, R., Moons, P., Jansen, A. et al. (2006b) Isolation and functional analysis of luxS in *Serratia plymuthica* RVH1. *FEMS Microbiology Letters*, 262(2), 201–209.

Waters, C.M. and Bassler, B.L. (2005) Quorum sensing: cell-to-cell communication in bacteria. *Annual Review of Cell and Developmental Biology*, 21, 319–346.

Wevers, E., Moons, P., Van Houdt, R. et al. (2009) Quorum sensing and butanediol fermentation affect colonization and spoilage of carrot slices by Serratia plymuthica. *Int. J. Food Microbiol.*, 134(1-2), 63–69.

Whiteway, M. and Oberholzer, U. (2004) Candida morphogenesis and host-pathogen interactions. *Current Opinion in Microbiology*, 7(4), 350–357.

Wickes, B.L., Mayorga, M.E., Edman, U., and Edman, J.C. (1996) Dimorphism and haploid fruiting in Cryptococcus neoformans: association with the alpha-mating type. *Proceeding of National Academic Science U S A*, 93(14), 7327–7331.

Williams, P., Winzer, K., Chan, W.C., and Camara, M. (2007) Look who's talking: communication and quorum sensing in the bacterial world. *Philos Trans R Soc Lond B Biol Sci*, 362(1483), 1119–1134.

Winzer, K., Hardie, K.R., and Williams, P. (2002) Bacterial cell-to-cell communication: sorry, can't talk now — gone to lunch! *Current Opinion in Microbiology*, 5(2), 216–222.

Winzer, K. and Williams, P. (2001) Quorum sensing and the regulation of virulence gene expression in pathogenic bacteria. *International Journal of Medical Microbiology*, 291(2), 131–143.

Withers, H., Swift, S., and Williams, P. (2001) Quorum sensing as an integral component of gene regulatory networks in Gram-negative bacteria. *Current Opinion in Microbiology*, 4(2), 186–193.

Wu, H., Song, Z., Hentzer, M. et al. (2004) Synthetic furanones inhibit quorum-sensing and enhance bacterial clearance in Pseudomonas aeruginosa lung infection in mice. *Journal of Antimicrobiol Chemotherophy*, 53(6), 1054–1061.

Wuster, A. and Babu, M.M. (2010) Transcriptional control of the quorum sensing response in yeast. *Molecular Biosystem*, 6(1), 134–141.

Yang, X. and Tsao, G.T. (1994) Mathematical modeling of inhibition kinetics in acetone-butanol fermentation by *Clostridium acetobutylicum. Biotechnology Progress*, 10(5), 532–538.

Zhu, J., Miller, M.B., Vance, R.E. (2002) Quorum-Sensing Regulators Control Virulence Gene Expression in Vibrio cholerae. *Proceeding of Natlural Academic Science U S A*, 99(5), 3129–3134.

Zupan, J., Avbelj, M., Butinar, B. (2013) Monitoring of Quorum-Sensing Molecules during Minifermentation Studies in Wine Yeast. *Journal of Agricultural and Food Chemistry*, 61(10), 2496–2505.

9

Biochemical and Cellular Events in Controlling Microbial Performance: A Comparative Account

Shadab Ahmed[1]*, Shreyas Niphadkar*[1]*, Somnath Nandi*[2]*, Satya Eswari*[3]*, Vishal Pandey*[1]*, Aishwarya Shankapal*[1]*, and Aishvarya Agrawal*[1]

[1] *Institute of Bioinformatics and Biotechnology, Savitribai Phule Pune University, Pune, India*
[2] *Department of Technology, Savitribai Phule Pune University, Pune, India*
[3] *Department of Biotechnology, National Institute of Technology, Raipur, Chhattisgarh, India*

9.1 Biochemical *vs.* Molecular Cues for Microbial Performances

The microbial performances or productions are always associated with biochemical or metabolic pathways of the microorganism employed for a particular metabolite production or pharmaceuticals. The most vital role is played by methods like statistical optimization, metabolic engineering, strain improvement, genetic manipulations etc., which is why industries like the healthcare, pharmaceutical, chemical, food and agriculture now rely on the latest advances in the field of biotechnology for enhancing the production scale (Figure 9.1). This figure shows how based on biochemical and molecular controls *in silico* designs or models and experimental set up can be created and desired product can produce in fairly high amounts. Essentially, the fact is that microbes serve as factories for the production of various biomolecules and their inherent biochemical pathways, cellular or molecular events provides us with clues to direct the production in certain schemes so as to get the targeted product (Figure 9.1).

9.1.1 Nutritional Parameters Optimization

The optimization of media composition can significantly affect microbial performance. The classical method of medium optimization involves changing one variable at a time, keeping the others at fixed levels. Being single dimensional, this painstaking and time-consuming method often does not necessary yield the desired optimal conditions for production. On the other hand, carrying out experiments with every possible factorial combination of the test variables is impractical because a large number of experiments will be needed to be carried out. Statistical optimization of the medium concentrations and physical factors plays a very

Microbial Sensing in Fermentation, First Edition. Edited by Satinder Kaur Brar, Ratul Kumar Das, and Saurabh Jyoti Sarma.

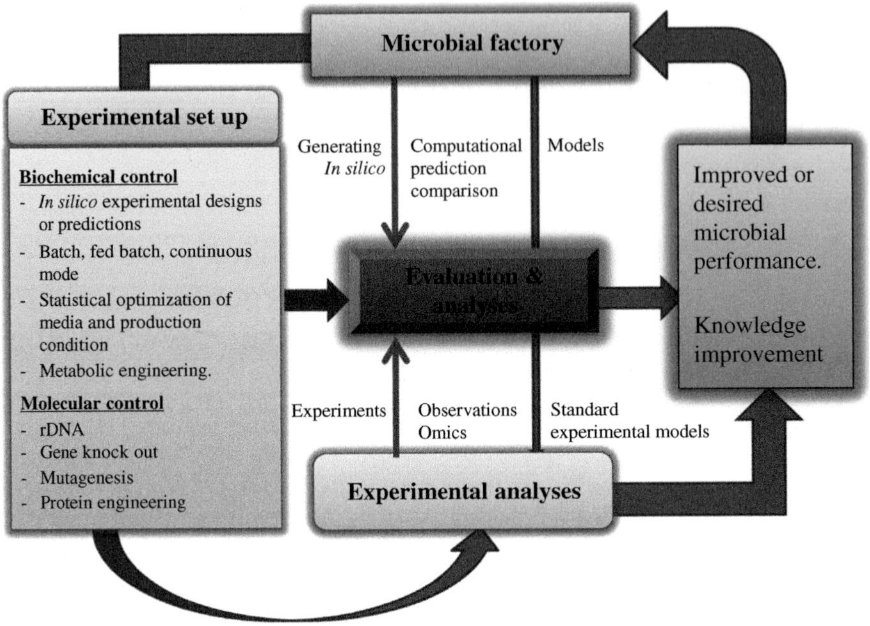

Figure 9.1 Schematic representation of biochemical and molecular events which can be explored for controlling or manipulating microbial performances. *Source:* Adapted from Lee et al. (2005). (*See insert for color representation of this figure.*)

significant role in enhancing the overall productivity. As early as in 1997, Lee et al. (Lee et al., 1997) discussed about physical conditions of fermentation for the production of exopolysaccharide (EPS). They explored various carbon sources for the production of EPS. They found that EPS concentration increased with nitrate salts and decreased with ammonium sulfate, whereas, inorganics salts limited feed were used as a medium but it didn't affect the production significantly. Hence, we can say how nutritional parameters and process conditions are altering the cellular metabolism for improving process performance and product improvement. Saraniya et al. (2014) also carried out statistical optimization of nutritional and non-nutritional factors, using Placket-Burman (1946) and Response Surface Methodology (Myers, 1974), for the maximal production of antimicrobial compounds from *Lactobacillus pentosus* SJ65. They could significantly increase the antibacterial production in this way by optimizing the medium as well as the production. Similarly, Rufino et al. (2014) reported RSM-based optimization process for the production of biosurfactant by *Candida tropicalis* grown on industrial waste substrates. Hence, we can say that statistical optimization of nutritional factor can significantly affect overall production of desired chemicals or metabolites.

9.1.2 Process Condition Optimization

A very well-organized way to improve the microbial performance or microbial production is choosing an appropriate experimental design procedure or method catering to particular needs. By designing the experiments, it is possible to vary all the investigational parameters simultaneously so as to understand their underlying effects on the final product. This can further guide the prospective biochemical investigation in right direction. Established investigational preparations markedly lessen the experimentation determinations and the wanted outcomes can be attained in a more cost-effective manner. In this we are illuminating the experimental design procedures for biological products or production optimization to lessen the tentative efforts at the procedure enlargement stage. The experimental data generated from such designs or models can be utilized to build response surface models for the bioprocess in order to investigate the influence of process conditions on the products. More often a particular or rather appropriate experimental design procedure is needed to be selected for achieving the required goal. All statistics-based experimental design procedures, namely, Plackett Burmen (PB), Taguchi designs and Central Composite Designs (CCD), response surface methodology (RSM), Genetic Algorithm (GA) based designs, etc. can be studied at the process development stage in order to pick the best design procedure (Plackett-Burman, 1946; Meyers, 1971; Box and Draper, 1987; Montgomery, 2001; Lee et al., 2005; Box and Norman, 2007). Bioprocess optimization plays a critical role in increasing productivity in bioprocesses.

9.1.3 Process Improvement by Using Batch and Fed-Batch via Process and Modeling

Bioprocess strategies involve mainly involves three modes which are run or operated in batch, fed-batch and continuous mode. The batch process involves normal culture maintenance. The input (nutritional parameters and process conditions) will be fixed at time zeroth hour and after incubation time the corresponding culture or product concentration could be analyzed. Whereas, fed-batch process involves the addition of input at regular intervals, which could be maintained and process output analyzed at the end of the process. The modes of input in fed-batch can be in two ways. In the first strategy there is a constant feed, same amount of feed given at regular intervals, whereas, in the second case there is a constant feeding but feed concentration is increased exponentially till the log phase, since in log phase utilization of substrates or nutritional parameters is more. A similar approach has been highlighted by Satya Eswari et al. (2013), where, they undertook rhamnolipids production by using the above feeding strategy employing glucose, nitrogen and phosphorous substrates. The data was used to construct the models and optimal control feeding strategies was developed by using tabu search. The exponential feeding strategy using glucose feeding resulted in improved rhamnolipid activity.

9.1.4 Metabolic Engineering for Improving Microbial Performance

9.1.4.1 Metabolic Flux Balance Analysis

9.1.4.1.1 Constraint Based Flux Balance Analysis

Metabolic flux balance analysis (FBA) is a powerful method for the ascertain biochemical networks, especially the metabolic reactions in an organism and the genes that encode each enzyme (Dmytruk et al., 2008; Durate et al., 2007). The constraint based metabolic flux analysis mathematically symbolizes metabolic reactions and the stoichiometric coefficients of each reaction are tabulated in the form of a numerical matrix (Covert et al., 2001; Covert et al., 2001b; Edwards, Ibarra and Palsson, 2001; Edwards, Covert and Palsson, 2002). The stoichiometry then imposes constraints on the course of metabolites through the metabolic network. The FBA calculates the direction of flow of metabolites from start to finish in this metabolic network, which allows us to predict the growth rate of an organism or the rate of production of a biotechnologically important metabolite. In FBA, the intracellular fluxes are calculated by making use of stoichiometric model (a kind of metabolic pathway signature or map) for the major intracellular reactions followed by applying mass balances around intracellular metabolites. A set of determined extracellular fluxes, usually uptake rates of substrates and secretion rates of metabolites, are used as input for the calculations. The outcome of flux calculation reveals a detailed metabolic flux scheme displaying a clear picture of the biochemical reactions considered in the calculations and also an estimate of the steady state rate (as flux) at which each reaction occurs. As of now metabolic models for as many as 104 organisms is already deduced (http://systemsbiology.ucsd.edu/InSilicoOrganisms/OtherOrganisms) and with the advent of advanced technologies the number is increasing every year.

9.1.4.1.2 *Defining Biological Objective to Optimize a Phenotype*

Defining a biological objective or target is the first task and it can vary from predicting growth or predicting yield or it could also be production condition. The defined objective is mathematically represented by an 'objective function' which clearly reflects how much each reaction contributes to the phenotype. The biomass reaction utilizes precursor metabolites from the system corroborating to stoichiometries which simulate biomass production. Furthermore, such a type of biomass reaction is automatically selected by 'objective function' to predict growth rates. Now, this particular biomass reaction is required to be scaled in order to keep the flux across it equal to the exponential growth rate (μ) of the organism (Edwards, Covert and Palsson, 2002; Falb et al., 2008; Orth, Theile and Palsson, 2010). The combination of mathematical illustration of the metabolic networks and the phenotype generates an array of linear equations (Edwards, Covert and Palsson, 2001; Oberhardt, Chavali and Papin, 2009). In FBA, these arrays of linear equations mentioned above are solved by making use of linear programming, for example the COBRA toolbox (category of constraint-based reconstruction and analysis) is a freely available under Matlab toolbox sections for performing these calculations (http://systemsbiology.ucsd.edu/Downloads/Cobra_Toolbox).

9.1.4.1.3 *Applications of Flux Analysis*

One of the common applications of FBA is to estimate the intracellular metabolic rate which is otherwise very difficult to measure. These intracellular metabolic rates are very difficult to measure. However, FBA overcomes this problem since one can easily measure the intracellular fluxes based on some measurable quantities, for example, the rate of formation of extracellular metabolites. The FBA can also be used for physiological studies, gap-filling efforts, and genome-scale synthetic biology by merely manipulating the restraints on certain reactions, growth on different media or multiple gene knockouts. FBA analysis can also be used to predict the yields of important cofactors like ATP or NADH or NADPH (Edwards, Covert and Palsson, 2002; Feist and Palsson, 2008).

9.1.5 Strain Improvement for Microbial Performance

Strain improvement is usually done with objectives like (i) developing strains that can utilize inexpensive and complex raw materials; (ii) developing strains which produce significantly less byproducts; (iii) strains with high extracellular secretion of product; (iv) strains which are less prone for product inhibition; (v) strains taking less time for fermentation; (vi) to improve morphology of the cells for better downstream processing potential (Parekh, Vinci and Strobel, 2000). Strain improvement becomes a necessity when the wild strains of microorganisms produce low quantities of desired compounds (Crook and Alper, 2012). It is also necessitated when we need to reduce or eliminate an unwanted property (Parekh, Vinci and Strobel, 2000; Crook and Alper, 2012; de Gerando et al., 2016). Such a technique can be exploited to sometimes engineer complex phenotypes in industrial strains (Crook and Alper, 2012; Johansen et al., 2014; Liu et al., 2011). Chemical mutation and recombination are the major strain improvement techniques where the product yield can increase up to 100 times and theoretically the scope for even higher titer exists (Vinci and Byng, 1999; Tu et al., 2016).

9.1.5.1 Mutagenesis for Strain Improvement

During the Second World War mutagenesis studies were primarily focused on producing permanent heritable change in the genetic makeup of *Penicillium chrysogenum* by using various physical (UV rays, X-rays, etc.) and chemical means (Ethyl methanesulfonate, N-methyl-W-nitro-N-nitrosoguanidine). This led to the development of high yielding [55-fold] strains (Karanam and Medicherla, 2008; Satya Eswari, Anand and Venkateshwarlu, 2013) of penicillin. Thereafter, UV mutagenesis and chemical mutagens were extensively used as strain improvement strategy for production of various metabolites like cephalosporin, bicyclomicin, candicidin etc., (Saraniya et al., 2014; Sasaki et al., 2014; Saxena, 2015). Nowadays, microbial strain improvement by various mutagenesis techniques is a frequently used advanced tool for industrial microbial fermentation processes as highlighted by the review of Saxena et al. (2015).

9.1.5.1.1 Physical Mutagenesis

Mutation can be artificially induced by physical means by employing ionizing (X-rays & γ-rays) or non-ionizing radiations (UV rays). The ionizing rays have short wavelength and have the ability to penetrate into deeper tissues resulting in ionization of the molecules along their way. The final effect of ionizing radiation is strand breakage in DNA which leads to deletions and structural changes in DNA. Finally, all this can lead to point mutations, and generally the rate of occurrence of such mutation increases with the increased exposure time. Now the damage due to the radiation occurs randomly and is both physiological and genetic. Karanam and Medicherla (2008) reported that non-ionizing UV ray although have longer wavelength and less penetration, enhanced the production of lipase from *Aspergillus niger* by random mutagenesis. They were able to produce the mutation by physical as well as chemical means (UV-irradiation, HNO_2 and N-methyl-N'-nitro-N-nitroso guanidine or NTG). The lipase production by the mutated strains was found to be 127%, 177% and 276%, respectively, higher than the parent strain when UV irradiation, HNO_2 and NTG were used as chemical mutagens.

9.1.5.1.2 Chemical Mutagenesis

Chemical mutation refers to the use of synthetic mutagens causing changes at DNA level of a gene resulting in structural change in the genome. The common chemical class of mutagens are alkylating agents (alkylation of nitrogenous bases either removes or modifies the base), DNA intercalating agents (produces frame shift mutation), and base analogues (structurally similar to nitrogenous bases and can be incorporated into the growing polynucleotide chain during replication). Many such chemical mutagenesis reports have been highlighted in Table 9.2.

9.1.5.1.3 Biological Mutagenesis

Another method of strain improvement is by genetic recombination wherein genetic information from two genotypes is combined to produce a recombinant organism. This can be brought about by transformation, transduction and conjugation. The process of transformation utilizes plasmids, phages and cosmids for recombination. Genomic shuffling allows recombination between genes of multiple parents. Wang et al. (2016), cloned pyruvate decarboxylase (PDC) gene into *Phanerochaete sordida* YK-624, a hyper lignin-degrading fungus to improve ethanol production. The improved mutant strain produced 1.41-fold higher ethanol than the wild-type strain. In summary, strain improvement of microorganisms provides the advantages of shorter fermentation time, reduced oxygen demand, ability of metabolizing low-cost substrates, decreased foam-formation, non-production of undesirable compounds, and tolerance to high concentrations of carbon or nitrogen sources and resistance to infections of bacteriophages. Liao et al. (2012) reported that site-directed mutagenesis resulted in improved thermostability and catalytic efficiency of *Aspergillus niger* N25 phytase mutated by I44E and T252R.

9.2 Sequential Evidences of Biochemical and Molecular Controlling Over Microbial Performances

There are incredible numbers of examples for evidence of enhanced microbial production or performances based on biochemical or molecular controlling or manipulations. Table 9.1 given below illustrates the processes where microbial performance was controlled by means of biochemical or molecular control or manipulations. Table 9.1 illustrates some of the examples as evidences of employing biochemical and molecular control for improving microbial performance

Table 9.1 Evidences of biochemical and molecular controlling and its direct or indirect effects on microbial performances.

Control	Technique	Strain	Description	References
BIOCHEMICAL	Physiological, Nutritional and culture condition	*B. subtilis* strain Rand	Higher protease activity presence of organic solvent.	Abusham et al., (2005)
		Brevundimonas sp. SGJ	Enhanced L-DOPA production	Surwase et al., (2012)
		S. fradiae NCIM 2418	Production of Neomycin	Vastrad & Neelagund, (2014)
		C.glutamicum MH 20-22	Enhanced the production of l-lysine.	Razak & Viswanath, (2015)
		A. terreus	Mevastatin production	Syed & Rajasimman, (2015)
		A. terreus (KM017963)	Maximal Lovastatin Production	Kamath et al. (2015)
		Xylaria sp. sof11	production of 19,20-epoxy-cytochalasin Q	Zhang et al. (2016)
		C. *thermocellum*/ *Thermoanaerobacterium*	Improved lignocellulosic biohydrogen production	Wang et al. (2016)
	Metabolic Engineering	*E. coli*	Autonomous induction of recombinant proteins by minimally rewiring native quorum sensing regulon	Tsao et al. (2010)
		S. elongatus PCC 7942	ATP drives direct photosynthetic production of 1-butanol in cyanobacteria	Lan and Liao (2012)
		E. coli	Metabolic engineering of microorganisms for the production of L-arginine and its derivatives	Shin & Lee (2014)
		E. coli	Metabolic engineering for the high-yield production of isoprenoid-based C5 alcohols	George et al. (2015)
MOLECULAR	Genetic manipulation, mutations, protein engineering etc.	Engineered *E. coli* Strains	Hyaluronic acid accumulation in recombinant *E. coli* transformed by libraries of engineered sigma factors.	Yu et al. (2008)
		L. delbrueckii 11842	Tuning *E. coli* for membrane protein overexpression	Wagner et al. (2008)
		E. coli	Programming cells by multiplex genome engineering and accelerated evolution	Wang et al. (2009)
		E. coli	Engineered d-lactate dehydrogenase utilizing NADPH and NADH as cofactors	Bastian et al. (2011)
		Recombinant *E. coli*	Plasmid biopharmaceutical production	Goncalves et al. (2014)
		E. coli	Engineered enzymes enable anaerobic 2-methylpropan-1-ol production.	Meng et al. (2016)

and consequently the production of biomolecules or pharmaceuticals. Various biochemical control or constraints were applied for example nutritional control, physiological conditions, culture conditions, metabolic engineering techniques for improving the microbial performance as illustrated in Table 9.1. Similarly, a few evidences for molecularly controlled microbial performances are also displayed in Table 9.2. More examples for both are further provided in Section 9.3, Table 9.2, though in different context.

9.3 Biochemically Influenced Molecular Events and Vice Versa

Now initially biochemically influenced molecular events were the usual approach for improving microbial performance but later on with the advent of high throughput technologies and recombinant DNA techniques the reverse is the more popular approach with researchers across the globe. Table 9.2 illustrates examples for biochemically influenced molecular events and vice versa. Table 9.2 illustrates examples for biochemically influenced molecular events and vice versa. The table 9.2 section A illustrates the biochemical influences like oxidative stress, adaptive evolution, stimulant addition, inhibitor addition, nutritional and statistical optimization etc. which caused change in molecular events leading to microbial performance getting manipulated towards higher production of certain targeted products. Further, example of events where the biochemical and molecular changes go side by side can be brought about by using adaptive evolution, short term evolution, and evolutionary strategy by selecting mutants growing in presence of valine and norvaline. Whereas, Table 9.2 section B highlights the molecular events leading to biochemical changes or process changes or output as a direct result of manipulations at molecular level. Genetic manipulations, site directed or random mutagenesis, gene knockout strategies etc., are some of the tools that were employed to bring about the changes at molecular level which further resulted in enhanced microbial performances.

9.4 Facts at the Interface of Biochemical and Molecular Controlling: Products vs Applied Parameters

9.4.1 Sulfur-Delivery into Biosynthetic Pathway

Kessler et al. (2006) reported enzymatic activation-based sulfur for incorporation into prokaryotic biosynthetic pathway. They added sulfideas a deciding factor to an activated acceptor during the biosynthesis of cysteine, which resulted in the formation of methionine, coenzyme A and a number of biologically important thiols. Similarly, the persulfides (R-S-SH) and thiocarboxylates groups on sulfur-containing proteins contribute to carbon and sulfur bond formation in primary metabolites and such sulfur incorporation biosynthetic pathways always

Table 9.2 Evidences for biochemically influenced molecular events and vice versa and the final benefit

A. Evidences for biochemically influenced molecular events

Strain	Technique	Result	References
Blakeslea trispora	Oxidative stress	Oxidative stress resulted in fivefold increase in carotene production	Nonau et al. (2010)
Saccharomyces cerevisiae	Evolutionary engineering	Increased consumption of arabinose and xylose under aerobic and anaerobic conditions. Also improves ethanol production at the expense of xylitol and glycerol and complete conversion of arabinose to arabitol.	Sanchez et al.(2010)
Escherichia coli	Adaptive Evolution	Increased ethanol tolerance, mutant able to grow at 40 g/l ethanol, also 48% faster volumetric ethanol productivity	Wang et al. (2011)
Escherichia coli	Evolutionary strategy by selecting mutants growing in presence of valine and norvaline	Increased isobutanol production (with yield of 76% of theoretical maximum)	Smith et al. (2011)
Saccharomyces cerevisiae	Adaptive evolution in xylose	Increased xylose utilizing capacity and ethanol production of engineered strain	Shen et al. (2012)
Saccharomyces cerevisiae	Short term evolution using hydrogen peroxide shocking schemes	Increased carotenoides production by 3-fold	Reyes et al. (2014)
Blakeslea trispora	Growth control	Enhanced Lycopene production in stirred-tank reactor when (+) mating type was inoculated after (-) mating type has been grown for certain period of time in nonsynchronous inoculation process	Wang et al. (2015)
Chlorella pyrenoidosa	Stimulant addition	70.9% increase in lutein content was observed on stimulation with Vitamin A	Li et al. (2015)

(*Continued*)

Table 9.2 (Continued)

A. Evidences for biochemically influenced molecular events			
Strain	**Technique**	**Result**	**References**
Blakeslea trispora	Addition of inhibitor to upregulate expression of gene of interest	Lycopene production increased from 1.7% to 90.1% in presence of cyclase inhibitor tripropylamine. Tripropylamine inhibits cyclase activity and upregulates the expression of genes associated with terpenoid biosynthesis.	Wang et al. (2015)
Blakeslea trispora	Nutritional optimization	Addition of sodium acetate to the medium leads to induction of 5 carotenogenesis gene expression and about 80% increment in β-carotene production was observed compared to control	Jing et al. (2016)
Bacillus licheniformis	Statistical optimization	Increased alkaline protease production through statistical optimization of media using response surface methodology	Kavitha et al. (2016)

B. Evidences for molecular level changes influencing biochemical events			
Strain	**Technique**	**Result**	**References**
Zymomonas mobilis	NTG random mutagenesis	Increased rates of levan (fructose polymer) synthesis	Ananthalakshmy et al. (1999)
Saccharomyces cerevisiae	Mutagenesis (UV)	Thermos-tolerance, ethanol tolerance and osmotolerance	Sridhar et al. (2002)
Escherichia coli	Recombinant technology	Enhanced L-Phenylalanine production	Gerigk et al. (2002)
Lactobacillus	genome shuffling	increased acid tolerance and strain produces 3-fold acid at pH 4.0	Patnaik et al. (2002)
Escherichia coli	Systematic and combinatorial methods for gene knockout	Recombinant strain overproducing lycopene by about 8.5-fold compared to recombinant wild type strain	Alper et al. (2005)
Saccharomyces cerevisiae	Global transcription machinery engineering	Increased glucose/ethanol tolerance	Alper et al. (2005)

Hansenula polymorpha	site directed mutagenesis	Increased ethanol yield up to 7.3-fold	Dmytruk et al. (2008)
Lactobacillus rhamnosus	Genome shuffling	Lactic acid production was about 70% higher than wild type	Yu et al. (2008)
Saccharomyces cerevisiae	Genome shuffling	Thermo-tolerance, ethanol tolerance and increased ethanol yields	Shi et al. (2009)
Saccharomyces cerevisiae	Genome shuffling	multiple stress tolerance to ethanol, glucose, and heat, also increased ethanol yield by 10%	Hou et al. (2009)
Saccharomyces cerevisiae	Genome shuffling	multi stress-tolerances (osmotic, heat, acid) and increased yield	Zheng et al. (2011)
Saccharomyces cerevisiae	Genome shuffling	increased ethanol yield (up to 15%) and VGH fermentation capacity	Liu et al. (2011)
Pichia stipitis	Genome shuffling	Enhanced xylose fermentation and 1.5-fold increase in ethanol production	Shi et al. (2014)
Brettanomyces anomalus	Genome shuffling	Eight-fold increase in production of extracellular β-glucosidase	Wu et al. (2014)
Saccharomyces cerevisiae	Genome shuffling	Expression of glutathion synthetase gene increased by 9.9-fold and resulted in enhanced glutathione production by 3.2-fold	Yin et al. (2016)
Clostridium acetobutylicum	Mutagenesis (UV) & genome shuffling	About 40% increase in production of butanol from cassava	Li et al. (2015)
Volvariella volvacea	genome shuffling	Low temperature resistant fungi (increased shelf life by 75%)	Zhu et al. (2016)
Clostridium beijerinckii	NTG mutagenesis & genome shuffling	increased isopropanol tolerance	de Gérando et al. (2016)

share enzymes (Mueller, 2006; Sasaki et al., 2014). Usually the genes encoding the catalyzing enzyme as well as the similar sulfur-carrier protein are often found in the same gene cluster (Burroughs et al., 2009). Further it has also been shown that such an enzyme can not only adenylate sulfur-carrier proteins but also catalyzes the subsequent thiolation. So, such controlled sulfur-delivery machinery of primary metabolism can be exploited for the production of 2-thioglucose or other value added sulfur-containing natural compounds.

9.4.2 Synthetic Biochemistry Platform for Production of Glucose

Synthetic biochemistry platform concept is the perfect example for use biochemical and molecular biology interface domain for production of commercial compounds. Under this, a cell-free system is designed to carry out biochemical conversions of biomass to biofuels or other value added products. Korman et al. (2017) reported such a design comprising of 27 enzymes for conversion of glucose into monoterpenes resulting in generation of cofactor NAD(P)H and ATP in a modified glucose breakdown module for building terpenes. They could produce different monoterpenes by changing or controlling which terpene synthase enzyme to employ. They illustrated that such a design has potential for continuous production of limonene, pinene and sabinene, for at least 5 days with just a single addition of glucose. There work encourages such exploitation of synthetic biochemistry approaches for production of bio-based chemicals.

9.4.3 Biochemical and Molecular Aspects of Metabolic Engineering Approaches

The recent advancement in Genome sequencing has opened many avenues in systems metabolic engineering for huge number of industrially important microorganisms (Covert, Schilling and Palsson, 2001a; Jacob, 2009; Jing et al., 2016; Tsao et al., 2010). Tsuge et al. (2016) extensively reviewed metabolic engineering approaches to enhance microbial strains for producing platform chemicals to synthesize aliphatic and high-performance aromatic polymers. Koirala et al. (2017) reported the production of isoflavonoid-7-O-methoxides from genetically engineered *Escherichia coli*. This engineered strain was reconstructed by incorporating *E. coli* K12-derived *metK* encoding *S*-adenosine-L-methionine (SAM) synthase to establish SAM as an efficient precursor for subsequent production of 7-*O*-methyl-genistein and 7-*O*-methyl-daidzein. After optimization at flask culture, the most suitable conditions were applied for fed-batch fermentation at bioreactor level (3 liters) to obtain the maximum yield of the products.

9.4.3.1 Engineering Regulatory Network

In this approach genes for enzymes involved in particular biosynthetic pathway along associated specific regulatory genes are grouped together on same chromosome as one cluster. These regulatory genes can be activators (positive) or repressor (negative) of gene expressions and are manipulated or engineered to overproduce particular metabolites by up and down regulation approach (Nakamura and Whited, 2003; Kumar and Prasad, 2011). The example of up regulation is the continuous expression of SARP (Antibiotic Regulatory Protein) to increase continuous production of different secondary metabolites such as actinorhodin in *Streptomyces coelicolor*. Similarly, inactivation of pathway specific repressors in *Streptomyces griseus* for increased production of chromomycin is an example of down regulation approach (Bekker et al., 2014).

9.4.3.2 Heterologous Expression of Entire Gene Cluster

Molecular cloning and heterologous expression of the complete gene cluster has been extensively exploited for the production of secondary metabolites. Pyeon et al., (2017) reported pSBAC-based heterologous expression of pikromycin biosynthetic gene cluster (60 kb) by making use of a single integration pSBAC vector. This insertion of pikromycin biosynthetic gene cluster into the non-producing strains ensued higher pikromycin production. Earlier, a similar approach was adopted by Tu et al. (2016) for genetic engineering and heterologous expression of the disorazol biosynthetic gene cluster via Red/ET recombineering (recombination facilitated genetic engineering) followed by heterologous expression *Myxococcus xanthus* DK1622. This further allowed for yield optimization of several biosynthetic products using promoter engineering. The best example for such technique is the insertion of an artificial synthetic promoter upstream to discrete acyl transferase encoding gene (disD) which ensured a 7-fold increase in disorazol production (Tu et al., 2016).

9.4.3.3 Rerouting Metabolic Pathway

This approach is based on multiple changes made in a single pathway so as to channel it for the synthesis of desired product. The latest approaches make use of integrated systems biology tools such as transcriptomics, proteomics, metabolomics, and fluxomics characterization of new mutants and the production of the desired fuel-grade products (Wang et al., 2011). Functional genomics approach is used for overproduction of folate and such genes are expressed in Lactobacillus plantarum WCFS1 (Wegkamp et al., 2010). Strain optimization process can be made very efficient by employing different models and computational analysis during fermentation process. Kim, Sang-Jeong and Ji-Sook et al. (2016) successfully carried out the redirection of pyruvate flux towards the desired metabolic pathways in *Saccharomyces cerevisiae* which led to an overall increase in 2,3-butanediol production (38%), with a simultaneous decrease in ethanol production without any genetic manipulation of the ethanol-producing pathway.

9.4.3.4 Integration of Metabolic Engineering and Process Engineering

This concept is based on the possibility that the entire pathway or at least a major portion of a pathway can be expressed introduced into a new host organism which will exhibit more desirable characteristics (Edwards, Ibarra, and Palsson, 2001; Falb et al., 2008; Feist and Palsson, 2008). This revolutionary concept has created an interface for integration of metabolic engineering and process engineering for enhanced production of a plethora of products. Bond-Watts et al. (2011) highlighted possibility of enzyme mechanism as a kinetic control component for designing synthetic biofuel pathways. Similarly, Zabala et al. (2012)

reported about protein engineering towards natural product synthesis and diversification. Pirie et al. (2013) previously analyzed the significance of protein engineering strategies for the advanced metabolic engineering of secondary metabolism pathways. This type of integrated approach can increase the collective toolkit of microbial engineering to shape the future of chemical manufacturing.

9.5 Conclusions

In conclusion we can say that a considerable advancement has been made in both biochemical strategies as well as cellular or molecular aspects, which has now allowed production of commercially important biomolecules to a scale never imagined before. The traditional physiological, nutritional and one variable optimization has been completely replaced with new and more robust statistical designs and model. Further boost to microbial manipulation occurred due the development of metabolic engineering as an important tool as this technique completely changes the understanding of metabolic pathway and its exploitation potential. Many hitherto difficult productions were made possible due to employment of metabolic engineering. The advent of DNA manipulation techniques like recombinant DNA, gene knockout, chemical or physical mutagenesis, protein engineering, etc., has further opened a new horizon for hyper expression of the desired therapeutic proteins and other pharmaceuticals. Overall, one can say deep understanding of biochemical as well as molecular schemes or mechanism could allow us to influence the production of target molecules in either direction (biochemical to molecule and vice versa) resulting in huge yield of active biomass. Currently the biotech industry market is projected to reach $228 billion in bioengineered proteins, $7.7 billion in genomics, proteomics, and epigenomics and $174 billion in global protein drugs (Yu et al., 2008a). Keeping this in mind in future the more and more such products are going to get approved for sale in the market and it could really boost the biotech industry and applied research.

References

Abusham, R.A., Rahman, R.N.Z.R., Salleh, A.B., and Basri, M. (2009) Optimization of physical factors affecting the production of thermo-stable organic solvent-tolerant protease from a newly isolated halo tolerant *Bacillus subtilis* strain Rand. *Microbial Cell Factories*, 8(1), pp.20.

Adsul, M., Khire, J., Bastawde, K., and Gokhale, D. (2007) Production of lactic acid from cellobiose and cellotriose by *Lactobacillus delbrueckii* mutant Uc-3. *Applied and Environmental Microbiology*, 73(15), pp.5055–5057.

Almeida, D.G., Soares da Silva R.C., Luna, J.M. et al. (2017) Response surface methodology for optimizing the production of biosurfactant by *Candida tropicalis* on industrial waste substrates.*Frontiers in Microbiology*, 8(157), pp. 1–13.

Alper, H., Miyaoku, K., and Stephanopoulos, G. (2005) Construction of lycopene-overproducing *E. coli* strains by combining systematic and combinatorial gene knockout targets. *Nature Biotechnology*, 23(5), pp.612–616.

Alper, H., Moxley, J., Nevoigt, E. et al. (2006) Engineering yeast transcription machinery for improved ethanol tolerance and production. *Science*, 314(5805), pp.1565–1568.

Ananthalakshmy, V.K. and Gunasekaran, P. (1999) Isolation and characterization of mutants from levan-producing *Zymomonasmobilis*. *Journal of Bioscience and Bioengineering*, 87(2), pp.214–217.

Bastian, S., Liu, X., Meyerowitz, J.T. et al. (2011) Engineered ketol-acid reductoisomerase and alcohol dehydrogenase enable anaerobic 2-methylpropan-1-ol production at theoretical yield in *Escherichia coli*. *Metabolic Engineering*, 13(3), pp.345–352.

bcc Research, In *Market Research Reports, Biotechnology* [2017] http://www.bccresearch.com/market-research/biotechnology.

Bekker, V., Dodd, A., Brady,D., and Rumbold, K. (2014) Tools for metabolic engineering in Streptomyces. *Bioengineered*, 5(5): pp.293–299.

Bond-Watts, B.B., Bellerose, R.J., and Chang, M.C.Y. (2011) Enzyme mechanism as a kinetic control element for designing synthetic biofuel pathways. *Nature Chemical Biology*,7, pp.222–227.

Box, G.P.E. and Norman, R.D. (2007) *Response Surfaces, Mixtures, and Ridge Analyses*. 2nd ed., John Wiley & Sons, Inc., pp 1–768.

Box, D.E.P. and Draper N.R. (1987) *Empirical Model-Building and Response Surfaces*, Wiley, New York, pp.1–669.

Burroughs, A.M., Iyer, L.M., and Aravind, L. (2009) Natural history of the E1-like superfamily: implication for adenylation, sulfur transfer, and ubiquitin conjugation. *Proteins*, 75, pp.895–910.

Chen, S.H. and Aitken, M.D. (1999) Salicylate stimulates the degradation of high-molecular weight polycyclic aromatic hydrocarbons by *Pseudomonas saccharophila* P15. *Environmental science & technology*, 33(3), pp.435–439.

Covert, M.W., Schilling, C.H., and Palsson, B. (2001a) Regulation of gene expression in flux balance models of metabolism. *Journal of Theoretical Biology*, 213(1), pp.73–88.

Covert, M.W., Schilling, C.H., Famili, I. et al. (2001b) Metabolic modeling of microbial strains in silico. *Trends in Biochemical Sciences*, 26(3), pp.179–186.

Crook, N. and Alper, H.S. (2012) Classical strain improvement. In: *Engineering Complex Phenotypes in Industrial Strains*, John Wiley & Sons Inc., pp.1–33.

de Gérando, H.M., Fayolle-Guichard, F., Rudant, L. et al. (2016) Improving isopropanol tolerance and production of *Clostridium beijerinckii* DSM 6423 by random mutagenesis and genome shuffling. *Applied Microbiology and Biotechnology*, pp.1-10.

Dmytruk, O.V., Dmytruk, K.V., Abbas, C.A. et al. (2008) Engineering of xylose reductase and overexpression of xylitol dehydrogenase and xylulokinase improves xylose alcoholic fermentation in the thermotolerant yeast *Hansenula polymorpha*. *Microbial Cell Factories*, 7(1), p.1.

Duarte, N.C., Becker, S.A., Jamshidi, N. et al. (2007) Global reconstruction of the human metabolic network based on genomic and bibliomic data. *Proceedings of National Academy of Sciences USA*,104(6), pp. 1777–1782.

Edwards, J.S., Covert, M., and Palsson, B. (2002) Metabolic modelling of microbes: the flux-balance approach. *Environmental Microbiology*, 4(3), pp.133–140.

Edwards, J.S., Ibarra, R.U., and Palsson, B.O. (2001) In silico predictions of *Escherichia coli* metabolic capabilities are consistent with experimental data. *Nature Biotechnology*, 19(2), pp.125–130.

Falb, M., Muller, K., Konigsmaier, L. et al. 2008. Metabolism of halophilic archaea. *Extremophiles*, 12, pp.177–196.

Feist, A.M. and Palsson, B.O. (2008) The growing scope of applications of genome-scale metabolic reconstructions using *Escherichia coli*. *Nature Biotechnology*, 26(6), pp. 659–67.

George, K.W., Thompson, M.G., Kang, A. et al. (2015) Metabolic engineering for the high-yield production of isoprenoid-based C_5 alcohols in *E. coli*. *Scientific Reports*, 5, pp. 11128.

Gerigk, M., Bujnicki, R., Ganpo-Nkwenkwa, E. et al. (2002) Process control for enhanced L-phenylalanine production using different recombinant *Escherichia coli* strains. *Biotechnology and Bioengineering*, 80(7), pp.746–754.

Gonçalves, G.A., Prather, K.L., Monteiro, G.A. et al. (2014) Plasmid DNA production with *Escherichia coli* GALG20, a pgi-gene knockout strain: fermentation strategies and impact on downstream processing. *Journal of Biotechnology*, 186, pp. 119–27.

Hou, L. (2010) Improved production of ethanol by novel genome shuffling in *Saccharomyces cerevisiae*. *Applied Biochemistry and Biotechnology*,160(4), pp. 1084–93.

Jacobs, P.P. (2009) Engineering complex type N-glycosylation in *Pichia pastoris* using GlycoSwitch technology. *Nature Protocol*, 4, pp. 58–70.

Jing, K., He, S., Chen, T. et al. (2016) Enhancing beta-carotene biosynthesis and gene transcriptional regulation in *Blakeslea trispora* with sodium acetate. *Biochemical Engineering Journal*, 114, pp.10–17.

Johansen, E., Øregaard, G., Sørensen, K.I., and Derkx, P.M.F. (2014) In: *Advances in fermented foods and beverages: Improving quality*, technologies and health benefits, Holzapfel W, (ed.) *Modern Approaches for Isolation, selection and improvement of bacterial strains for fermentation applications*, Woodhead Publishing Ltd., Cambridge, UK, pp. 227–248.

Kamath, P.V., Dwarakanath, B.S., Chaudhary, A., and Janakiraman, S. (2015) Optimization of culture conditions for maximal lovastatin production by *Aspergillus terreus* (KM017963) under Solid State Fermentation. *HAYATI Journal of Biosciences*, 22(4), pp.174–180.

Karanam, S.K. and Medicherla, N.R. (2008) Enhanced lipase production by mutation induced *Aspergillus japonicus*. *African Journal of Biotechnology*, 7(12), pp. 2064–2067.

Kardos, N. and Demain, A.L. (2011) Penicillin: the medicine with the greatest impact on therapeutic outcomes. *Applied Microbiology and Biotechnology*, 92, pp.677–687.

Kavitha, M.D., Kathiresan, S., Bhattacharya, S., and Sarada, R., 2016. Culture media optimization of *Porphyridium purpureum*: production potential of biomass, total lipids, arachidonic and eicosapentaenoic acid. *Journal of Food Science and Technology*, pp.1–9.

Kessler, D. (2006) Enzymatic activation of sulfur for incorporation into biomolecules in prokaryotes. *FEMS Microbiology Reviews*, 30, pp.825–840.

Kim, S., Bae, S.J., and Hahna, J.S. (2014) Metabolic flux redirection from a central metabolic pathway toward a synthetic pathway using a metabolic toggle switch. *Metabolic Engineering*, 23, pp.175–84.

Kim, S., Bae, H.J., and Hahna J.S. (2016) Redirection of pyruvate flux toward desired metabolic pathways through substrate channeling between pyruvate kinase and pyruvate-converting enzymes in *Saccharomyces cerevisiae. Scientific Reports*, 6, pp. 24145.

Koirala, N., Thuan, N.H., Ghimire, G.P. et al. (2017) Metabolic engineering of *E. coli* for the production of isoflavonoid-7-O-methoxides and their biological activities. *Biotechnology and Applied Biochemistry*, DOI: 10.1002/bab.1452.

Korman,T.P., Opgenorth, P.H., and Bowie, J.U. (2017) A synthetic biochemistry platform for cell free production of monoterpenes from glucose. *Nature Communication*, DOI: 10.1038/ncomms15526

Kumar, R.P. and Prasad, S. (2011) Metabolic engineering of bacteria. *Indian Journal of Microbiology*, 51(3), pp. 403–409.

Lan, E.I. and Liao, J.C. 2012. ATP drives direct photosynthetic production of 1-butanol in cyanobacteria. *Proceedings of National Academy of Sciences USA*, 109(16), pp. 6018–23.

Ledolter, J. and Swersey, A.J. (2007) In *Testing 1-2-3: Experimental design with applications in marketing and service operations*, Stanford University Press, Chicago, pp. 1–312.

Lee, I.Y., Seo, W.T., Kim, G.J. et al. (1997) Optimization of fermentation conditions for production of exopolysaccharide by *Bacillus polymyxa. Bioprocess engineering*, 16(2), pp.71–75.

Lee, S.Y., Lee, D.Y., and Kim T.Y. (2005) Systems biotechnology for strain improvement. *Trends in biotechnology*, 23(7), pp. 349–358.

Li, S.B., Qian, Y., Liang, Z.W. et al. (2016) Enhanced butanol production from cassava with *Clostridium acetobutylicum* by genome shuffling. *World journal of Microbiology and Biotechnology*, 32(4), pp.1–10.

Li, T., Bai, D., Tian, L. et al. (2015) Effects of stimulators on lutein and chlorophyll biosynthesis in the green alga *Chlorella pyrenoidosa* under heterotrophic conditions. *In*: Zhang, TC., Nakajima M. (eds) *Advances in Applied Biotechnology. Lecture Notes in Electrical Engineering*, 332, Springer, Berlin, Heidelberg, pp 389–398.

Liao, Y., Zeng, M., Wu, Z. F. et al. (2012) Improving phytase enzyme activity in a recombinant phyA mutant phytase from *Aspergillus niger* N25 by error-prone PCR. *Applied Biochemistry and Biotechnology*, 166, 549–562.

Liu, J.J., Ding, W.T., Zhang, G.C., and Wang, J.Y. (2011) Improving ethanol fermentation performance of *Saccharomyces cerevisiae* in very high-gravity fermentation through chemical mutagenesis and meiotic recombination. *Applied Microbiology and Biotechnology*, 91(4), pp.1239–1246.

MacLean, C.R., Torres-Barceló, C., and Moxon, R. (2013) Evaluating evolutionary models of stress-induced mutagenesis in bacteria. *Nature Reviews Genetics*, 14, pp. 221–227.

Mahadevan, R. and Schilling, C.H. (2003) The effects of alternate optimal solutions in constraint-based genome-scale metabolic models. *Metabolic Engineering*, 5(4), pp.264–276.

Meng, H., Liu, P., Sun, H. et al. (2016) Engineering a D-lactate dehydrogenase that can super-efficiently utilize NADPH and NADH as cofactors. *Scientific Reports*, 6, 24887; doi: 10.1038/srep24887.

Montgomery, D. (2001) *Design and Analysis of Experiments*, John Wiley & Sons, New York, NY, USA.

Mueller, E.G. (2006) Trafficking in persulfides: delivering sulfur in biosynthetic pathways. *Nature Chemical Biology*, 2, pp.185–194.

Meyers R.H. (1971) Response Surface Methodology, *Boston Allyn & Bacon*, 31(2), pp. 159–171.

Nakamura, C.E. and Whited, G.M. (2003) Metabolic engineering for the microbial production of 1,3-propanediol. *Current Opinion in Biotechnology*, 14, pp. 454–459.

Nanou, K. and Roukas, T. (2010) Oxidative stress response and morphological changes of *Blakeslea trispora* induced by butylated hydroxytoluene during carotene production.*Applied Biochemistry and Biotechnology*, 160(8), pp. 2415–23.

Oberhardt, M.A., Chavali, A.K., and Papin, J.A. (2009) Flux balance analysis: interrogating genome-scale metabolic networks.*Methods in Molecular Biology*, 500, pp. 61–80.

Orth, J.D., Thiele, I., and Palsson, B.O. (2010) What is flux balance analysis? *Nature Biotechnology*, 28(3), pp.245–248.

Parekh, S., Vinci, V.A., and Strobel, R.J. (2000) Improvement of microbial strains and fermentation processes. *Applied Microbiology and Biotechnology*, 54, pp.287–301.

Patnaik, R., Louie, S., Gavrilovic, V. et al. (2002) Genome shuffling of Lactobacillus for improved acid tolerance. *Nature Biotechnology*, 20(7), pp.707–712.

Pirie, C.M., Mey, M.D., Prather, K.L.J., and Ajikumar, P.K. 2013. Integrating the protein and metabolic engineering toolkits for next-generation chemical biosynthesis. *ACS Chemical Biology*, 8, pp.662–672.

Plackett, R.L. and Burman, J.P. (1946) The design of optimum multifactorial experiments. *Biometrika*, 33(4), pp.305–325.

Pyeon, H.R., Nah, H.J., Kang, S.H. et al. (2017) Heterologous expression of pikromycin biosynthetic gene cluster using Streptomyces artificial chromosome system. *Microbial Cell Factories*, 16(96), DOI 10.1186/s12934-017-0708-7.

Razak, M.A. and Viswanath, B. (2015) Optimization of fermentation upstream parameters and immobilization of *Corynebacterium glutamicum* MH 20-22 B cells to enhance the production of l-lysine. *3 Biotech*, 5(4), pp.531–540.

Reyes, L.H., Gomez, J.M., and Kao, K.C. (2014) Improving carotenoids production in yeast via adaptive laboratory evolution. *Metabolic Engineering*, 21, pp.26–33.

Rufino, R. D., Luna, J. M., Campos–Takaki, G. M., and Sarubbo, L. A. (2014) Characterization and properties of the biosurfactant produced by Candida lipolytica UCP 0988. *Electronic J. Biotech.* 17, 34–38.

Sanchez, R.G., Karhumaa, K., Fonseca, C. et al. (2010) Improved xylose and arabinose utilization by an industrial recombinant *Saccharomyces cerevisiae* strain using evolutionary engineering. *Biotechnology for Biofuels*, 3(1), p.1.

Saraniya, A. and Jeevaratnam, K. (2014) Optimization of nutritional and non-nutritional factors involved for production of antimicrobial compounds from

Lactobacillus pentosus SJ65 using response surface methodology. *Brazilian Journal of Microbiology*, 45(1), pp.81–88.

Sasaki, E., Zhang, X., Sun, H.G. et al. (2014) Co-opting sulphur-carrier proteins from primary metabolic pathways for 2-thiosugar biosynthesis. *Nature*, 510(7505), pp.427–431.

Satya Eswari, J., Anand, M., and Venkateswarlu Ch. (2013) Optimum culture medium composition for rhamnolipid production by pseudomonas aeruginosa AT10 using a novel multi-objective optimization method. *Journal of Chemical Technology And Biotechnology*, 88(2), 271–279.

Saxena S. (2015) Strategies of Strain Improvement of Industrial Microbes. In *Applied Microbiology*, Springer, Part of Springer Nature, New Delhi.

Shen, Y., Chen, X., Peng, B. et al. (2012) An efficient xylose-fermenting recombinant Saccharomyces cerevisiae strain obtained through adaptive evolution and its global transcription profile. *Applied Microbiology and Biotechnology*, 96(4), pp.1079–1091.

Shi, D.J., Wang, C.L., and Wang, K.M. (2009) Genome shuffling to improve thermotolerance, ethanol tolerance and ethanol productivity of *Saccharomyces cerevisiae*. *Journal of industrial microbiology and biotechnology*, 36, pp. 139–147.

Shi, J., Zhang, M., Zhang, L. et al. (2014) Xylose-fermenting *Pichia stipitis* by genome shuffling for improved ethanol production. *Microbial biotechnology*, 7(2), pp.90–99.

Shin, J.H. and Lee, S.Y. (2014) Metabolic engineering of microorganisms for the production of L-arginine and its derivatives. *Microbial Cell Factories*, 13, pp.166.

Smith, K.M. and Liao, J.C. (2011) An evolutionary strategy for isobutanol production strain development in *Escherichia coli*. *Metabolic Engineering*,13(6), pp.674–681.

Sridhar, M., Sree, N.K., and Rao, L.V. (2002) Effect of UV radiation on thermotolerance, ethanol tolerance and osmotolerance of *Saccharomyces cerevisiae* VS 1 and VS 3 strains. *Bioresource Technology*, 83(3), pp.199–202.

Surwase, S.N., Patil, S.A., Jadhav, S.B., and Jadhav, J.P. (2012) Optimization of L-DOPA production by *Brevundimonas* sp. SGJ using response surface methodology. *Microbial Biotechnology*, 5(6), pp.731–737.

Syed, M.B. and Rajasimman, M. (2015) Fermentative production and optimization of mevastatin in submerged fermentation using *Aspergillus terreus*. *Biotechnology Reports (Amst.)*, 6, pp.124–128.

Thiele, I. and Palsson, B.O. (2010) A protocol for generating a high-quality genome-scale metabolic reconstruction. *Nature Protocols*, 5(1), pp.93–121.

Tsao, C.Y., Hooshangi, S., Wu, H.C. et al. (2010) Autonomous induction of recombinant proteins by minimally rewiring native quorum sensing regulon of *E. coli*. *Metabolic Engineering*, 12(3), pp. 291–297.

Tsuge, Y., Kawaguchi, H., Sasaki, K., and Kondo, A. (2016) Engineering cell factories for producing building block chemicals for bio-polymer synthesis. *Microbial Cell Factories*, 15 (19), pp.1–12.

Tu, Q., Herrmann, J., Hu, S. et al. (2016) Genetic engineering and heterologous expression of the disorazol biosynthetic gene cluster via Red/ET recombineering. *Scientific Reports*, 6:21066. doi: 10.1038/srep21066.

Vastrad, B.M. and Neelagund, S.E. (2014) Optimization of medium composition for the production of neomycin by *Streptomyces fradiae* NCIM 2418 in solid state fermentation. *Biotechnology Research International*, 2014.

Vinci, V.A. and Byng, G. (1999) Strain improvement by nonrecombinant methods. *Manual of Industrial Microbiology and Biotechnology*, 2, pp.103–113.

Wagner, S., Klepsch, M.M., Schlegel, S. et al. (2008) Tuning Escherichia coli for membrane protein overexpression. *Proceedings of the National Academy of Sciences USA*, 105(38), pp.14371–14376.

Wang, H.H., Isaacs, F.J., Carr, P.A. et al. (2009) Programming cells by multiplex genome engineering and accelerated evolution. *Nature*, 460(7257), pp. 894–8.

Wang, J., Hirabayashi, S., Mori, T. et al. (2016) Improvement of ethanol production by recombinant expression of pyruvate decarboxylase in the white-rot fungus *Phanerochaete sordida* YK-624. *Journal of Bioscience &Bioengineering*, 122(1), pp.17–21.

Wang, M., Zhao, Q., Li, L. et al. (2016) Contributing factors in the improvement of cellulosic H2 production in *Clostridium thermocellum/Thermoanaerobacterium* co-cultures. *Applied Microbiology and Biotechnology*, 100(19), pp.8607–8620.

Wang, Q., Feng, L.R., Luo, W. et al. (2015) Effect of inoculation process on lycopene production by *Blakeslea trispora* in a stirred-tank reactor. *Applied Biochemistry and Biotechnology*, 175(2), pp.770–779.

Wang, Y., Chen, X., Hong, X. et al. (2016) Cyclase inhibitor tripropylamine significantly enhanced lycopene accumulation in *Blakeslea trispora*. *Journal of Bioscience and Bioengineering*, 122(5), pp.570–576.

Wang, Y., Manow, R., Finan, C. et al. (2011) Adaptive evolution of nontransgenic *Escherichia coli* KC01 for improved ethanol tolerance and homoethanol fermentation from xylose. *Journal of Industrial Microbiology & Biotechnology*, 38(9), pp.1371–1377.

Wegkamp, A., Mars, A.E., Faijes, M. et al. (2010) Physiological responses to folate overproduction in *Lactobacillus plantarum* WCFS1. *Microbial Cell Factories*, 9:100.

Wesolowski, J. (2010) Antifungal compounds redirect metabolic pathways in yeasts: metabolites as indicators of modes of action. *Journal of Applied Microbiology*, 108, pp.462–471

Wu, P., Zhao, X., and Pan, S. (2014) Intraspecific protoplast fusion of *Brettanomyces anomalus* for improved production of an extracellular β-glucosidase. *Biotechnology &Biotechnological Equipment*, 28(5), pp.878–881.

Yin, H., Ma, Y., Deng, Y. et al. (2016) Genome shuffling of Saccharomyces cerevisiae for enhanced glutathione yield and relative gene expression analysis using fluorescent quantitation reverse transcription polymerase chain reaction. *Journal of Microbiological Methods*, 127, pp.188–92.

Yu, H., Tyo, K., Alper, H. et al. (2008a) A high-throughput screen for hyaluronic acid accumulation in recombinant Escherichia coli transformed by libraries of engineered sigma factors. *Biotechnology and Bioengineering*, 101(4), pp.788–796.

Yu, L., Pei, X., Lei, T. et al. (2008b) Genome shuffling enhanced L-lactic acid production by improving glucose tolerance of *Lactobacillus rhamnosus*. *Journal of Biotechnology*, 134(1), pp.154–159.

Zabala, A.O., Cacho, R.A., and Tang, Y. (2012) Protein engineering towards natural product synthesis and diversification. *Journal of Industrial Microbiology and Biotechnology*, 39, pp.227–241.

Zhang, Y., Cai, J., Huang, L. et al. (2016) Improving the productivity of 19,20-epoxy-cytochalasin Q in *Xylaria* sp. sof11 with culture condition optimization. *Preparative Biochemistry and Biotechnology*, 46(5), pp. 461–466.

Zheng, D.Q., Wu, X.C., Tao, X.L. et al. (2011) Screening and construction of *Saccharomyces cerevisiae* strains with improved multi-tolerance and bioethanol fermentation performance. *Bioresource Technology*, 102(3), pp.3020–3027.

Zhu, Z., Wu, X., Lv, B. et al. (2016) A new approach for breeding low-temperature-resistant *Volvariella volvacea* strains: Genome shuffling in edible fungi. *Biotechnology and Applied Biochemistry*, 63(5), pp. 605–615.

10

Qualitative vs. Quantitative Control Over Microbial Products

Rachna Goswami[1], *Vijay Kumar Mishra*[1*], *and Radhika Pilli*[2]

[1] *Biosciences Department, Rajiv Gandhi University of Knowledge Technologies, India*
[2] *Microbiology Department, Bharathiar University, Coimbatore, TN, India*

10.1 Introduction

Microorganisms can be used by humans to produce foods and other useful substances. They are suitable for this role because they grow rapidly, may have simple nutrient requirements and can produce a wide range of products. Humans have unknowingly made use of microorganisms for centuries to carry out fermentation processes to make wine, beer, yoghurt, cheese and bread. Products such as bread, beer, wine, distilled spirits, vinegar, cheese, pickles, and other fermented materials have been with us for centuries, being provided by bacteria and fungi (Bennett, 1998). Since the 1860s we have known about the existence of bacteria and since the 1970s we have been able to genetically modify them, increasing the scope of biotechnology.

Since prebiblical times, fungi, including both true filamentous fungi and yeasts, have been used to produce products such as beer, wine, bread, and cheese. The fungi as food and feed are very nutritious, since they contains essential and non-essential amino acids. The fungi are used as dietary sources and in fermented beverages. Archaeological evidence traces association of edible wild mushroom to the inhabitants of Chile, almost 13,000 years ago (Ghorai et al., 2009; Rojas & Mansur, 1995); However, it was in China where the consumption of wild fungi was first reliably noted, several hundred years before the birth of Christ (Aaronson, 2000; Ghorai et al., 2009). The first evidence of fermentation and the use of fermented foods were found in Sumeria and Babylon (Elander and Lowe, 1994). In spite of this very old practice of fungi consumption and their utilization in production of other food materials, their full potential was not explored until the latter half of the twentieth century when it was boosted by the advent of the golden age of industrial microbiology. Since then, diverse community of fungi

**This author equally contributed*

Microbial Sensing in Fermentation, First Edition. Edited by Satinder Kaur Brar, Ratul Kumar Das, and Saurabh Jyoti Sarma.

represented by yeasts, mushrooms and filamentous fungi have been exploited in a myriad of food products both for human and livestock consumption.

The fungal kingdom possesses some natural advantages in terms of their dietary supremacy over the rest of the vegetarian platter. These advantages are: (a) a good protein content (20–30% of dry matter) having all the essential amino acids (yeasts are especially enriched in lysine) thus capable of substituting meat; (b) chitinous wall to act as a source of dietary fiber; (c) high vitamin B content; (d) low in fat; and (e) virtually free of cholesterol (Ghorai et al., 2009). The lucrativeness of mushroom cultivation is further enhanced by their low cost of production, since most of mushrooms are cultivated on agro-wastes or other industrial waste products. The ability of certain *Pleurotus* sp. to grow on many lignocellulose agricultural wastes has been exploited both in bioremediation and production of a consequent cash crop in the form of mushroom. Moreover, the harvested mushrooms (spent compost) can be used as a useful animal feed because of its high protein content, soil conditioner with its high content nutrients and polymeric components that enhance soil structure and even used to digest pollutants (like polychlorinated phenols) on land-fill waste sites by virtue of its population of microorganisms able to digest the natural phenolic components of lignin (Chiu et al.,1998). Also, the multi-species amenability and subsequent ease of upgradation to large-scale cultures have substantially aided the use of fungi commercially. The twentieth century, a golden age of industrial microbiology, yielded a myriad of products made by fermentation processes: solvents, antibiotics, enzymes, vitamins, amino acids, polymers, and many other useful compounds (Adrio and Demain, 2003; Demain, 1999). The development of molecular biology techniques provided new ways to use yeasts and molds as microbial cell factories for the production of homologous and heterologous (especially mammalian) proteins as well as other metabolites, such as antibiotics, pigments, and fatty acids. The choice of the strain is made on the basis of production yields and regulatory issues, especially for fungi used in the food industry. Several species of fungi have that status and are currently being used for large-scale production of recombinant proteins and metabolites (Adrio and Demain, 2003; Punt et al., 2002).

In this chapter discussed about the qualitative and quantitative role over microbial products. Mostly, fungal based products (organic acid, enzymes secondary metabolites etc.) are considered for the contributory effects at morphological angle. Discussion with some leading examples with an emphasis on the preferential outcome from qualitative and quantitative studied. The chapter has enlightened the core factors of making microbial processes happen (qualitative) and the potential of speeding up (quantitative) the same.

10.2 Qualitative vs. Quantitative Control Over Microbial Products/Fungal Products

Fungal usage for production of wine to torrent of several metabolites for various applications started with Biblical and Vedic period to modern age through stream of deviations. The vague studies of qualitative and quantitative role of fungal products were started at old days itself. With the age of industrial microbiology, the unique interest on study of qualitative and quantitative role over microbial products found initial place for commercial and industrial evolution (Adrio and Demain, 2003; Molitoris, 1995; Bennett, 1998).

Qualitative control is a behavioral approach. It focuses on behavior of fungi in fermentation over the productions of primary and secondary metabolites which are desirable. It emphasizes the significance of understanding various factors that influence the fungal behavior in fermentation. The core factors include selection of fungal product which is of interest, screening of the fungi which produce the product of interest, growth and production factors, environmental factors and other factors. It provides insights into the divergence or helps to develop ideas for potential quantitative studies. Qualitative studies are also used to uncover the concealed characteristics of fungi and dive deeper directions for microbial process to happen. The sample size of qualitative control studies is typically small, and the outcomes will lead to the quantitative control studies (Kavanagh, 2005).

Quantitative control is used to quantify the fungal product in the fermentation process. This emphasizes the use of mathematics, assays and optimization methods. It is used to optimize growth and production and environmental conditions. The sample size is typically large and mainly focuses on the way to large scale industrial production. Quantitative studies formulate the conditions and uncover patterns in research. Quantitative control methods are much more structured than qualitative control methods.

10.2.1 Qualitative Control and Fungal Product

The qualitative or behavioral approach of fungi in industrial microbiology point of view is that they are tremendously vital saprophytes (Kavanagh, 2005). The primary metabolic pathways of fungi are active, produce energy and produce intermediate chemicals and end products. So, fungi produce various primary metabolites for digesting the substrate or dead matter and absorb the nutrients for their growth. This digestion process can be called as decomposition, spoilage, mineralization, digestion, bioremediation and fermentation. Along with the growth, it also produces mainly secondary metabolites for multiplication or sporulation and protection. Either primary metabolites which are produced in initial phases of fungal growth curve, i.e. lag and log phases or secondary metabolites which are produced in stationary or declining phases of growth curve are helpful for the fungal metabolism and are produced usually into the environment. The metabolites which are produced outside the fugal cells are called extra cellular metabolites, for, e. g. enzymes and antibiotics. But some metabolites are produced inside the cells are called intracellular metabolites, for, e.g. fatty acids. The industrially important primary metabolites are enzymes, alcohol, organic acids, etc., whereas secondary metabolites are antibiotics, vitamins, pharmaceuticals, growth factors, etc. (Molitoris, 1995; Bennett, 1998; Adrio and Demain, 2003; Kavanagh, 2005; Norouzian, 2008).

Several species of fungi have their unique properties. The evaluation of the isolated fungi for the fungal product of interest can be done through various methods of screening based on the type of fungal product. This qualitative control of screening offers several advantages like simplicity, low cost, ability to test enormous number of microorganisms and even ease to interpret results provided. Additionally, numerous studies have revealed the fact that there is a good correlation between the qualitative and quantitative studies (Kavanagh, 2005; Balouiri et al., 2016).

There is limitation for this qualitative technique that it cannot determine the minimum inhibitory concentration (MIC), because it is impossible to measure the amount of the test compound diffused into the agar. This can be achieved through quantitative techniques. However, approximate MIC of fungal product can be calculated by comparing the inhibition zones with stored algorithms (Balouiri et al., 2016). There are several techniques for screening of fungal product i.e. Diffusion technique, thin layer chromatography (TLC), chromatography-bioautography and High-performance liquid chromatography (HPLC). Among these diffusion techniques chromatography-bioautography is very common for screening of antimicrobial activity.

10.2.1.1 Diffusion Techniques

Simple diffusion technique for pigments is the easiest and rapid technique. The fungal strain of interest is screened by inoculating in different fungal media and incubated for growth and production of the pigments. The change of the color of the agar media around the colony or a zone of pigmentation in the agar media indicates the strain produces pigment as fungal product in the specific media (Figure 10.1). The zone of pigmentation is measured to refer its effectiveness (Lopes et al., 2013).

The preliminary qualitative screening of citric acid is done by diffusion technique. The fungal media is supplemented with Bromo cresol green (1%) as an indicator at pH 6. The isolated fungal strain is inoculated aseptically and incubated at suitable conditions. After incubation, observe for yellow color development in the media around the colony, indicates the citric acid production (Figure 10.2). The yellow zone diameter measured and analyzed for fungal product efficiency. Similarly, the screening of Gluconic acid production from fungi is achieved through diffusion. The fungal culture media is supplemented with Calcium carbonate (5%) for detection of gluconic acid. The fungal media is autoclaved without $CaCO_3$ and $CaCO_3$ is sterilized separately and added to the medium at the time of plating. The fungus of interest is inoculated and incubated under desirable conditions. The development of a clear zone around the fungal

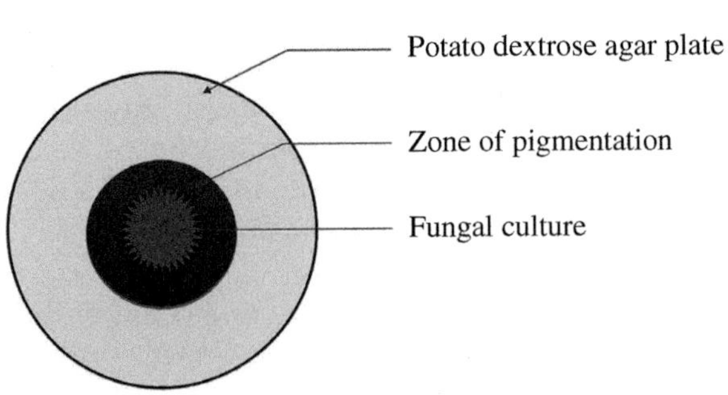

Figure 10.1 Screening of pigment producers. (*See insert for color representation of this figure.*)

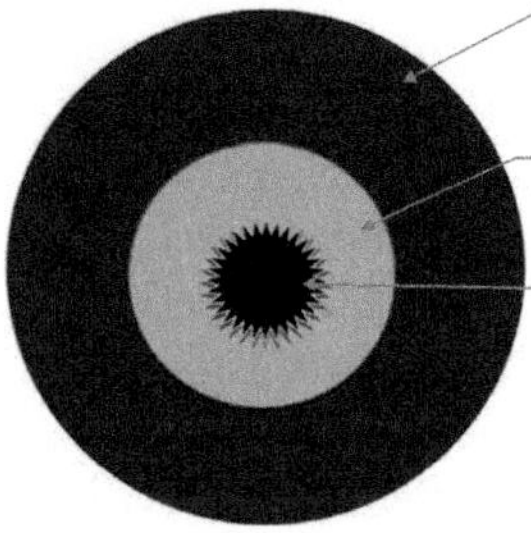

Figure 10.2 Screening of citric acid producers. (*See insert for color representation of this figure.*)

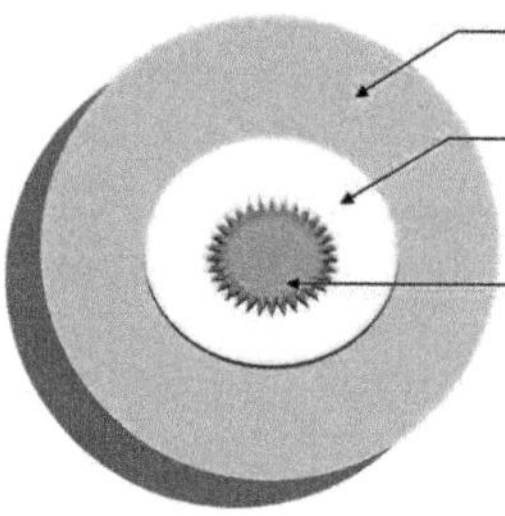

Figure 10.3 Screening of organic acid producers. (*See insert for color representation of this figure.*)

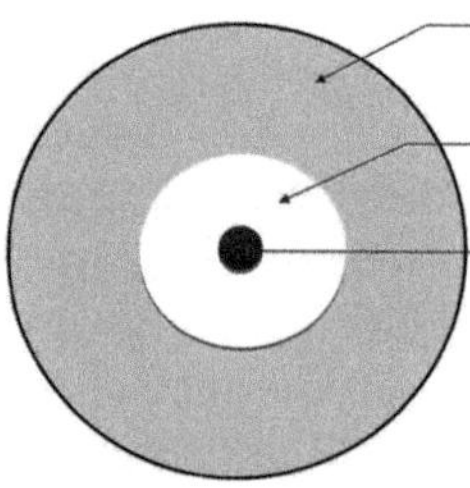

Figure 10.4 Screening of antimicrobial agent producers – Agar Disc diffusion method. (*See insert for color representation of this figure.*)

colony indicates the production of gluconic acid which is developed due to reaction of gluconic acid with $CaCO_3$ and formation of Calcium gluconate (Figure 10.3). The diameter of the clear zone against the colony diameter will be measured and this indicates the acid zone formed by the fungal product (Singh et al. 1999).

Agar disc-diffusion technique for antimicrobial activity is the conventional and well-known method for testing the antimicrobial susceptibility of fungal products (Figure 10.4). This technique will help to study the fungal product against many bacteria, yeast and even filamentous fungi, though some fastidious bacteria cannot be tested accurately. There are few criteria such as specific cultural media, pH, temperature, incubation period and interpretation of the inhibition zone will reveal the antimicrobial properties of the fungal product.

This technique is mainly applicable for extracellular fungal products (Balouiri et al., 2016). The agar plates are inoculated with the test microorganism (bacteria or fungi). Then sterile filter paper discs of diameter about 6mm are dipped into the crude extract containing the fungal product of interest at desire concentration. The discs are placed aseptically into a freshly inoculated test microbial agar plate and incubated at suitable conditions for the test organism. The fungal crude extract which contains the antimicrobial compound diffuses into the agar and inhibits the growth of the test microorganism or group of test microorganisms. This leads to the formation of a halo inhibition growth zone around the disc. The diameter of the inhibition zone is measured to discover the efficiency of the fungal product. Hence, it is a typing tool for antimicrobial property of fungal product against susceptibility of microorganism (Balouiri et al., 2016).

Agar well diffusion technique for antimicrobial activity is more widely used compared to the disc diffusion technique. The procedure of agar plate preparation, the crude extract of fungal product preparation and spread plate method of inoculating the test microorganism are similar to disc diffusion technique. Then, a well or a hole is made with a sterile cork borer or a sterile tip with a diameter of 6–8 mm aseptically instead of introducing the disc. The fungal crude extract of 20–100 μL or 2–3 drops is introduced into the well and incubated at suitable conditions for the test microorganism. The extract containing antimicrobial compound diffuses into the agar and inhibits the germination and growth of the microorganism (Figure 10.5). The diameter of inhibition zone is measured for fungal product efficacy (Magaldi et al., 2004; Valgas et al., 2007; Balouiri et al., 2016).

Agar plug diffusion technique for antagonistic procedure is also similar to disc diffusion technique with respect to steps like agar plate preparation for the test microorganism to grow and spreading the test microorganism in the agar plate. But the well or disc is replaced by an agar plug. The fungus of interest is inoculated on the specific agar plate for its growth and secretion of fungal product and incubated. The exuded fungal product is diffused into the culture medium. After incubation, an agar plug is cut aseptically using a sterile tip or a cork borer from the diffused agar plate. The agar plug is introduced aseptically on the surface of the agar plate freshly inoculated with test microorganism, instead of using the crude extract for the fungal product testing. The deposited fungal product in the plug diffuses into below agar plate and effects test microorganism (Figure 10.6). The inhibition zone shows the efficiency of product (Balouiri et al., 2016).

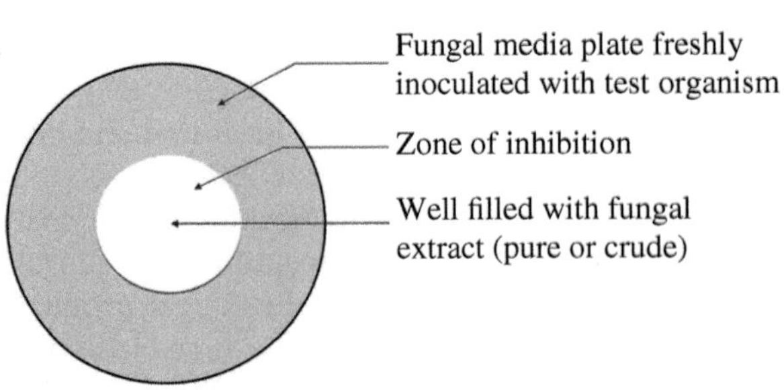

Figure 10.5 Screening of antimicrobial agent producers – Agar well diffusion method. *(See insert for color representation of this figure.)*

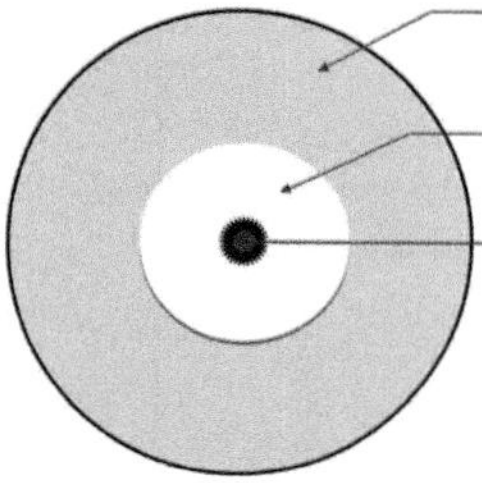

Figure 10.6 Screening of antimicrobial agent producers – Agar plug diffusion method. (*See insert for color representation of this figure.*)

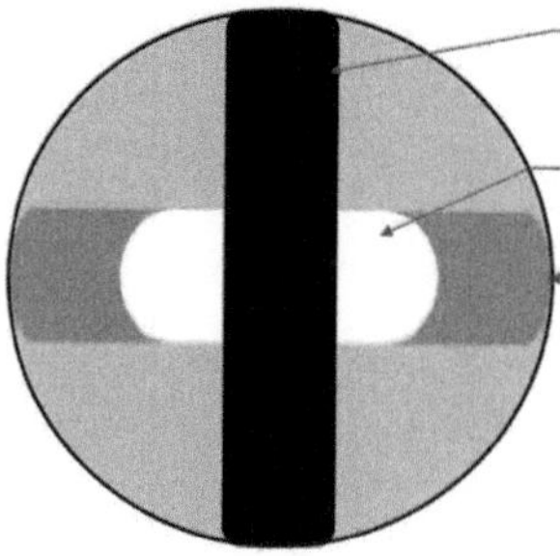

Figure 10.7 Screening of antagonistic agent producers – Cross streak method. (*See insert for color representation of this figure.*)

Cross streak technique for antagonistic activity method is simple and used for rapid screening. The fungal strain of interest is seeded by a single streak in the center of agar plate. After an incubation period of fungal strain, the same plate is seeded with test microorganism by single streak perpendicular to the central streak. Further incubation, the antimicrobial activities are analyzed by observing the inhibition zone size (Balouiri et al., 2016) (Figure 10.7).

The screening of enzyme producing fungi can be done by diffusion plate assay method. Enzymes are substrate specific. The substrate specific media plates are used for the screening. If the fungi utilize the substrate for its growth, then there will be a clear zone around the colony. This interprets that the fungi produced the specific enzyme as product. This is the easy and economic method of selecting the fungi which are of interest in producing desired enzymes. Qualitative assay for determination of some of the industrial and pharmaceutical important enzymes are listed in Table 10.1.

10.2.1.2 Thin Layer Chromatography (TLC)

Thin Layer Chromatography is used for the separation of biological and chemical compounds in the mixture or crude extract of fungal product. TLC is used for the rapid screening confirmation of fungal product. The crude extract is subjected to thin layer chromatography (TLC). Here the separation of compound in the mixture of crude extract with the help of solvent or mixture of solvents with respect to the product of interest on stationary phase (e.g. silica gel). The crude

Table 10.1 Qualitative assay for determination of enzymes.

Enzyme	Common substrate added to fungal culture media for screening	Principle involved (after incubation)	Reference
Protease	Protein (Casein/skimmed milk/ Gelatin)	Protein hydrolysis make clear zone formation around the colony.	Sharma et al. 2015
α-amylase	Starch	Starch hydrolysis causes hydrolytic zone around the colony visible by flooding the plate with 1% iodine solution	Khan and Yadav 2011; Karri et al. 2014
Lipase	Tributyrin/acrylglycerols/Tween 80/rhodamine-B	Appearance of hydrolytic zone	Griebeler et al. 2011; Pacheco et al. 2015; Narasimhan and Bhimba 2015
Pectinase	Pectin in Yeast extract Pectin agar medium/Citrus pectin in Mineral salt agar medium	Appearance of hydrolytic zone	Reddy and Sreeramulu 2012
Chitinase	Colloidal chitin	Appearance of hydrolytic zone	Thiagarajan et al. 2011; Natsir et al. 2010
Laccase	Guaiacol/Syringalazine	In Guaiacol agar media, Guaiacol is oxidized results reddish brown color appear around the colony; In Syringaldazine agar media, Syringaldazine (pale yellow color) is oxidized to purple color compound results appearance of purple color zone around the colony.	Kumar et al. 2011
L-asparaginase	L-asparagine + pH indicator (phenol red or Bromothymol blue)	appearance of color change of medium around the colony	Shirazian et al. 2016
L-glutaminase	L-glutamine + pH indicator (phenol red or Bromothymol blue)	appearance of color change of medium around the colony	Shirazian et al. 2016

sample is run against the standard compound (e.g. antibiotic or enzyme digest). Rf values were calculated to determine the product.

10.2.1.3 Chromatography-bioautography for Screening of Antimicrobial Activity

The combination of TLC-bioautography was introduced by Fischer and Lautner (1961), after the introduction of paper chromatography with contact bioautography by Goodall and Levi (1946) to detect different strains of penicillium This technique introduces a complex separation and detection system, called Bio-Arena by combination of TLC with bioautography. The procedure in bioautography is similar to agar diffusion technique. In diffusion, the fungal product is diffused from culture exudates to agar media, whereas in TLC-bioautography, the fungal product which produced in TLC plate or paper chromatogram diffuses from the chromatogram to agar media.

In Contact bioautography technique, the TLC plate or paper chromatogram is dipped into or sprayed with fungal crude extract of fungal product or fungal product itself. Then, TLC plate or paper is placed on freshly prepared test microbial agar medium for some minutes or hours to allow diffusion. After, the fungal product diffuses into the agar medium, TLC plate or paper is removed and agar plate is incubated. The zone of inhibition of growth is observed and is measured for analyzing the antibiotic property of fungal product (Choma and Grzelak, 2011; Balouiri et al 2016).

In immersion bioautography or agar-overlay bioautography technique, the TLC plate is first immersed in or cover with agar medium. This plate is dipped or sprayed or inoculated with spore suspension of fungus to be screening for the fungal product of interest. Then, the TLC plate is placed on the agar medium which is inoculated with test microorganisms and incubated. TLC plates are kept at low temperature for few hours before incubation enables the diffusion. This technique is a combination of contact and direct bioautography. The transfer of fungal product from TLC plate diffuse into the microbial agar medium like in contact bioautographic method and the agar layer of TLC remains onto the chromatogram surface during the incubation and visualization like in direct bioautography (Choma and Grzelak, 2011; Balouiri et al., 2016).

Direct bioautography is most widely applied among all bioautographic techniques due to its rapidity and versatility. The developed TLC plate is dipped or sprayed in microbial broth which has well grown microbial suspension or standardized microbial suspension to be screened for antimicrobial activity of the fungal product of interest which was run on developed TLC plate and incubated overnight in dark at 100% relative humidity. Because the silica of TLC plate absorbed the nutrients of broth, they in turn support growth of microorganisms on TLC plate itself. There will be formation of growth inhibition zone of microbes at antimicrobial compound (ran fungal product) is present. For better visualization of zone, TLC plate is sprayed with dehydrogenase activity-detecting reagents like tetrazolium salts such as *p*-iodonitrotetrazolium violet (INT - 2mg/mL) and further incubate the plate under suitable conditions. White bands or spots appear at inhibition zone against purple background where no microbial growth. Because the dehydrogenase produced by living microorganisms on TLC plate converted INT to formazan gives purple color. This is extensively used for testing antimicrobial activity (Suleiman et al., 2010; Choma and Grzelak, 2011; Balouiri et al., 2016).

10.2.1.4 High-performance Liquid Chromatography (HPLC)

High-performance liquid chromatography (HPLC; formerly known as high-pressure liquid chromatography), is used to separate, identify, and quantify each component in a mixture. It relies on pumps to pass a pressurized liquid solvent containing the sample mixture through a column filled with a solid absorbent material. HPLC is an advanced technique of column liquid chromatography. The solvent usually flows through the column with the help of gravity but in HPLC technique the solvent will be forced under high pressure up to 400 atmospheres so that sample can be separated into different constituents with the help of a difference in relative affinities. HPLC are different types like partition chromatography, normal phase chromatography, displacement chromatography, reverse phase chromatography, which are used for the separation or identification of the fungal primary of secondary metabolites (Frisvad, 1989). HPLC separations have theoretical parameters and equations to describe the separation of components into signal peaks when detected by instrumentation such as by a UV detector or a mass spectrometer.

10.2.2 Quantitative Control of Fungal Products

Quantitative control studies on fungal product start with the end of qualitative control which provides adequate information of which product to be studied for quantitative control, gives provision for the selection of fungal strain to be studied and even provides information about elementary nutritional and physiological needs to optimize them. Quantitative control is quantification of qualitative control for the optimization of fungal product yield in the fermentation process. Quantitative control methods are much more structured than qualitative control methods.

After a series of qualitative studies, the array of strains is selected for various fungal products. It is necessary to select the product of our interest for further studies based on its applications and to avoid time and cost consumption. The product selected is produced by an array of strains to segregate optimum producer by quantitative assay. Dilution methods (Choma and Grzelak, 2011; Balouiri et al., 2016) is one of the most common technique for the determination of MIC values, since they offer the possibility to estimate the concentration of the tested antimicrobial agent in the agar (agar dilution) or broth medium (macrodilution or micro-dilution). Selection of fungus producing optimum production (i.e. assay) to be assayed like filtrate or cell harvest.

In submerged fermentation, if the product is produced in to the media, we go for the filtration process and making filtrate. This can be achieved by centrifugation at 3000 to 5000 rpm, followed by filtration through Whatmann filter paper. The filtrate is collected and used for assay of quantitation of fungal product of interest (Emeko et al., 2015). Some examples are discussed below:

- Oxalic acid assay by spectrophotometric method is the technique used for determination of the fungal oxalic acid carried out by Jiang et al. 1996. 1ml of the filtrate was added to 0.5mL of 0.06M potassium dichromate. Then, 0.2mL of 2.5M H_2SO_4 and 0.1 mL of 3.28 x 10^{-4}M rhodamine B were combined and brought up to 10mL in a test tube and mixed thoroughly. The mixture was placed in a water bath at 90°C for 8 min. The reaction was quenched by cooling under tap water. The absorbance of the mixture was read at 555 nm against the blank solution. The oxalic acid produced was quantitated using a standard curve prepared with synthetic oxalic acid (Emeko et al., 2015);
- lipase titration is based on NaOH titration of fatty acids. The fatty acids release by the action of lipase extract from filtrate on fat (e.g. triglycerols of olive oil emulsifies in Arabic gum). One unit of lipolytic activity was defined as the amount of enzyme that releases 1 μmol of fatty acid per minute per mL of enzyme extract (1 U = 1 μmol/min/mL) under the test conditions (Burkert et al., 2004; Colla et al., 2016);
- phytase activity is determined by estimating the amount of liberated inorganic phosphate. The reaction mixture composes 0.9 mL of acetate buffer (0.2M, pH 5.5), 1mM phytate and 0.1 mL of the enzyme extract. After incubation for 30 min at 37oC, add 1 mL of 10% TCA (Trichloro acetic acid). The aliquot is analyzed for inorganic phosphate (Singh and Satyanarayana, 2006; Awad et al., 2014); and
- assay of mycophenolic acid: The filtrate has collected, and its pH was adjusted to 2.0. 50mL of filtrate was extracted twice with ethyl acetate (100mL/each extract). The organic phase is recovered, pooled and evaporated at reduced pressure in a rotary evaporator. The residue was dissolved in methanol (1mL). Then filtered through 0.45μm filter and measured by HPLC (High Performance Liquid Chromatography) (Patel et al., 2016);

The fungal biomass is harvested (cell harvest) after fermentation by centrifuging the fermentation broth and the supernatant is filtered through Whatmann filter paper NO. 1. The mycelia are washed twice with distilled water. They are transferred to lyophilizer*in vacuo* for drying to a constant weight. They are homogenized to powder for determination of various contents (Yousufi, 2012, Yang et al., 2014). Some examples for quantitative way of determination of fungal product by cell harvest methodology are discussed below:

- Quantitative determination of polysaccharides: The extraction of dry mycelial powder is prepared by taking mycelial powder and distilled water in a ratio of 1:5 for 3.5 hours in water bath at 80°C. Then the extract is centrifuged at 7000g for 10 min. Phenol- sulfuric acid method is used for the determination of polysaccharides (Yang et al., 2014);
- quantitative determination of protein: The protein content of single cell protein (SCP) produced by filamentous fungi and various extracts protein content is studied through conventional method like AOAC method, flow cytometry, mass cytometry, affinity arrays, mass spectrometry and even through genetic and chemical probes. The extraction is prepared, and the amount of protein content was estimated by above methods (Yousufi, 2012, Wu and Singh, 2012).
- Quantitative determination of Adenosine: The extract is prepared by 0.1g of dry mycelial powder with 5mL of distilled water for 3 hours in water bath at

45°C. Then the extract is centrifuged at 12000g for 10 min. the supernatant is filtered with 0.45 μm membrane. The filtrate is analyzed for adenosine by HPLC (Yang et al., 2014); and

- quantitative determination of mannitol: The extract is prepared by 0.1g of dry mycelial powder with 5mL of distilled water for 3 hours in water bath at 45°C. Then the extract is centrifuged at 8000g for 10 min. The supernatant is collected and diluted 50 times. Then the amount of mannitol is determined by spectrophotometric method (Yang et al., 2014).

10.2.3 Speeding Up Fungal Product

One of the most important concerns for industry is the speeding up of fungal. It can be achieved by the optimization of fermentation media. Optimization of fermentation media processes is accomplished either by one-factor-at-a-time approach (OFAT) or multiple factor or parameter by statistical methods. Some common biostatics methods are central composite design (Patel et al., 2016), response surface methodology (Emeko et al., 2015, Colla et al., 2016), artificial neural network (Emeko et al., 2015) and box-Behnken factorial experimental design (Awad et al., 2014).

Optimization of the fermentation media composition for classically, optimization of fermentation media processes is accomplished by one-factor-at-a-time approach (OFAT) (Emeko et al., 2015). Optimization of fermentation variables in productivity of fungi is critical to exploiting its yield.

OFAT is conventional method of optimization of fermentation medium. These fermentations are carried out in duplicate or triplicate.

Carbon source is a major nutrient for fungal growth and production of metabolites. There are list of specific carbon sources available for optimization. They are glucose, sucrose, fructose, maltose, lactose, galactose, glycerol, xylose, myo-inositol, glycerol, sorbitol, mannitol, starch, cellulose, cellobiose, and carboxymethyl cellulose, etc. They are studied usually at an initial concentration of 1% of w/v for optimizing the carbon source. Then the carbon source that yields the best productivity is selected to examine at various initial concentrations. The concentrations usually studied are 0.25% to 1.0% of w/v with variance of 0.25% or 0.1% to 1.0 % with variance of 0.1%. In some cases, the optimization of the carbon source and the concentrations are done simultaneously for optimum production of fungal product. The optimization of carbon source reveals the best source and optimization of concentrations reveals the best minimal concentration of the carbon source for productivity. The remaining media composed ingredients other than carbon source; their physiological conditions and fermentation duration are selected according to selected fungal media at qualitative studies. The carbon source and its concentration that yield high productivity in all the variable carbon sources and variable concentrations is selected for further studies (Lata et al., 2013; Patel et al., 2016).

Nitrogen source is the major ingredient in nutritional composition after carbon source. The optimization of nitrogen source is carried out using both synthetic inorganic nitrogen sources like ammonium chloride, ammonium phosphate, ammonium sulphate, ammonium acetate, ammonium oxalate, ammonium ferrous sulphate and sodium nitrate and organic sources like casein, urea, peptone, yeast extract, beef extract, malt extract, soybean meal, tryptone and baker's yeast. For optimizing the nitrogen sources, the carbon source is considered as glucose in some cases and in other cases the optimized carbon source with its optimized concentration. The initial concentration of various nitrogen sources is of 6% of w/v. The optimized nitrogen source that may be organic or inorganic is selected for optimizing its concentrations. The concentrations usually studed are 0.25% to 1.0% of w/v with variance of 0.25% or 1.0% to 6.0% with a variance of 1.0% in case of organic nitrogen source and 0.025% to 0.1% of w/v with variance of 0.025% in case of inorganic nitrogen source. The remaining ingredients of the media, fermentation conditions and duration are as specified in qualitative studies of fermentation media (Lata et al., 2013; Awad et al., 2014; Patel et al., 2016).

Phosphorus is also important nutrient for fungus. There are different sources of phosphorus available like KH_2PO_4, K_2HPO_4, Na_2HPO_4, NaH_2PO_4 for optimizing the fermentation media. The initial concentration of various phosphates is 0.6% of w/v. The optimized phosphate source is selected for optimizing the best concentration. The concentrations usually studied are 0.1% to 0.6% with a variance of 0.1%. Optimum concentration of optimized phosphate source is used for fermentation. The other components of media, fermentation conditions and duration are followed as specified in the qualitative studies of fermentation media (Patel et al., 2016).

Different metal ions are required for the fermentation process to happen. The common metal ions usually involved in the growth of fungi and production of specific fungal products are Fe^{+2}, Mg^{+2}, Zn^{+2}, Cu^{+2}, Mn^{+2}, Ca^{+2}, Na^{+}, K^{+}, Co^{+2}, Ni^{+2}, Ba^{+2}, Pb^{+3}, Hg^{+2} and Ag^{+2} as chlorides or sulphates or nitrates occurring to availability and requirement. They are optimized for fungal fermentation by supplementing them each with initial concentrations of 0.1% of w/v along with trace metals composition of selected media through quantitative control of fungal product. The optimized metal ion is selected for optimizing the concentration. The concentrations are 0.025% to 0.1% with a variance of 0.025%. The rest of the media components are as specified in the qualitative study for the selection of fermentation media for definite fungal product (Lata et al., 2013).

Optimization of fermentation kinetics with One-factor-at-a-time method is also a very important factor for speeding up of fungus product. This is conventional method as described at optimization of fermentation media (Irfan et al., 2014). The factors of fermentation kinetic studies for optimizing are initial pH, incubation temperature, inoculum size, aeration and agitation and time course of fermentation. These factors are each studied one at a time for optimizing. So, this is an easy method. These fermentations are carried out in duplicate or triplicate.

Initial pH of fermentation media is the most important factor for growth and product output. So it's very important to optimize its pH for fermentation. The composition of media for fermentation kinetics is selected from the designed media through optimization of fermentation media, but in the,case of the fermentation media not being optimized, then the selected fungal media through qualitative control of fungal product is used for these studies. The initial pH is adjusted in a range of 4-9 with variance 1. The rest of the fermentation conditions are followed with the qualitative control reports of the fungal product (Siddalingeswara and Lingappa, 2010; Lata et al., 2013).

The optimum incubation temperature for production of fungal product is determined in a range of 25°C to 50°C with variance of 5°C (Siddalingeswara and Lingappa, 2010; Lata et al., 2013).

Optimum productivity of inoculum size is investigated by selecting different inoculum concentration of 1,2,3,4, and 5% of v/v or 1×10^5 to 1×10^6 at different levels of variance. The remaining conditions are followed as above (Siddalingeswara and Lingappa, 2010; Lata et al., 2013).

The aeration and agitation is achieved by shaking an incubator for laboratory studies or for batch fermentation. The agitation speed also affects the availability of oxygen for fungal growth and production. So, the optimization is carried out by studying aeration at 0.5, 1.0 and 1.5 vvm at different agitation speeds of 100, 200 and 300 rpm correspondingly (Bakri et al., 2011).

The incubation period of fermentation is very much reflected on the type of production. Primary metabolites are produced at log phase, so the harvest of the product can be optimized at early days of fermentation whereas secondary metabolites are produced at stationary and declining phase so the harvest can be optimized late days of fermentation. Usually the optimization of time period of fungal products study is investigated on fourth, fifth, sixth, seventh and eighth days of incubation. In some case, Maximum yield of xylanase enzyme was achieved from 14 to 17 days fermentation time period of *Trichoderma viride*. Increased fermentation time period results the decrease of fungal product yield due to nutrient depletion and further alters the physiology of fungi and inactivation of secretary system. (Rengeet al., 2012; Irfan et al., 2014).

Solid state fermentation is an alternative method of fermentation to submerged fermentation where the substrate is in solid form. The composition of media is not defined. The fermentations are mainly preferred on solid states because they are cheap, waste, natural products and have mixed nutrients. Solid state fermentations customs culture substrates with reduced water activity. Optimization of substrate for fungal product is very much required because it's undefined for application in fermentation.

Advantages of solid state (SSF) over submerged fermentation and factors effecting solid state fermentation (Hölker and Jurgen, 2005; Irfan et al., 2014) are below:

- High volumetric productivity;
- relatively high concentration of products;
- less effluent generation;
- requirement of simple fermentation equipment;
- lower capital investment; and
- lower operating cost.

Factors influencing SSF (Pandey, 2003; Irfan, 2014) are listed below:

- Selection of suitable substrate;
- selection of suitable microorganism;
- pre-treatment of substrate;
- particle size of substrate;

- water content and water activity of substrate;
- relative humidity;
- type and size of inoculum;
- control of temperature of fermenting matter/ removal of metabolic heat;
- period of cultivation;
- maintenance of uniformity in the environment; and
- gaseous atmosphere (O_2 consumption and CO_2 evolution rate).

Screening for optimization of solid substrate is carried out based on the fungal product of interest. Some very common solid substrates are listed in Table 10.2. Each solid dry substrate (5 grams) with fungal media composition which was dissolved in calculated volume of water in 250 ml flask. Fungal media composition and fermentation conditions are followed as per qualitative control studies of that fungal product. The substrate media which produce optimum yield of the product is selected.

The percent of moisture content is calculated as volume of medium divided by total weight of dry solid substrate and medium volume (Awad et al., 2014).

Nutrient value of the solid substrate and fungal product in consideration, wide combinations of solid substrates (for e.g. Corn cob and corn bran in a ratio of 1:4, 2:3, 2.5:2.5, 3:2 and 4:1 were studied for phytase production) are used. The optimization can be achieved by using different individual substrates or mixture of different substrates and at different concentration ratios. These substrates can be wetted with water or selected fungal broth with respect to the product of interest and productivity of yield (Pandey et al., 2008; Awad et al., 2014).

If the yield with nutrients is better than without the fungal media nutrients, the optimization of the fermentation media in solid state fermentation also required. The optimization nutrients for solid state fermentation procedure are as same as optimization of fermentation media for submerged fermentation. Except, the supplementation of optimized solid substrate and moisture content are as given at solid substrate optimization (Pandey et al., 2008; Awad et al., 2014).

In addition, the moisture content of the media is greatly involved in solid state fermentation. The moisture is most important thing in fungal fermentation. The amount moisture utilization is very less compared to submerged fermentation. But the requirement varies with the fungi and its product. The initial amounts of moisture used for optimization are 33%, 50%, 58%, 66% and 75%. The optimized moisture content at which fungi produced high yield of product is used for further studies (Awad et al., 2014).

The process of optimization of fermentation kinetic studies (initial pH, incubation temperature, and fermentation period and inoculum size) in solid state fermentation is as same as submerged fermentation.

10.3 Fungal Morphology and Product Spectrum: a Representative Theme

Fungi are eukaryotic multicellular living organisms as like plants and animals. Unlike these other groups, however, fungi have filaments called hyphae (singular: hypha); their cells are long and thread-like and connected end-to-end. Because of

Table 10.2 Solid substrates used for fermentation.

Product		Solid substrate for fermentation	Reference
Cell harvest	SCP	Inedible parts of rapeseed, Corn cobs, Sorghum hydrolysate, cellulosic waste, Prawn shell wastes, citrus fruit peel, sugar cane bagasse	Rhishipal and Philip, 1998; Anupama and Ravindra 2000
Primary metabolite	Alcohol	Grape pomace, maize starch, sugarcane, sugar beet, Tapioca, sweet sorghum, apple	Hang et al., 1982; Yu et al., 2008; Thomas et al., 2013
Primary metabolite-Organic acids	Citric acid	Peat Moss (with glucose), Treated cassava Bagasse, fruits waste,sugar cane bagasse, corn husk, pineapple waste, apple pomace, carob pod, coffee husk, grape pomace, Mussel processing wastes, rice bran	Pandey et al., 2008; Thomas et al., 2013
	Lactic acid	sugar cane bagasse and cassava bagasse hydrolysate	
	Oxalic acid	sweet potato	
	Gluconic acid	fig	
Primary metabolite-Enzymes	Xylanase	Eucalyptus pulp, Bagasse pulp, corn cobs, Wheat bran, Rice bran, Rice straw and wheat bran, sorghum flour, wheat straw	Pandey et al., 2008; Thomas et al., 2013,
	Neutral protease	Wheat bran, coconut oil cake and wheat bran	
	Alkaline protease	Wheat bran and soy protein, Green gram husk,	
	Pectinase	Deseeded sunflower head, grape pomace, wheat bran and orange bagasse,	
	Lipase	wheat bran and olive oil, soy cake, Babassu oil cake, sugar cane bagasse, almond meal	
	Lovastatin	Impregnated liquid media with artificial inert support, Rice with only nutrients in the seed culture	Pandey et al., 2008; Thomas et al., 2013
	Cephalosporin C	Wheat rawa with soluble starch, yeast extract and minerals	
	Pencillin G	Impregnated liquid media with natural inert support (Sugarcane bagasse)	
	Griseofulvin	Rice bran with glucose, yeast extract, minerals and precursor, Millet	
Antibiotics and other secondary metabolites	Gibberellic acid	Wheat bran with soluble starch and minerals	Pandey et al. 2008, Thomas et al. 2013
	Nigerloxin	Wheat bran with trisodium citrate	
	Cyclosporin A	Wheat bran with millet flour and jowar flour	
	Mycophenolic acid	Wheat bran with Mannitol, ragi flour, millet flour, $(NH_4)2HPO_4$, casein hydrolysate	
	Monacolin K	Rice	

this diffuse association of their cells, the body of the organism is given the special name mycelium, a term which is applied to the whole body of any fungus (Byrne and Ward, 1989). When reproductive hyphae are produced, they form a large organized structure called a sporocarp, or mushroom. This is produced solely for the release of spores, and is not the growing portion of the fungus. A hyphal element consists of a main hypha, usually with a number of branches, branches of branches, etc. that originates from one spore. A flock is a loosely packed, temporary agglomerate of hyphal elements (Pazouki and Panda, 2000). A pellet or layer is a dense and, under normal process conditions, almost permanent configuration of hyphae or hyphal elements (see Figure. 10.8). In addition to being filamentous, fungal cells often have multiple nuclei. In the chytrids and Zygomycetes, the cells are coenocytic, with no distinction between individual cells (Nielsen, 1992). Rather, the filaments are long and tubular, with a cytoplasm lining and large vacuole in the center. By contrast, the ascomycetes and basidiomycetes are septate; their filaments are partitioned by cellular cross-walls called septa. The structure of these septa varies, and is taxonomically useful. Another feature of fungi is the presence of chitin in their cell walls. This is a long carbohydrate polymer that also occurs in the exoskeletons of insects, spiders, and other arthropods. The chitin adds rigidity and structural support to the thin cells of the fungus, and makes fresh mushrooms crisp. Several fungi can grow in submerged culture with different forms ranging from dispersed filaments to pellets (Byrne and Ward, 1989; Nielsen, 1992; Pazouki and Panda, 2000). Cellular growth and morphology of fungi are affected by several factors which are level and type of inoculum, medium shear, medium constituents and the pH (Blakebrough and Hamer, 1963; Atkinson and Daoud, 1976). Numerous fungal fermentation requires a morphology for optimum product yield. Pelleted growth is preferred for itaconic acid production by *Aspergillus terreus* and citric acid by *Aspergillus niger* (Metz et al., 1979), whereas pelleted plus filamentous growth have been found suitable for penicillin production (Metz Kossen, 1977; Pazouki and Panda, 2000).

Most members of the kingdom Fungi lack flagella; the structures are completely absent in all stages of their life cycle. The only exception is the chytrids, which produce flagellated gametes. The absence of flagella then, is a synapomorphy which unites all the remaining groups of fungi. This has had a tremendous impact on fungal biology, because it means that no fungus can produce motile

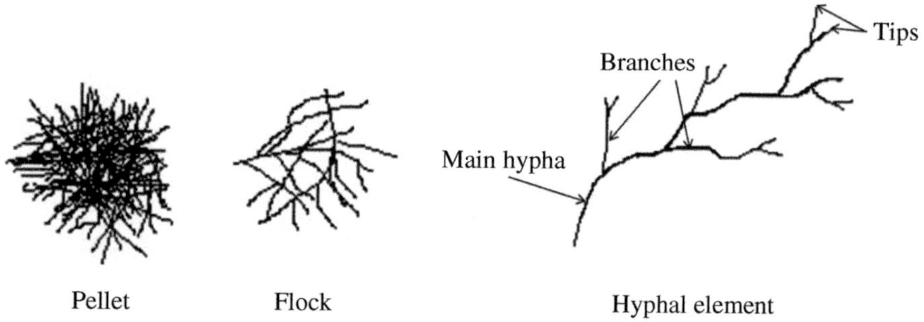

Figure 10.8 Fungi morphology: hyphal elements.

gametes, and two organisms must therefore come into direct physical contact to effect sexual reproduction (Walker and White, 2005). The protoplast of fungal cells is typically eukaryotic containing membrane-bound nucleus and other cell organelles, like mitochondria, rough and smooth endoplasmic reticulum, microtubules, Golgi bodies etc. Ribosomes are of 80S type. Vacuoles are often present occupying the major part of the cells, pushing the cytoplasm to the periphery. Cytoplasmic streaming is also observed. As reserve material, fungi generally accumulate glycogen which is a branched polymer of glucose. Fats and oils are also often present (Walker and White, 2005).

One of the industrially important morphological characters of filamentous fungi in submerged fermentation is fungal pellets. Filamentous fungal pellets are generally globular or spherical appearance with short, fluffy, loose, swollen filaments with swollen tips. They are coagulative and non-coagulative pellets based on their process of formation caused by driving forces. Coagulative pellets are formed by the cultivation of spore aggregate within 6 to 8 h, followed by germination and hyphal growth or by already aggregated spores germinate and form hyphae are pellets. Several spores form one coagulative pellet. Coagulative pellets are generally used in citric acid production. The non-coagulative pellets are formed by initial spore germination followed by pellet formation. One spore forms one non-coagulative pellet. The advantages of fungal pellets are low viscosity of fermentation media, easy to harvest, high productivity These pellets are extensively used in citric acid production especially from *Aspergillus niger* (*A. niger* pellets), yield 99% of global citric acid production which counts about 1.5 million tonnes per year. The factors influencing the fungal pellet formation to optimize the product yield are inoculum size, aeration, agitation, pH, temperature, trace metals, additives, nucleating agents, type of reactor, etc. These factors cause organism morphology from productive pellet form to unproductive filamentous form. (Gomez et al., 1988; Papagianni and Mattey, 2006; Berovic and Legisa, 2007; Zhang and Zhang, 2015).

Microbes are the leading producers of important useful natural products (Table 10.3). Natural products from microbes and plants make excellent drugs. Significant portions of the microbial genomes are devoted to the production of several useful secondary metabolites. A single microbe can make a number of secondary metabolites, as high as 50 compounds (Demain, 2014). The most useful products include antibiotics, anticancer agents, immunosuppressants, but products for many other applications, e.g. antivirals, anthelmintics, enzyme inhibitors, nutraceuticals, polymers, surfactants, bioherbicides, and vaccines have been commercialized (Demain, 2014).

Fungi are morphologically microscopic and macroscopic. Mushrooms and truffles come under macroscopic fungi which consist of several edible species used for delicacies. Molds or filamentous fungi and yeasts come under microscopic and have industrially elaborate applications in food, pharmaceuticals, industries and research (Molitoris, 1995; Adrio and Demain, 2003; Kavanagh, 2005).

Fungi are used in many industrial processes, such as the production of enzymes, vitamins, polysaccharides, polyhydric alcohols, pigments, lipids, and glycolipids. Some of these products are produced commercially while others are potentially valuable in biotechnology (Table 10.4). Fungal secondary metabolites are extremely important to our health and nutrition and have tremendous

Table 10.3 Example of microbial product and process.

Category	Sub-category	Examples
Commodities	Food additives	Organic acid, amino acids, polyunsaturated fatty acids, thickening agents, vitamins etc.
	Solvents	Acetone, butanol, ethanol etc.
	Bulk enzymes	Serine proteases, amylases, lipases, cellulases, chymosin
	Whole cells	Bakers' yeast
	Biofuels	Ethanol, biodiesel, methane, hydrogen
	Agrochemicals	Feed additives, biopesticides
Processes	Waste water treatment	Sewage treatment
	Bioremediation	Detoxification of hazardous wastes including hydrocarbons, halogenated hydrocarbons and metals
	Biobleaching	Removal of lignins from wood pulp in paper industries
	Biofiltration	Removal of volatile organic solvents from air
	Biocatalysis	Production of optically pure chiral molecules
Fine chemicals	Pharmaceutical	Antibiotics, antifungal, chemotherapeutic agents
	Laboratory reagents	Enzymes, biochemicals, whole cells
	Diagnostic reagents	Purified antigen, nucleic acids, proteins
Emerging technology	DNA arrays	Research and diagnostic
	Biosensor	
	Specialized research services	

economic impact. In addition to the multiple reaction sequences of fermentations, fungi are extremely useful in carrying out biotransformation processes (Adrio et al., 2003).

Based on the commercial importance, industrial application and critical need, the fungal product is selected from the database of fungal metabolites. The product may be primary metabolite or secondary metabolite for studying the qualitative role in fermentation. The better understanding about the applications of the product, economic strategies and essentiality for the human welfare leads proper selection of fungal product.

10.4 Effectiveness of Qualitative Domain for Different Microorganisms

Microbes have been very important in the production of natural product drugs. Of 23,000 active compounds from microorganisms, i.e. antimicrobials, antivirals, cytotoxic, and immunosuppressive compounds, 42 % are made by fungi and

Table 10.4 Fungal product spectrum*.

Era	Process or product	Organisms
Premodern' fungal products (traditions of the 19th century)	Asian food fermentations	
	Ang-kak	*Monascuspurpurea*
	Miso	*Aspergillus oryzae*
	Ontjam	*Neurospora crassa*
	Soy sauce	*Aspergillus oryzae, A. sojae*
	Tempeh	*Rhizopus niveus*
	Brewing and baking	*Saccharomyces cereviseae, S. carlbergensis*
	Mold-ripened cheeses	*Penicillium roqueforti, P. camembetii*
	Mushroom cultivation	*Agaricusbisporus, Auriculariasp., Flammulinavelutipes, Lentinusedodes, Pleurotussp., Volvariellavolvacea*
'Modern' fungal products (traditions of the 20th century)	Antibiotics and other pharmaceuticals	
	Penicillins	*Penicillium chrysogenum*
	Cephalosporin	*Cephalosporium acremonium*
	Cyclosporin (immunosuppressive)	*Tolypocladiuminflatum*
	Ergot alkaloids	*Clavicepspurpurea*
	Griseofulvin	*Penicillium griseofulvin*
	Mevalonin	*Aspergillus terreus*
	Enzymes	
	a-Amylases & Invertase	*A. niger, A. oryzae*
	Cellulase	*Humicolainsolens, Penicillium funiculosum, Trichoderma viride*
	Glucoamylases	*Aspergillus phoenicis, Rhizopus delemar, R. niveus*
	Glucose oxidase	*A. niger*
	Laccase	*Coriolusversicolor*
	Pectinase	*A. niger, A. oryzae, Humicolainsolens*
	Proteinases	*A. oryzae, Aspergillus melleus, R. delemar*
	Rennin (microbial)	*Mucor miehei, M. pusillus*
	Organic acids	
	Citric acid	*A. niger*
	Itaconic acid	*A. terreus*

**Source*: Bennett, 1998.

32 % by filamentous bacteria, the actinomycetes. Out of about 1 million natural products, approximately 25 % are biologically active, i.e. show positive activities or toxicity. About 60 % of these are from plants and most of the rest from microbes; some are from animal sources. We already know the structures of over

160,000 natural products, about half from plants and half from microbes. This figure grows at about 10,000 per year. So effectiveness of qualitative domain for different microorganism will be achieved generally by the strain improvement and with suitable fermentation media composition and their fermentation kinetics.

Strain improvement leads to the improvement and manipulation, results of boosting metabolic activities of microbial strain for various industrial and biotechnological applications. The need of strain improvement is found to be increased productivity with limited substrate leads resulting in manufacturing cost reduction. Advantages of strain improvement are increased quantity of metabolites, revertants of non-producing strains showed the productivity, resistant to nutritional repression, improved production on agar, enhance the permeability leads to metabolite excretion, qualitative changes in the mixture of fermentation products leads elimination of undesirables in the mixture, helps to elucidate biosynthetic pathways. Strain improvement includes natural recombination, mutagenesis and genetic engineering (Adrio and Demain, 2006; Pathak et al., 2015).

Natural recombination deals with transfer of genetic material. This is carried out in fungi mainly by transformation and through protoplast fusion. Transformation of fungal strain achieves the modification of existing properties, introducing new characteristics without adversely affecting other desirable properties or eliminating of unwanted properties. Increased yield of inositol independent transformants were achieved through transformation (Mishra and Tatum, 1973; Pathak et al., 2015).

Protoplast fusion is used mainly to improvement industrial important yeasts and filamentous fungi. This is proficient technique to induce heterokaryon formation and recombination with high incidence. Medically important antibiotics were achieved through protoplast fusion and further recombination. This technique is carried out mainly between different mutant lines. Penicillin production yield was improved from 12g/L to 18g/L through protoplast fusion of *P. chrysogenum* (Adrio and Demain, 2006; Pathak et al., 2015).

Mutagenesis simple, conventional and effective method of improving industrially important microbes. This is carried out as induced mutagenesis by repeated mutations, followed by strain selection. The mutagens are two based on its nature, physical mutagens and chemical mutagens.

Physical mutagens are ultraviolet (UV) rays, gamma rays and X-rays. UV rays are extensively used because of more effectiveness due to shorter wave length (200-300 nm) and low penetration power avoids damage to genome. UV irradiation of *Aspergillus niger* led 2 fold increase of CMCase activity, *Pencilliumchrysogenum* led high cellulose activity, *Rhizopus sp.* yielded high lipase and glucoamylase activites (Pathak et al., 2015). Whereas gamma irradiation on *Aspergillus niger* also yielded 2 to 2.5 fold increase in production of glucoamylase (GA).

More effective chemical mutagens are ethyl methyl sulphonate (EMS), methyl methane sulphonate (MMS), nitrosoguanidine (NTG), 4-nitroquinolone-1-oxide, hydroxylamine (HA) and ethidium bromide. The combined treatment of chemical mutagens and UV irradiation on *Aspergillus awamori* resulted increase yield of amyloglucosidase and on *Aspergillus niger* yielded 2 to 4 fold increase productivity of glucoamylase.

Genetic engineering is another most effective tool for improvement of strain and enhances the fungus production (Adrio and Demain, 2006).

Production of fungal product is mainly based on the nutrient medium it grows. It is the major factor in quality of fungal product. Common media for isolation of fungi are Potato Dextrose Agar (PDA), CzapekDox Agar (CZA), Sabouraud's Dextrose Agar(SDA), Malt extract Agar(MEA), etc. The given media without agar is called broth which is used as fermentation media. This type of fermentation is called submerged fermentation. Based on the media composition, quality and the yield of the fungal product change. So fungal product is not only specific with organism, but it is also specific with nutrient media. In this step, the media for the product fermentation will be selected by trial and error method (Kavanagh 2005). The media which showed better respond and quality will be selected for the quantitative fermentation studies.

The study of environmental factors and their causes in fermentation is called fermentation kinetics. Microorganisms grow, reproduce and harvest active products for industrial and commercial exploitation when they are present in physiologically friendly environment. Number of factors affect productivity of the fungus, regardless of nature of fermentation process or type of fungal cells used (free or immobilized). The requirements of the process lead to the selection of fermentation system. The major factors comes under fermentation kinetics are pH, temperature, aeration and agitation, time period and inoculum size. Variation in this diversity and the factors that cause it will impact on quality of the fungal product (Kavanagh, 2005).

One of the main factors influencing the environment of fungal fermentation is pH or hydrogen ion concentration. pH of fermentation media influence the growth of the fungi and production rate. Therefore, changes in pH of the medium can shift metabolite product. In case of organic acid production, shift of pH occurs due to control and balance of NADH, a key in the metabolic pathways of fungal cells in fermentation. The fermentation media have low pH (more acidic) or high pH (more alkaline) does not support growth and enzyme production. Fungi are well-known to grow and secrete the desired products in acidic pH. Even though, *Fusarium globulosum* grows at pH range of 3 to 10, it produces lipase at pH 7 i.e. neutral pH (Kavanagh, 2005).

Usually the fungi are mesophilic which grows at 25 to 35°C of temperature. However, each fungus shows its optimum growth at specific temperature and its productivity to specific fungal product at specific temperature. The temperature of the fermentation affects the growth of fungus, type of product, effectiveness of the product and product yield of fungi. Temperature traffics the sensitivity of yeast to product concentration, growth rate, rate of fermentation, viability, length of lag phase and membrane function. Hence, the temperature of the incubation during fermentation dictates the achievement of good yield in harmony with growth of fungal biomass (Kavanagh, 2005).

For better aeration or oxygen transfer, agitation of the fermentation media is required which is achieved through shake flasks. They are agitated by shaker incubator as orbital or linear shaking which provides surface aeration. This aeration and agitation of shake flasks are influenced by flask size, fill volume, construction material, shaking frequency and type of sealing plug. Shake flasks are easy to operate and are in use up to 90% of industrial and research processes of fungal products. The usual agitation for the shaker flasks is in the range of 30 to 200 rpm per 100mL. This is adequate enough for batch fermentation without constraining microbial growth (Kavanagh, 2005; Betts and Baganz, 2006).

For fermentation process is also influenced by initial inoculum. Freshly grown cultures have the vital cells to attain the optimum yield. The inoculum preparation is achieved by

- Unfiltered inoculum consisting of hyphae and microspores.
- Filtered inoculum consisting only microspores filtered in Whatmann filter.

Some other methods which are in regular practice are spore suspension technique, spectrophotometric technique and haemocytometer technique. All these techniques used to accomplish the specific concentrations required for the fermentation (cfu/mL). In spore suspension method, the homogenous inoculum attained by volume of the media for inoculum preparation makes it constant for acquiring the required concentration. In spectrophotometric method, the required concentration is attained by reading the optical density of inoculum at 530nm to obtain standardized preparation. In haemocytometer method, the standard concentration is obtained by counting the spores or conidia against haemocytometer (Kavanagh, 2005).

10.5 Emphasizing the Need: Qualitative and Quantitative Importance

Any microorganism can be symbolized as a factory where thousands of responses and reactions occur in or outside of the cell. Those extracts have enormous applications for human welfare. So the qualitative study employs detection of the interested product, effect of factors like nutritional, physiological and environmental conditions for reaction to happen. The quantitative study employs high volumetric production, relatively higher concentration of fungal products, and generation of less effluent, requires simple fermentation, lowers the capital investment and even lowers the operating cost (Norouzian, 2008). The qualitative and quantitative control of fungal products is significantly increased due to high need for potential bioactive compounds and their elaborative applications, simple to culture and GRAS (Generally Regarded as Safe) (Kavanagh, 2005). The nature is sole and rich platform of all the potent microorganisms which are unrevealed. These can only can be brought about through qualitative and quantitative studies.

Currently, interest is growing in the research and development of new natural agents which can counterattack diseases, microbial resistance; industrial applications to reduce or avoid chemical usage and develop an eco-friendly environment; to screen biodegradable agents, etc. Have become of universal concern to the public (Balouiri et al., 2016). This can only be achieved through qualitative and quantitative studies.

These are basic, reliable, explorative, preliminary, introductory, cheap economically, time saving, requires no specific sophisticated equipment and are easy to handle (Choma, 2011). Some equipment like GC-MS and HPLC can also be used for screening and quantification. While qualitative studies disclose the potent fungal products through screening and detection, quantitative studies brings blooming through optimization and strain improvements.

The qualitative study reveals the characteristics of fungal products and leads the way to need for quantitative studies. The quantitative study reveals the conditions required for the best production and leads to the way for impending large scale production to success. Different laboratory methods are in use to evaluate or detect and quantify the nature of fungal product of extract or pure compound. The most common and known method of qualitative studies is diffusion method and for quantitative studies is through spectrophotometry.

Review of literature revealed that the results of comparison between microbial products became difficult. This is due to the use of diverse non-standardized methods in qualitative and quantitative studies of different microbial products like inoculum preparation, inoculum size, growth medium, pH, temperature and time of incubation and end points determination. (Balouiri et al., 2016). Thus the need prompted the studies to standardize qualitative and quantitative control.

Optimization of productivity of fungal product in both submerged and solid state fermentation leads to the path of high yield at economic conditions.

Strain improvement through mutations and recombinant DNA technologies also plays a greater role in qualitative and quantitative studies for yield of fungal product. Mutant strain screening studies revealed that mutations alter secretion properties causes improvement of yield of final product. Successive metabolic engineering also provides increased yield of final product (Adrio and Demain, 2003).

Owing to the new charm of the properties of new fungal products either antibiotics, enzymes, pigments, organic acids, fatty acids or hormones, etc., it is substantial to develop better understanding of the current procedures available for screening and/ or quantifying the effect of fungal products of an extract or a pure compound for its applications in medicine, food, beverage, agriculture, industry and ecosystem (Balouiri et al., 2016).

Therefore, a greater devotion has been paid to fungal products screening, evaluating and quantifying methods.

10.6 Conclusions

Microbes are the leading producers of useful natural products. A single microbe can make a number of secondary metabolites, as high as 50 compounds. Fungi are used in many industrial processes, such as the production of enzymes, vitamins, polysaccharides, polyhydric alcohols, pigments, lipids, and glycolipids. Fungal secondary metabolites are extremely important to our health and nutrition and have tremendous economic impact. In addition to the multiple reaction sequences of fermentations, fungi are extremely useful in carrying out biotransformation processes. Several factors influence the qualitative and quantities control for the fungal product. Among the qualitative prospect diffusion method, thin layer chromatography HPLC and Chromatography-bioautography. Quantitative enhancement of fungal product can be achieved by strain improvement by mutation and recombinant methods and by the optimization of fermentation media and kinetics.

References

Aaronson, S. (2000) Fungi. In: K.F. Kiple and K.C. Ornelas (Eds), *The Cambridge World History of Food* (pp. 313–336). Cambridge, UK: Cambridge University Press.

Adrio, J.L., Demain, A.L. (2003) Fungal biotechnology. *International microbiology*, 6:191–199.

Adrio, J.L., Demain, A.L. (2006) Genetic improvement of processes yielding microbial products. *Federation of European Microbiological Societies Microbiology Review*, 30:187–214.

Anupama, R.P. (2000) Value-added food: Single cell protein. *Biotechnology Advances*, 18: 459–479.

Atkinson, B., Daoud, L.S.(1976) Microbial focs and focculation in fermentation process engineering. *Adv. Biochem. Eng.* (Ghose, T.K.; Fiechter, A.; Blakebrough, N. eds), 4, 41–124, Springer-Verlag, Berlin.

Awad, G.E.A., Mohamed, M.I., Helal, E.N. et al. (2014) Optimization of phytase production by GE1 under solid state fermentation by using Box–Behnken design, *Saudi Journal of Biological Sciences*, 21(1): 81–88.

Bakri, Y., Mekaeel, A., and Koreih, A. (2011) Influence of agitation speeds and aeration rates on the Xylanase activity of Aspergillus niger SS7. *Brazilian Archives of Biology and Technology*, 54(4): 659–664.

Balouiri, M., Sadiki, M., and Ibnsouda, S.K. (2016) Methods for in vitro evaluating antimicrobial activity: Areview. *Journal of Pharmaceutical Analysis*, 6: 71–79.

Bennett, J.W. (1998) Mycotechnology: the role of fungi in biotechnology. *Journal of Biotechnology*, 66: 101–107.

Betts, J.I. and Baganz, F. (2006) Miniature bioreactors: current practices and future opportunities. *Microbial cell factories*, 5(21): 1–14.

Berovic, M. and Legisa, M. (2007) Citric acid production. *Biotechnology Annual Review*, 13: 303–343.

Blakebrough, N. and Hamer, G. (1963) Resistance to oxygen transfer in fermentation broths. *Biotechnol. Bioeng.*, 5: 59–74.

Burkert, J.F.M, Maugeri, F., and Rodrigues, M.I. (2004) Optimization of extracellular lipase production by sp. using factorial design. *Bioresource Technology*, 91(1): 77–84.

Byrne, G.S. and Ward, O.P. (1989) Effect of nutrition on pellet formation by *Rhizopus arrhizus. Biotechnol. Bioeng.*, 33:912–914.

Chiu, S.W., Ching, M.L., Fong, K.L., and Moore, D. (1998) Spent oyster mushroom substrate performs better than many mushroom mycelia in removing the biocide pentachlorophenol. *Mycological Research*, 102, 1553–1562.

Choma, I.M., Grzelak, E.M. (2011) Bioautography detection in thin-layer chromatography. *Journal of Chromatography A*, 1218: 2684–2691.

Colla, L.M., Primaz, A.L., Benedetti, S., Loss, R.A., Lima, M. de, Reinehr, C.O., Bertolin, T.E., Costa, J.A.V. (2016) Surface response methodology for the optimization of lipase production under submerged fermentation by filamentous fungi. *Brazilian Journal of Microbiology*, 47 (2): 461–467.

Demain, A. (2014) *The business of biotechnology.* Gen publishing inc., a maryannliebert inc. Company 3(3): 270–283.

Demain, A.L. (1999) Metabolites, primary and secondary. In: M.C. Flickinger and S.C. Drew (eds) *Encyclopedia of Bioprocess Technology: Fermentation, Biocatalysis and Bioseparation*. Wiley, New York, pp 1713–1732.

Elander, R.P. and Lowe, D.A. (1994) Fungal biotechnology: An overview. In: *Handbook of Applied Mycology* (pp. 1–34). New York, NY: Marcel Dekker.

Emeko, H.A., Olugbogi, A.O., and Betiku, E. (2005) Appraisal of artificial neural network and response surface methodology in modeling and process variable optimization of oxalic acid production from cashew apple juice. *BioResources*, 10(2): 2067–2082.

Fischer R., Lautner, H., (1961) Zum papierchromatographischen Nachweis von Penicillinpräparaten. *Archiv der Pharmazie*, 294:1–7.

Frisvad, J.C. (1989) The use of high-performance liquid chromatography and diode array detection in fungal chemotaxonomy based on profiles of secondary metabolites. *Botanical Journal of the Linnean Society*, 99: 81–95. doi:10.1111/j.1095-8339.1989.tb00393.x

Gomez, R., Schnabei, I., Garrido, J. (1988) Pellet growth and citric add yield of *Aspergillus niger* 110. *Enzyme and Microbial Technology*, 10: 188–191.

Ghorai, S., PrBanik, S., Verma, D., Chowdhury, S., Mukherjee, S., and Khowala, S. (2009) Fungal biotechnology in food and feed processing, *Food Research International*, 42(5): 577–587.

Griebeler, N., Polloni, A., Remonatto, D. et al. (2011) Isolation and screening of lipase-producing fungi with hydrolytic activity. *Food Bioprocess Technol.*, 4:578–586.

Hang, Y.D., Lee, C.Y., and Woodams, E.E. (1982) A Solid State Fermentation System for Production of Ethanol from Apple Pomace. *Journal of Food Science*, 47: 1851–1852.

Hölker, U. and Lenz, J. (2005) Solid-state fermentation — are there any biotechnological advantages?, *Current Opinion in Microbiology*, 8(3): 301–306.

Irfan, M., Nadeem, M., and Syed Q. (2014) One-factor-at-a-time (OFAT) optimization of xylanase production from Trichoderma viride-IR05 in solid-state fermentation. *Journal of Radiation Research and Applied Sciences*, 7: 317–326.

Zhi-Liang, J., Mei-Xiu, Z., and Lin-Xiu, L. (1996) Catalytic spectrophotometric methods for the determination of oxalic acid, *Analytica Chimica Acta*, Volume 320(1): 139–143.

Karri, S., Talla, S.G., and Renuka, D.S. 2014. Screening and production optimisation of alpha amylase from *Aspergillus* strains by using solid state fermentation. *International Journal of Current Microbiology and Applied Science*, 3(4): 623–631.

Kavanagh, K., (2005) Fungal Fermentation Systems and Products. In: K. Kavanagh (ed.) *Fungi Biology and Applications*. John Wiley & Sons Ltd., England, 89–112.

Khan, J.A. and Yadav, S.K. (2011) Production of alpha amylases by *Aspergillus nigerusing* cheaper substrates employing solid state fermentation. *International Journal of Plant, Animal and Environmental Science*, 1(3): 100–108.

Kumar, V.V., Kirupha, S.D., Periyaraman, P., and Sivanesan, S. (2011) Screening and induction of laccase activity in fungal species and its application in dye decolorization. *African Journal of Microbiology Research*, 5(11):1261–7.

Lata, S., Rastogi, S., Kapoor, A., and Imran, M. (2013) Optimization of culture conditions for the production of phytase from *Aspergillus heteromorphus*MTCC 10685. *International Journal of Advanced Biotechnology and Research*, 4(2): 224–235.

Lopes, F.C, Tichota, D.M, Pereira, J. Q. et al. (2013) Pigment production by filamentous fungi on agro-industrial byproducts: an eco-friendly alternative. *Appl. Biochem. Biotechnol.*, 171(3):616–25.

Magaldi, S., Mata-Essayag, S., Hartung De Capriles, C. et al. (2004) Well diffusion for antifungal susceptibility testing. *International Journal of Infectious Diseases*, 8(1), 39–45.

Metz, B. and Kossen, N.W.F. (1977) Biotechnology Review: The growth of molds in the form of pellets-A literature review. *Biotechnol. Bioeng.*, 19: 781–799.

Metz, B., Kossen, N.W.F., and van Suijdam, J.C. (1979) The rheology of mould suspensions. *Adv. Biochem. Eng.* (Ghose, T.K., Fiechter, A., Blakebrough, N. eds.), 11:104–156, Springer- Verlag, Berlin.

Molitoris, H.P. (1995) Fungi in biotechnology. Past, present, future; *Czech Mycol.*, 48(1), 53–65.

Narasimhan, B. and Bhimba, V. (2015) Screening of extracellular lipase releasing microorganisms isolated from sunflower vegetable oil contaminated soil for bio-diesel production. *Asian Journal of Pharmaceutical and Clinical Research*, 8(2):427–430.

Natsir, H., Patong, A.R., Suhartono, M.T., and Ahmad, A. (2010) Production and characterization of chitinase enzymes from Sulili hot spring in south Sulawesi, *Bacillus* sp. HSA,3-1a. *Indo. J. Chem.*, 10 (2), 263–267.

Nielsen, J. (1992) Modeling the growth of filamentous fungi. *Adv. Biochem. Eng. Biotechnol.* (Fiechter, A. ed.), 46, 187–223, Springer-Verlag, Berlin.

Papagianni, M. and Mattey, M. (2006) Morphological development of *Aspergillus niger* in submerged citric acid fermentation as a function of the spore inoculum level. Application of neural network and cluster analysis for characterization of mycelial morphology. *Microbial Cell Factories*, 5: 3.

Norouzian, D. (2008) Effect of different factors on fermentative production of enzyme by fungi. *Dynamic biochemistry, process biotechnology and molecular biology*, 2(1), 14–18.

Pacheco, S.M.V., Júnior A.C., Morgado, A.F. et al. (2015) Isolation and screening of filamentous fungi producing extracellular lipase with potential in biodiesel production. *Advances in Enzyme Research*, 3:101–114.

Pandey, A. (2003) Solid-state fermentation. *Biochemical Engineering Journal*, 13(2-3): 81–84.

Pandey, A., Soccol,C.R., and Larroche, C. (2008) *Current Developments in Solid-state Fermentation*. Springer, Asiatech Publishers, Inc. New Delhi; PP 3–12.

Patel, G., Patil, M.D., Soni, S. et al. Production of mycophenolic acid by —A comparison of two methods of optimization. *Biotechnology Reports*, 11: 77–85.

Pathak, S.S., Sandhu, S.S., and Rajak, R.C. (2015) Mutation studies on fungal glucoamylase: a review. *International Journal of Pharmacy and Biological Sciences*, 5(2): 297–308.

Pazouki, M. and Panda, T. (2000) Understanding the morphology of fungi. *Bioprocess Engineering*, 22: 127–143.

Punt, P.J, van Viesen, N., Conesa, A. et al. (2002) Filamentous fungi as cell factories for heterologous protein production. *Trends Biotechnol*, 20: 200–206.

Reddy, P.L. and Sreeramulu, A. (2015) Isolation, identification and screening of pectinolytic fungi from different soil samples of chittoor district. *International Journal of Life Science Biotechnology and Pharma Research*, 1(3): 186–193.

Renge, V.C., Khedkar, S.V., and Nandurkar, N.R. (2012) Enzyme synthesis by fermentation method: A Review. *Scientific Reviews and Chemical Communications*, 2(4): 585–590.

Rhishipal, R. and Philip, R. (1998) Selection of marine yeasts for the generation of single cell protein from prawn-shell waste. *Bioresource Technology*, 65: 255–256.

Rojas, C. and Mansur, E. (1995) Ecuador: informacionesgeneralessobreproductos non madererosen Ecuador. In Memoria, consulta de expertossobreproductosforestales no madereros para America Latina y el Caribe (pp. 208–223). Santiago, Chile: FAO Regional Office for Latin America and the Caribbean (Serie Forestal#1).

Sharma, A.K., Sharma, V., Saxena, J. et al. (2005) Isolation and screening of extracellulaṙ protease enzyme from bacterial and fungal. *International Journal of Scientific Research in Environmental Sciences*, 3(9):0334–0340.

Shirazian, P., Asad, S., and Amoozegar, M.A. (2016) The potential of halophilic and halotolerant bacteria for the production of antineoplastic enzymes: L-asparaginase and L-glutaminase. *EXCLI Journal*, 15:268–279.

Siddalingeshwara, K.G. and Lingappa, K. (2010) Screening and Optimization of L-Asparaginase- A Tumour Inhibitor from *Aspergillus terreus* Through Solid State Fermentation. *Journal of Advanced Scientific Research*, (1): 55–60.

Singh, B. and Satyanarayana, T. (2006) Phytase production by thermophilic mold *Sporotrichum thermophile* in solid-state fermentation and its application in dephytinization of sesame oil cake. *Appl. Biochem. Biotechnol.*, 133: 239–250.

Singh, O.V., Preira, B.M.J., and Singh R.P. (1999) Isolation and characterization of a potent fungal strain *Aspergillus niger* ORS-4 for gluconic acid production. *Journal of Scientific and Industrial Research*, 58: 594–600.

Suleiman, M.M., McGaw, L.J., Naidoo, V., and Eloff, J.N. (2010) Detection of antimicrobial compounds by bioautography of different extracts of leaves of selected south african tree species. *Afr. J. Traditional Complementary and Alternative Medicines*, 7(1): 64 – 78.

Thiagarajan, V., Revathi, R., Aparanjini, K. et al. (2011) Extra cellular chitinase production by *Streptomyces* sp. PTK19 in submerged fermentation and its lytic activity on *Fusarium oxysporum*PTK2 cell wall. *International Journal of Current Science*, 1: 30–44.

Thomas, L., Larroche, C., and Pandey, A. (2013) Current developments in solid-state fermentation. *Biochemical Engineering Journal*, 81:146–61.

Valgas, C., Souza, S.M. de, Smânia, E.F.A., and Smânia Jr., A. (2007) Screening methods to determine antibacterial activity of natural products. *Brazilian Journal of Microbiology*, 38(2): 369–380.

Walker, G.M. and White, N.A. (2005) Introduction to Fungal Physiology, in: K. Kavanagh (ed.) *Fungi Biology and Applications*, John Wiley & Sons Ltd, England, pp 89–112.

Wu, M. and Singh, A.K. (2011) Single-Cell Protein Analysis. *Curr Opin Biotechnol*, 23(1): 83–88.

Yang, S., Jin, L., Ren, X. et al. (2014) Optimization of fermentation process of Cordyceps militaris and antitumor activities of polysaccharides *in vitro*. *Journal of Food and Drug Analysis*, 22: 468–476.

Yousufi, M.K. (2012) To determine protein content of single cell protein produced by using various combinations of fruit wastes and two standard food fungi. *International Journal of Advanced Biotechnology and Research*, 3(1): 533–536.

Yu, J. and XuZhang, T.T. 2008. Ethanol production by solid state fermentation of sweet sorghum using thermotolerant yeast strain. *Fuel Processing Technology*, 89: 1056–1059.

Zhang, J. and Zhang, J. (2015) The filamentous fungal pellet and forces driving its formation. *Critical Reviews in Biotechnology*. Early Online: 1–12.

11

Microbes and Their Products as Sensors in Industrially Important Fermentations

Ritu Raval[1] *and Keyur Raval*[2]

[1] *Department of Biotechnology, Manipal Institute of Technology, Manipal, Karnataka, India*
[2] *Department of Chemical Engineering, National Institute of Technology, Surathkal, Karnataka, India*

11.1 Introduction

Fermentation reactions have an ever-increasing role to play in the biotechnological processes involved in major industrial and environmental sectors. These applications include food, pharmaceuticals, water treatment and energy. Thus to yield an optimal result, precise control of the process conditions are important. The need of the hour for the industrial level fermentations is always rapid and sensitive on-line monitoring and control. Though the methods of spectrophotometery and chromatography are generally made for assessing the quality of the product but it is unsuitable for on-line measurements. It is in these situations that electrochemical monitoring using sensors is convenient. One such type of sensors called biosensors are used to a large extent in the industrial level fermentations wherein the transducer for temperature, pH, optical activity is in conjunction with a biologically active system (Hikuma et al., 1979a; Karube et al., 1977; Turner, Karube, and Wilson, 1987)

The advances in the sensor technology started with the use of the biocatalysts in conjunction with the transducers, the first reports hinting this development came from Updike (Updike and Hicks, 1967). The highly specific biocatalysts coupled with the high sensitivity of the transducer systems ameliorated the entire regime of process control. The first generation of biosensors using biocatalysts paved the way to the second generation of biosensors where in whole microbe cells were utilized. The pioneer studies by Divies in 1974 reported the use of whole microbial cells for the detection of ethanol in the year 1974. The first industrial application was reported in 1977 to determine the BOD in waste water by Karube (Karube et al., 1977). Even though the first generation sensors using biocatalysts have an edge of specificity and response time over the microbe sensors, the latter score over the former in terms of its higher tolerance to the measuring conditions, a long lifetime and cost performance. In addition, the second

Microbial Sensing in Fermentation, First Edition. Edited by Satinder Kaur Brar, Ratul Kumar Das, and Saurabh Jyoti Sarma.

generation of sensors using whole microbe cells do not require downstream processes of extraction and purification and co-factors. Thus the cost of the second generation sensors could be economized. Over and above, these sensors could be recharged by soaking in the nutrient broth. This makes them active for a longer period with the added advantage of reusability. The main idea of the use of the microbe as sensors revolves around their characteristics to adapt to the analyte. In the present chapter, we would discuss the different aspects of the microbes as sensors in the fermentation of major commercially viable products and environmental monitoring.

11.2 Sensors

The first microbial sensor was developed by Divies in 1974 for determination of ethanol using cells of *Acetobacter xylinum* (Divies, 1974). The working of the microbe sensors are based on the last phase of the aerobic respiration involving the electron transport chain (López-Barneo, Pardal, and Ortega-Sáenz, 2001). Another improvisation of these sensors are the mediator based microbe sensors. In these type of sensors, the electron produced as a result of the respiration process is sequestered by the mediator, and this mediator finally passes the electron to the anode which is one of the two transducers in the system. The signal can be detected either by the dissolved oxygen sensor or electron transfer measuring systems (Du, Li, and Gu, 2007). The main basis of a biosensor is the intimate contact between microbe and transducer element. Microbes are immobilized to improve the contact. This immobilization can either involve physical or chemical methods (Cassidy, Lee, and Trevors, 1996). The physical methods are largely utilized for the immobilization as it facilitates microbes to retain their biological activity. The sensitivity of the physically immobilized microbe sensors also surpasses in detection sensitivity as compared to the other methods of immobilization.

11.3 Transducers in Conjunction With Microbe Sensors

11.3.1 Dissolved Oxygen (DO) Electrode

A DO electrode is the most general transducer for the microbial sensor. The sensors used today are the improved micro-oxygen electrode developed based on semiconductor fabrication technology given by Karube et al. (Suzuki, Tamiya, and Karube, 1988). The microbial sensors are dipped in the solution saturated with DO; this causes an increase in the respiratory activity of the microorganisms which in turn results in a decrease in the DO concentration near the membrane of the polarographic electrode. A suitable polarization voltage between the anode and cathode selectively reduces oxygen at the cathode. The results of these chemical reactions are shown as a current which is proportional to the DO concentration. Using a DO electrode, substrate concentration can be measured from the oxygen decrease.

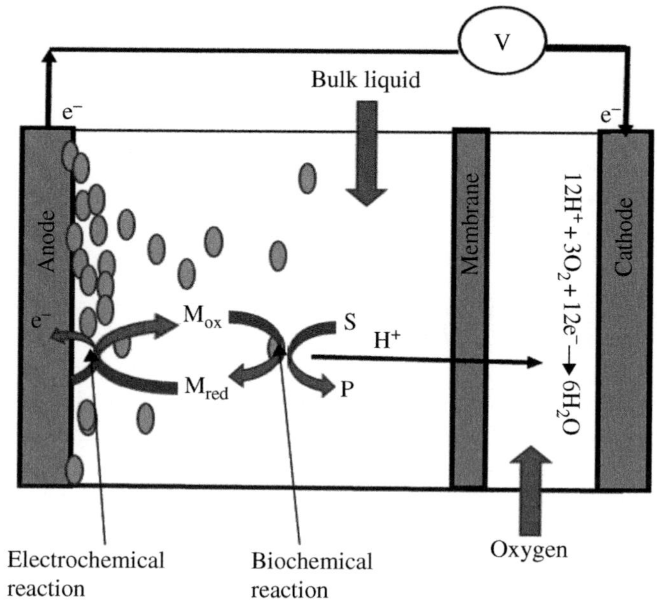

Figure 11.1 Principle of the Oxygen Sensor based on the respiration activity. (*See insert for color representation of this figure.*)

The principle in this sensor type is shown in Fig. 11.1. The DO electrode is also used to measure the amount of the substrate as well as the toxic compounds present in the media. An improved version of the oxygen sensor based on the optical sensing device for biochemical oxygen demand (BOD) was first reported by Preininger et al. in 1994 (Preininger, Klimant, and Wolfbeis, 1994). This sensor based on an oxygen-sensitive ruthenium complex [(Ru) complex] relied on the oxygen quenching of luminescence. Since the initial work, several BOD biosensors have been developed using different kinds of microbes (Chee et al., 2000; Li et al., 1994; Lin et al., 2006). Even though the addition of mediators in these biosensors can enhance the electron transfer, these biosensors have poor stability because of the toxicity of mediators. Recently, mediator-less microbial fuelcells have been exploited to fabricate novel BOD sensors forcontinuous and real-time monitoring (Chang et al., 2004). Furthermore, Kim et al. reported that the performance of a microbial fuelcell as BOD sensor was improved using respiratory inhibitors (Chang, Moon, Jang, & Kim, 2005).

11.3.2 Electron Transfer Measuring Systems

In these types of sensors, the electron shuttles or mediators or color indicators are used between the microbes and the electrode. The principle of this sensor type is shownin Fig. 11.2. The mediator most commonly used for the microbial sensors is potassium hexacyanoferrate (III) [HCF (III)]. In the aerobic respiration, the end point acceptor for anelectron is oxygen. However competition is imposed by the HCF (III) when present in the reaction medium. Thus it acts as the electron

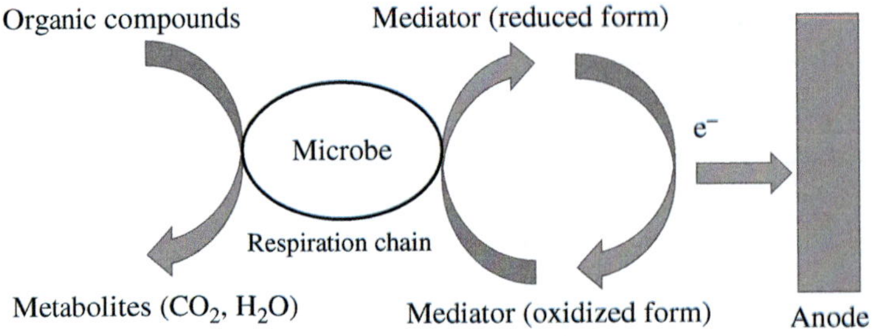

Figure 11.2 Principle of the Oxygen Sensor based on the Electron Transfer Measuring Systems. (*See insert for color representation of this figure.*)

acceptor and is preferentially reduced to HCF (II) during the metabolic oxidation of organic substances. The reduced HCF (III) is then reoxidized at a working electrode. This was initially tried with Pseudomonas sp. (Yoshida et al., 2001) with similar report on with a consortium of microbes (Catterall et al., 2003). The sensor system utilizing electron transfer has many advantages over the DO electrode. Firstly, the solubility of mediator or color indicator is much higher than that of DO. Secondly, the sensor system does not require an aeration system and can be greatly simplified to a mobile type device. Thirdly, the detectable potential of the mediator is lower compared tothat of DO. A further improvement of the electron transfer system sensor is a combination of microbe and the redox color indicators (RCI). This sytem allows simultaneous measurements simultaneous spectroscopic measurement using 96 or 364 microwells (Yoshida et al., 2002).

11.4 Metabolite Measuring Systems

In addition to the whole cell DO electrode and electron transfer systems, electrically active metabolites, such as H_2, CO_2, NH_3 and organic acidssecreted from microorganisms, can also be used as the microbialsensors. This type of sensor mostly uses a gas-permeable membrane todetect gaseous compounds in aqueous or gaseous samples. The principle of this sensor type is shown in Figure 11.3. The microbes employed in this sensor can be either aerobes or anaerobes.

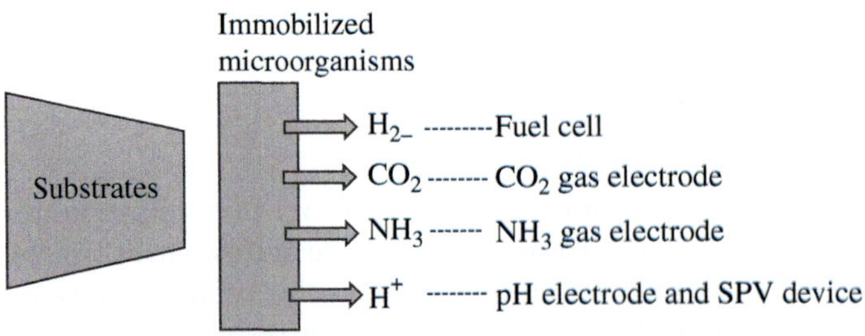

Figure 11.3 The metabolites produced by the microbes can also use as sensors. (*See insert for color representation of this figure.*)

The transducers for H_2 detection have been fuel cell type, CO_2 electrodes, NH_3 electrodes, or pH electrodes (including ion-sensitive field-effect transistors (ISFETs)). Most of these sensors are based on potentiometry. Though they have a large measurable range, there have a limitation of lower detection limits

11.5 Other Measuring Systems

In addition to the metabolites that the microbes secrete, several other parameters can also be used of for the sensors. For example,to measure the metabolic heat evolved by the immobilized microorganisms, microbial sensors can be constructed by placing them in proximity to a thermistor. Another method of detecting is to use photobacteriain conjunction with a photodetector (e.g. a photomultiplier (PMT) or photodiode (PD). The luminescence intensity of photobacteria also known as luminobacteria is dependent on its metabolic activity. Therefore, nutrients like, glucose, amino acids) supplied to the bacteria can easily be detected. Also, the inhibitors, eg. toxins, heavy metals which deter the metabolic rate of the bacteria could also be detected using this type of device. In general, luminescence intensity is a more sensitive parameter for metabolic activity than respiration activity or heat generation. Obviously, only photobacteria can be used for this purpose. The modulation in optical properties such as UV–vis absorption, bio- and chemi-luminescence, reflectance and fluorescence brought by the interaction of the biocatalyst with the target analyte is the basis for optical microbial biosensor (Mulchandani and Rogers, 1998; Tran-Minh, 1993; Turner et al., 1987). Optical based biosensors offer advantages of compactness, flexibility, resistance to electrical noise, and a small probe size.

11.5.1 Bioluminescence Biosensor

Bioluminescence is associated with the emission of light by living microorganisms, and it plays a very important role in realtime process monitoring. The bacterial luminescence *lux* gene has been widely applied as a reporter either in an inducible or constitutive manner. In an inducible manner, the reporter *lux* gene is fused to a promoter regulated by the concentration of a compound of interest. As a result, the concentration of the compound can be quantitatively analyzed by detecting the bioluminescence intensity (Rensing and Maier, 2003; Sagi et al., 2003). In a constitutive manner, the reporter gene is fused to promoters that are continuously expressed as long as the organism is alive and metabolically active (Rensing and Maier, 2003). This kind of reporter is good for evaluating the total toxicity of contaminant. Both types of reporters have been shown to be useful for biosensor development. Heavy metal-mediated toxicity in the environment is dependanton bioavailable metal concentrations. Bioluminescent microbial biosensors have been extensively investigated to monitor bio available metal. *Ralstonia eutropha* AE2515 was constructed by transcriptionally fusing *cnr* YXH regulatory genesto the bioluminescent *lux* CDABE report system to fabricate a whole cell biosensor for the detection of the bioavailable concentration of Ni^{2+} and Co^{2+} in soil (Ivask, Rõlova, and Kahru, 2009). Several optical biosensors consisting of bacteria that contain gene fusion between the regulatory region of the *mer* operon (*mer*R) and *lux* CDABE havebeen developed to quantitatively response to Hg^{2+}. The *mer* promoter is activated when Hg^{2+} binds to *Mer* R, then result inthe transcription of the *lux* reporter gene and subsequent lightemission (Hansen and Sørensen, 2001; Omura, Kiyono, and Pan-Hou, 2004). Bioavailable copper in soil is also monitoredby using engineered *P. fluorescens* through mutagenesis of *P. fluorescens* containing copper-induced gene and *Tn5::lux*AB Promoter probe transposon (Bereza-Malcolm, Mann, and Franks, 2014). To monitor nutrients in an aquatic ecosystem, a biosensor for monitoring phosphorus bioavailability to *Cyanobacteria* (*Synechococcus* PCC 7942) was developed (Schreiter et al., 2001). The reporter strain *Synechococcus* harbors the gene coding the reporter protein luciferase under the control of aninducible alkaline phosphatase promoter, which can be induced under phosphorous limitation and shows improvement to conventional phosphorus detection methods (Schreiter et al., 2001). Bioluminescent microbial biosensors using the inducible reporter gene have also been developed for the measurement of bioavailable NAPH

Thus from the above applications, we can generalize that the microbial species chosen for biosensor development must fulfil at least one of the two criteria:

i) aerobic uptake of oxygen in the respiratory process for assimilation of the substrates (in this case the microbial sensor is constructed by coupling the microorganisms with an amperometric oxygen electrode); or
ii) electrode-active products liberated derived from reactions of the microbial metabolism, e.g. protons, ammonium ions, H_2S, CO,and H,O_2,which can be detected by potentiometric or amperometric electrodes.

11.6 Applications of Microbe Sensors in Some Commercially Important Products

The microbial biosensors based on DO and Electron transfer mediators have been utilized for the food analysis and other fermentation processes like brewing and fuel production. During the process of brewing, the alcohol concentration is important. The works of Hikuma et al., in 1979 showed that using Trichosporon brassicae, the detection of the analyte i.e. ethanol in the range of 2–22.5 mg/L was possible. In another study with *Clostridium butyricum* the levels of formic acid in the range of 10–1000 mg/L could be detected (Matsunaga, Karube, and Suzuki, 1980). In another study, the gaseous ethanol was determined using by *T. brassicae* or *Gluconobacteroxydans* with ferricyanide mediation (Vostiar et al., 2002). The details are listed in Table 11.1. In all these examples from the brewing and the food industry, simply DO electrodes have been employed to detect the analyte. In recent times many improvements of the DO sensors have been realized. One such improvisationis the inclusion of magnetite-functionalized *Bacillus subtilis* that can be immobilized and regenerated on an ultramicroelectrode array (UMEA). Modification and regeneration are magnetically controlled. The assay can be performed within 5 min owing to the fast mass transfer of the magnetite-functionalized microbes on the surface of the UMEA. These sensors

Table 11.1 A list of the sensors and their applications in the fermentation industry.

Analyte	Microbe Used as Sensor	Electrode Type	Measuring time(min)	Stability (in days)	Reference
Ethanol	*Trichosporon brassicae*	Amperometric	<10	>21	(Hikuma, Kubo, Yasuda, Karube, and Suzuki, 1979b)
Formic acid	*Clostridium butyricum*	Amperometric	20	>20	(Matsunaga et al., 1980)
Acetic acid	Trichosporon brassicae.	Amperometric	8	>20	(Hikuma et al., 1979a)
Glutamic acid	Escherichia coli	Potentiometric	5	>21	(Hikuma, Obana, Yasuda, Karube, and Suzuki, 1980)
Aspartic acid	Pseudomomas				
Glucose	*Pseudomonas fluorescens*	Amperometric	10	>14	(Karube, Mitsuda, and Suzuki, 1979)
Ammonia	*Nitrifying bacteria*	Amperometric	4	>10	(Ikeda, Hachiya, Ito, Asano, and Imato, 1998)
Vitamin B12	*E.coli* 215	Amperometric	120	25	(Karube, Wang, Tamiya, and Kawarai, 1987)

have been successful applied to the determination of BOD in spiked water samples (Wang et al., 2017).

11.6.1 Red Wine

Wines, particularly red wines, contain numerous biologically active compounds, the most important of which are phenolic compounds. The nutritional importance of phenolic compounds is attributed to their antioxidant properties. In particular, flavonoids and related phenolic compounds which are naturally found in red wines have gained increasing interest (Campanella et al., 2004). Red wines have been reported to be preventive of many ailments, and they play a possible role in reducing thrombotic and anthrogenic processes. Phenolic compounds also contribute substantially to the quality of wines and affect their color, flavor, stability and aging behavior (Luximon-Ramma et al., 2005). See Figure 11.4. In one such study, *Alcaligense sp.*-immobilized biosensor based on CdS-MWNT and Cu2S-MWNT supports prepared by a one-step radiation reaction were fabricated. The sensing range of the *Alcaligense sp.*-immobilized biosensor based on CdS-MWNT and Cu2S-MWNT supports for phenol was in the range of 0.5 ~ 5.0 mM and 0.7 ~ 10 mM, respectively. Both biosensors exhibited a wide linear range, high sensitivity, and good stability. The prepared biosensors were used to the determination of phenolics in commercial red wines. The results showed that the amount of phenolic. Compounds in commercial red wines were in the range of 926.1 ~ 1,018 mg/L, calculated from the calibration curve of phenol measured by the *Alcaligense sp.*-immobilized biosensor based on QD-MWNT supports The relatively high amounts of phenolic compounds in Blue Nun are responsible for the bitter taste of the red win (Kim, Kwen, and Choi, 2011).

11.6.2 Fermentation of Cereal Products

The main volatile compounds reported by studies on cereal-based liquid fermentation have been mainly associated with the carbohydrate (mainly carboxylic acids and aldehydes, ketones, esters) and amino acid metabolism (mainly aldehydes and alcohols). Diacetyl (butane-2,3-dione) is a ketone responsible for a butter-

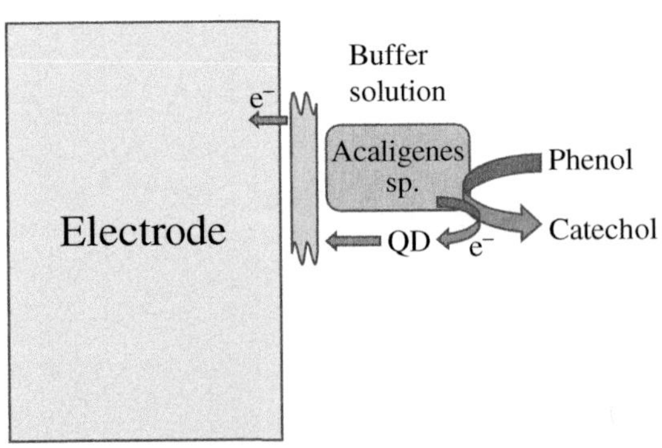

Figure 11.4 Application of microbe sensor in the quality control of red wine. *Source:* Adapted from Kim et al., 2011. (*See insert for color representation of this figure.*)

scotch-like aroma, and the verylow odor threshold (0.005 mg/l in water) is often exceeded during LAB fermentation of cereal substrates ((Burdock, 2002). This volatile is mainly formed from the oxidative decarboxylation of α-acetolactate (α-AL), an intermediate metabolite formed mainly duringsugar, citrate and amino acid catabolism (Hugenholtz et al., 2000). Metabolically related todiacetyl are the less flavoursome acetoin, formed by the reduction of diacetyl or afterenzymatic decarboxylation of α-AL, and 2,3-butanediol, which results from the reduction ofacetoin (Montel, Masson, and Talon, 1998). Although considered as being off-flavours in beer (Bokulich & Bamforth, 2013), these low-molecular-weight compounds are also responsible for mellowing theflavour during cereal fermentation (Mugula et al., 2003). Only some LAB strains, e.g. *L. lactis, Lb. Plantarum, Oenococcusoeni*, can metabolize citrate to pyruvate that can be eventually re-directed into the acetoin/diacetyl pathway. This pathway is responsible for the high accumulation of α-AL during LAB fermentation of milk-based matrices (citrate concentration of ca. 1500 mg/l)(Garde et al., 2002). Although citrate in cereal malt gruels has been reported as being rather low (170 mg/l) (Owuama, Chiangi, and Adeyemo, 2011), higher levels (871 mg/kg) have been reported in a sorghum-based liquid substrate (Acetaldehyde is a highly volatile aldehyde formed from pyruvate or threonine catabolism(Ardö, 2006). It has been described as delivering a pungent, fruity (green apples) flavor withsweet notes (odor threshold of 0.027 mg/l in water) (Salmerón et al., 2015)) showed that the high level of acetaldehyde (1.36 mg/l) found after fermentation of a malt-based beverage with *Lb.Plantarum* NCIMB 8826 positively contributed to the high acceptance of the beverage.

11.6.3 Mevalonate Production

In whole-cell biosensing, changes in cellular metabolism, pH, and gene expression have been quantified as a response of the sensing elements to the presence of target molecules (Ivask et al., 2009; Pfleger et al., 2007; Schultheiss et al., 2008). Microbial auxotrophy has been used used to monitor growth-limiting small molecules. For example, Pfleger et al. constructed an autotrophic *E. coli* strain for the detection and quantification of mevalonate,an intermediate in the biosynthesis of isoprenoids, a large class of industrially important secondary metabolites that includes flavor, fragrance, anti-oxidants, steroids, and the anti-malarial drug artemisinin. Since mevalonate is a key precursor whose production must be increased toimprove the production of isoprenoids, optimizing the level of mevalonate is important in developing recombinant strains for enhanced isoprenoid production. The authors deleted the native pathway for the production of isopentenyl pyrophosphate (IPP) and dimethylallyl pyrophosphate (DMAPP) that are necessary for growth, and incorporating the mevalonate-utilizing pathway Hence, using simple growth monitoring, the concentrations of mevalonate could be determined.

11.6.4 Bioaerosols

The recent advances in analytical techniques open a new door for the chemical characterisation of bioaerosol. Specifically, chemical analysis of microbial volatile organic compounds (MVOCs) can be a reliable and rapid assessment of the nature of ambient bioaerosols (Lemfack et al., 2014) as microbial communities express different MVOCs profiles depending in which environment they are in Ref.(Konuma et al., 2015). Further to this, it has been shown that species-specific volatiles may serve as marker compounds for the selective detection of pathogenic microbial species in indoor and outdoor environments (Lemfack et al., 2014). MVOCs are secondary metabolites produced by fermentation and are volatile due to their physicochemical properties (low molecular weight, low boiling point, and high vapor pressure)(Schenkel et al., 2015). Characterizing and quantifying MVOCs can also be used as a proxy approach to estimate microbial concentration (Araki et al., 2012)and to detect invasive *Aspergillus fumigatus* from breath samples (Chambers et al., 2009).

11.6.5 Aptamers

Metabolites and sensitivity and have a long history as detectors in complex samples in the food, environmental and medicinal sectors. The design of new biosensors is now facilitated by the availability of increasingly sophisticated methods for the recruitment of aptamers that comprise RNA(Ozer, Pagano, and Lis, 2014). Similar to antibodies or enzymes, aptamers feature a high affinity for both small molecules and larger structures (Garde, 2002, Hansen, 2001, He, 2016, Hikuma, 1979a,b) such as proteins. Aptamers can, however, can be converted into genetic actuators called riboswitches by combining them with another RNA sequence referred to as an expression platform (van der Meer and Belkin, 2010).These switches undergo conformational changes on ligand binding that result in a switch between states that are OFF (no translationof an adjacent open reading frame) and ON (translation occurs of an adjacent open reading frame encoding, for example,a green fluorescent protein (GFP)). The combination of highly specific ribo switches with the appropriate microbial sensor chassis thus allows the generation of highly specific microbial sensors. Taking advantage of this sensing principle, here we identify *Bacillus subtilis* strains that secrete particularly high levels of vitamin B2 when using cellobiose as a carbon source. Cellobiose is one of the main components of cellulosic biomass hydrolysates,

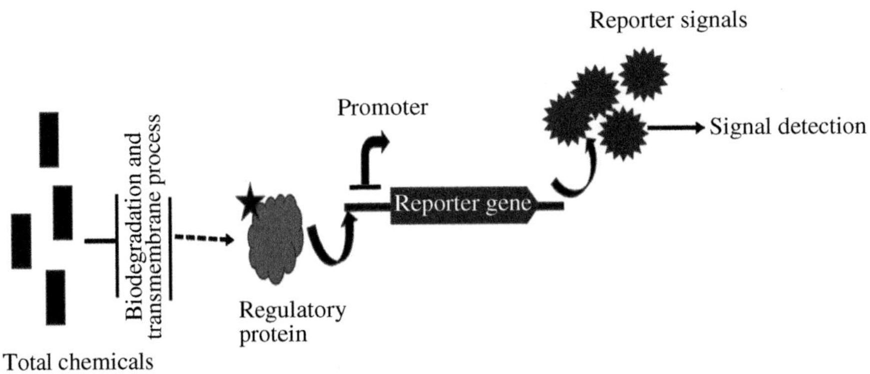

Figure 11.5 The regulation, construction and organization of the Genetically engineered microbial whole-cell biosensors. *Source*: Adapted from He et al. (2016). (*See insert for color representation of this figure.*)

which makes it a potential future feedstock for industrial biotechnology products. B2 secreted by a *B. subtilis* cell is taken up by metabolically active, but growth-arrested, E. coli sensor cells. The sensor cells then transform B2 into flavin mononucleotide (FMN) which binds with high affinity to a specifically developed hammer head riboswitch, switching its conformation to ON and triggering GFP production in the sensor cells (Fig. 11.5) in a concentration-dependent manner (Meyer et al., 2015).

11.7 Conclusions

In this chapter, an attempt has been made to introduce various transducer systems available at an industrial scale. In addition, major applications of these sensors in various industries have been highlighted. The main motive of the chapter was to help in analyzing the sensitivity of each sensor type to help extend these sensors in the field of research.

References

Araki, A., Kanazawa, A., Kawai, T. et al. (2012) The relationship between exposure to microbial volatile organic compound and allergy prevalence in single-family homes. *Science of the total environment*, 423, 18–26.

Bereza-Malcolm, L.T., Mann, G.L., and Franks, A.E. (2014) Environmental sensing of heavy metals through whole cell microbial biosensors: a synthetic biology approach. *ACS synthetic biology*, 4(5), 535–546.

Burdock, G.A. (2002) Regulation of flavor ingredients. *Nutritional Toxicology (2nd ed.)*, Taylor and Francis, New York, NY, 316–339.

Campanella, L., Bonanni, A., Finotti, E., and Tomassetti, M. (2004) Biosensors for determination of total and natural antioxidant capacity of red and white wines: comparison with other spectrophotometric and fluorimetric methods. *Biosensors and Bioelectronics*, 19(7), 641–651.

Cassidy, M., Lee, H., and Trevors, J. (1996) Environmental applications of immobilized microbial cells: a review. *Journal of Industrial Microbiology*, 16(2), 79–101.

Catterall, K., Zhao, H., Pasco, N., and John, R. (2003) Development of a rapid ferricyanide-mediated assay for biochemical oxygen demand using a mixed microbial consortium. *Analytical Chemistry*, 75(11), 2584–2590.

Chambers, S.T., Syhre, M., Murdoch, D.R. et al. (2009) Detection of 2-pentylfuran in the breath of patients with Aspergillus fumigatus. *Medical Mycology*, 47(5), 468–476.

Chang, I.S., Jang, J.K., Gil, G.C. et al. (2004) Continuous determination of biochemical oxygen demand using microbial fuel cell type biosensor. *Biosensors and Bioelectronics*, 19(6), 607–613.

Chang, I.S., Moon, H., Jang, J.K., and Kim, B.H. (2005) Improvement of a microbial fuel cell performance as a BOD sensor using respiratory inhibitors. *Biosensors and Bioelectronics*, 20(9), 1856–1859.

Chee, G.-J., Nomura, Y., Ikebukuro, K., and Karube, I. (2000) Optical fiber biosensor for the determination of low biochemical oxygen demand. *Biosensors and Bioelectronics*, 15(7), 371–376.

Divies, C. (1974). [*Remarks on ethanol oxidation by an "Acetobacter xylinum" microbial electrode (author's transl)*]. Paper presented at the Annales de microbiologie.

Du, Z., Li, H., and Gu, T. (2007) A state of the art review on microbial fuel cells: a promising technology for wastewater treatment and bioenergy. *Biotechnology advances*, 25(5), 464–482.

Garde, A., Jonsson, G., Schmidt, A.S., and Ahring, B.K. (2002) Lactic acid production from wheat straw hemicellulose hydrolysate by Lactobacillus pentosus and Lactobacillus brevis. *Bioresource technology*, 81(3), 217–223.

Hansen, L.H. and Sørensen, S.J. (2001) The use of whole-cell biosensors to detect and quantify compounds or conditions affecting biological systems. *Microbial ecology*, 42(4), 483–494.

He, W., Yuan, S., Zhong, W.-H. et al. (2016) Application of genetically engineered microbial whole-cell biosensors for combined chemosensing. *Applied microbiology and biotechnology*, 100(3), 1109–1119.

Hikuma, M., Kubo, T., Yasuda, T. (1979a) Amperometric determination of acetic acid with immobilized Trichosporon brassicae. *Analytica chimica acta*, 109(1), 33–38.

Hikuma, M., Kubo, T., Yasuda, T. et al. (1979b) Microbial electrode sensor for alcohols. *Biotechnology and bioengineering*, 21(10), 1845–1853.

Hikuma, M., Obana, H., Yasuda, T. et al. (1980) A potentiometric microbial sensor based on immobilized Escherichia coli for glutamic acid. *Analytica chimica acta*, 116(1), 61–67.

Hugenholtz, J., Kleerebezem, M., Starrenburg, M. et al. (2000) Lactococcus lactis as a cell factory for high-level diacetyl production. *Applied and Environmental Microbiology*, 66(9), 4112–4114.

Ikeda, M., Hachiya, H., Ito, S. et al. (1998) Development of long-term stable ammonium ion sensor in conjunction with a microbial membrane. *Biosensors and Bioelectronics*, 13(5), 531–537.

Ivask, A., Rõlova, T., and Kahru, A. (2009) A suite of recombinant luminescent bacterial strains for the quantification of bioavailable heavy metals and toxicity testing. *BMC biotechnology*, 9(1), 1.

Karube, I., Matsunaga, T., Mitsuda, S., and Suzuki, S. (1977) Microbial electrode BOD sensors. *Biotechnology and bioengineering*, 19(10), 1535–1547.

Karube, I., Mitsuda, S., and Suzuki, S. (1979) Glucose sensor using immobilized whole cells of Pseudomonas fluorescens. *Applied microbiology and biotechnology*, 7(4), 343–350.

Karube, I., Wang, Y., Tamiya, E., and Kawarai, M. (1987) Microbial electrode sensor for vitamin B12. *Analytica chimica acta*, 199, 93–97.

Kim, S.-K., Kwen, H.-D., and Choi, S.-H. (2011) Fabrication of a microbial biosensor based on QD-MWNT supports by a one-step radiation reaction and detection of phenolic compounds in red wines. *Sensors*, 11(2), 2001–2012.

Konuma, R., Umezawa, K., Mizukoshi, A. et al. (2015) Analysis of microbial volatile organic compounds produced by wood-decay fungi. *Biotechnology letters*, 37(9), 1845–1852.

Lemfack, M.C., Nickel, J., Dunkel, M. et al. (2014) mVOC: a database of microbial volatiles. *Nucleic acids research*, 42(D1), D744–D748.

Li, X.-M., Ruan, F.-C., Ng, W.-Y., and Wong, K.-Y. (1994). Scanning optical sensor for the measurement of dissolved oxygen and BOD. *Sensors and Actuators B: Chemical*, 21(2), 143–149.

Lin, L., Xiao, L.-L., Huang, S. et al. (2006) Novel BOD optical fiber biosensor based on co-immobilized microorganisms in ormosils matrix. *Biosensors and Bioelectronics*, 21(9), 1703–1709.

López-Barneo, J., Pardal, R., and Ortega-Sáenz, P. (2001) Cellular mechanism of oxygen sensing. *Annual Review of Physiology*, 63(1), 259–287.

Luximon-Ramma, A., Bahorun, T., Crozier, A. et al. (2005) Characterization of the antioxidant functions of flavonoids and proanthocyanidins in Mauritian black teas. *Food research international*, 38(4), 357–367.

Matsunaga, T., Karube, I., and Suzuki, S. (1980) A specific microbial sensor for formic acid. *European journal of applied microbiology and biotechnology*, 10(3), 235–243.

Meyer, A., Pellaux, R., Potot, S. et al. (2015) Optimization of a whole-cell biocatalyst by employing genetically encoded product sensors inside nanolitre reactors. *Nature chemistry*, 7(8), 673–678.

Montel, M., Masson, F., and Talon, R. (1998) Bacterial role in flavour development. *Meat Science*, 49, S111–S123.

Mugula, J., Nnko, S., Narvhus, J., and Sørhaug, T. (2003) Microbiological and fermentation characteristics of togwa, a Tanzanian fermented food. *International Journal of Food Microbiology*, 80(3), 187–199.

Mulchandani, A. and Rogers, K. R. (1998) *Enzyme and microbial biosensors*: Humana Press.

Omura, T., Kiyono, M., and Pan-Hou, H. (2004) Development of a specific and sensitive bacteria sensor for detection of mercury at picomolar levels in environment. *Journal of health science*, 50(4), 379–383.

Owuama, C., Chiangi, J., and Adeyemo, M. (2011) Effect of Discorea Cayensis and Ipomoea Batatas Extracts On Sugar Content of Sorghum Mash. *Nigerian Journal Of Microbiology*, 25.

Ozer, A., Pagano, J.M., and Lis, J.T. (2014) New technologies provide quantum changes in the scale, speed, and success of SELEX methods and aptamer characterization. *Molecular Therapy—Nucleic Acids*, 3(8), e183.

Pfleger, B. F., Pitera, D. J., Newman, J. D. et al. (2007) Microbial sensors for small molecules: development of a mevalonate biosensor. *Metabolic engineering*, 9(1), 30–38.

Preininger, C., Klimant, I., and Wolfbeis, O.S. (1994) Optical fiber sensor for biological oxygen demand. *Analytical Chemistry*, 66(11), 1841–1846.

Rensing, C. and Maier, R. M. (2003) Issues underlying use of biosensors to measure metal bioavailability. *Ecotoxicology and environmental safety*, 56(1), 140–147.

Sagi, E., Hever, N., Rosen, R. et al. (2003) Fluorescence and bioluminescence reporter functions in genetically modified bacterial sensor strains. *Sensors and Actuators B: Chemical*, 90(1), 2–8.

Salmerón, I., Loeza-Serrano, S., Pérez-Vega, S., and Pandiella, S. S. (2015) Headspace gas chromatography (HS-GC) analysis of imperative flavor compounds in Lactobacilli-fermented barley and malt substrates. *Food Science and Biotechnology*, 24(4), 1363–1371.

Schenkel, D., Lemfack, M.C., Piechulla, B., and Splivallo, R. (2015). A meta-analysis approach for assessing the diversity and specificity of belowground root and microbial volatiles. *Frontiers in plant science*, 6.

Schreiter, P.P.-Y., Gillor, O., Post, A. et al. (2001) Monitoring of phosphorus bioavailability in water by an immobilized luminescent cyanobacterial reporter strain. *Biosensors and Bioelectronics*, 16(9), 811–818.

Schultheiss, E., Weiss, S., Winterer, E. et al. (2008) Esterase autodisplay: enzyme engineering and whole-cell activity determination in microplates with pH sensors. *Applied and Environmental Microbiology*, 74(15), 4782–4791.

Suzuki, H., Tamiya, E., and Karube, I. (1988) Fabrication of an oxygen electrode using semiconductor technology. *Analytical Chemistry*, 60(10), 1078–1080.

Tran-Minh, C. (1993). *Biosensors*. Chapman Hall and Masson, New York, 87.

Turner, A., Karube, I., and Wilson, G.S. (1987). *Biosensors: fundamentals and applications*. Oxford University Press.

Updike, S., and Hicks, G. (1967) The enzyme electrode. *Nature*, 214, 986–988.

van der Meer, J.R. and Belkin, S. (2010) Where microbiology meets microengineering: design and applications of reporter bacteria. *Nature Reviews Microbiology*, 8(7), 511–522.

Vostiar, I., Tkac, J., Sturdik, E., and Gemeiner, P. (2002) Amperometric urea biosensor based on urease and electropolymerized toluidine blue dye as a pH-sensitive redox probe. *Bioelectrochemistry*, 56(1), 113–115.

Wang, J., Li, Y., Bian, C. et al. (2017) Ultramicroelectrode array modified with magnetically labeled Bacillus subtilis, palladium nanoparticles and reduced carboxy graphene for amperometric determination of biochemical oxygen demand. *Microchimica Acta*, 184(3), 763–771.

Yoshida, N., Hoashi, J., Morita, T. et al. (2001) Improvement of a mediator-type biochemical oxygen demand sensor for on-site measurement. *Journal of biotechnology*, 88(3), 269–275.

Yoshida, N., McNiven, S.J., Morita, T. et al. (2002) A simple, multiple simultaneous spectrophotometric method for bod determination using DCIP as the redox color indicator. *Analytical letters*, 35(9), 1541–1549.

12

Practical Aspects and Case Studies of Industrial Scale Fermentation

Sara Magdouli[1], *Thana Saffar*[2], *Tayssir Guedri*[1], *Rouissi Tarek*[1], *Satinder Kaur Brar*[1], *and Jean François Blais*[1]

[1] *Institut national de la recherche scientifique, Centre - Eau Terre Environnement, Québec, Canada*
[2] *Université du Québec en Abitibi-Témiscamingue, Rouyn-Noranda (Québec), Canada*

12.1 Introduction

For a long time, fermentation technology has been used for the production and conservation of food; recently, it was widely used for the production of bulk chemicals, fine chemicals, and pharmaceuticals as well. The concept of "white biotechnology" (i.e. industrial biotechnology) is newly developed and is known to be one of the important technologies that aimed to reduce the energy dependence on fossil fuels. Among value added products (VAP), produced by fermentation processes, organic acids, alcohols, antibiotic, fuels are very common:

i) Organic acids (citric acid, fumaric acid etc.) are mainly produced by filamentous fungi such as *Aspergillus niger* (Shu and Johnson, 1948) and *Rhizopus* (Das and Brar, 2014);
ii) alcohols are generally produced by bacterial species such *E.coli*, 1,3-propanediol is one of the important alcohol found in industrial sector, this latter is produced through the DuPont process by genetically modified *Escherichia coli* strain (Adkesson et al., 2004);
iii) antibiotics are generally produced by fungi such as *Penicillium chrysogenum* (Moyer, 1948);
iv) recombinant proteins such as insulin are again produced by fermentation process at large scale by *E. coli* (Johnson, 1983); and
v) biofuels by appropriate strains (Magdouli et al., 2014). The production of listed compounds in the industrial sector needs to be controlled and optimized. Generally, fermentation process optimization is carried out to reach higher volumetric production rates (= units of product formed per unit reactor volume per time) as well as higher by-product titers or (VAP). Both

Microbial Sensing in Fermentation, First Edition. Edited by Satinder Kaur Brar, Ratul Kumar Das, and Saurabh Jyoti Sarma.

of these parameters are important to achieve an economical process since high product titers will reduce the cost of downstream processing. Various points must be considered during the fermentation process, but they can be summarized in two key factors (strain improvements and process improvements).

For the strain improvement, diverse genetic tools are carried out (Ostergaard et al., 2000; Lütke-Eversloh and Bahl, 2011; Woolston et al., 2013). On the other side, for the process improvement, many issues and strategies can be done that include the variation of feeding strategy during the fermentation process, the manipulation of the media composition and other operational factors such pH, oxygen etc. To achieve the success of any biotechnological bioprocess, upstream and the downstream studies must also be well understood and optimized (Figure 12.1). Although the progress and successful results achieved in fermentation processes, the control of the suitable process recipe production is still in empirical way. Generally, experiments were performed at bench scale, then moved to pilot-scale and finally transferred to full-scale production fermenters, and further optimized there if needed. In this regard, many studies and diverse models are proposed to facilitate the transfer line from bench scale to large scale. The process analytical technology (PAT) guidance was developed to provide a better knowledge and details on the process, However, this guidance doesn't result in improved scaling-up strategies. So, other tools and equipment such as near infrared (NIR) or Raman spectroscopy were given. Again, these models were invalid and far from being fully developed, due to many limitations (e.g. variability and heterogeneity of a fermentation broth (i.e. complex media, cell morphology, and medium rheology), interference of gas bubbles and biomass particles with the liquid phase of aerobic fermentations). Besides, the investment in these tools is again costly compared to the low VAP desired during the fermentation process (Gernaey et al., 2012; Cervera et al., 2009; Scarff et al., 2006).

This chapter focused on some of the challenges related to the process improvement and the scale up efficiency. Afterwards, the chapter aimed to discuss some of the major issues that hurdle the bioprocess performance. Finally, some case studies were considered.

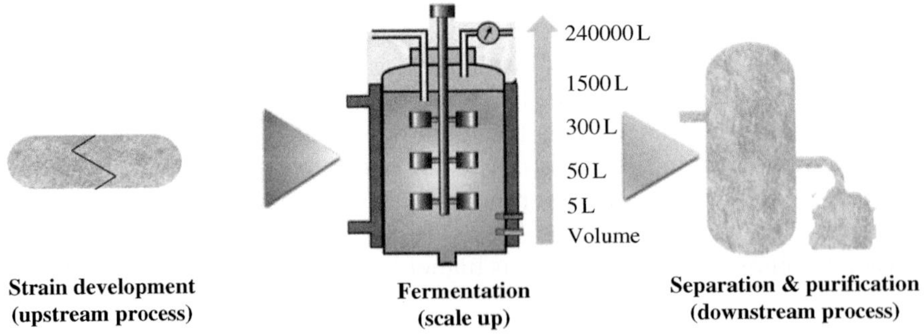

Figure 12.1 General schemes of bioprocess stages. (*See insert for color representation of this figure.*)

12.2 Scale Up Challenges

The scale-up of fermentation processes is a multidisciplinary approach of a very complex nature. Many fermentation process-related challenges that hurdle the fermentation performance are given in this section. Accordingly, scaling up a fermentation process aimed to transform optimal operating conditions found in laboratory- or pilot-scale bioreactors to the production-scale bioreactor in attempt to obtain maximal volumetric productivity of the full-scale process. Meanwhile, as discussed earlier, other tasks need to be considered before scaling up a process which include the screening and selection of the efficient strain, the inquiry of suitable tools for strain improvements and the manipulation of other culture conditions, for instance media composition, and various other parameters (Posch et al., 2013). Generally, these tasks are fulfilled in lab scale while using microtiter plate cultivations and then shake flasks. Microtiter plate cultivations are designed for screening purposes. When potential strain candidates are selected, experiments will be performed in bench-scale fermenter, generally (stirred bioreactors with a volume capacity between 0.5 to 20 L). Latter, the process parameter optimization is carried out in pilot-scale equipment.

A major drawback related to the use of bench-scale bioreactors and the large pilot-scale bioreactors (20–2000 L) is the difficulty to compare experimental data from small to large bioreactors. Efficient and fast scaling up of a process was reviewed and detailed by Schmidt (Schmidt, 2005).

To date, scale-up is based on an empirical approach that relied on oxygen transfer correlations.

Practically, scale up is done iteratively, by comparing the results of pilot plant experiments with key data obtained from the industrial-scale processes. The scale-up is considered successful if the pilot obtained results can be reproduced at industrial scale. And if it is not the case, additional experiments should be performed to investigate the effect of additional parameters on process performance.

12.2.1 Agitation

Accordingly, various physical differences between scales impacts the reproducibility of results. For instance, in bench-scale bioreactor, perfect mixing of the broth can be assumed, while in a large bioreactor the mixing time for achieving 95% homogeneity (θ_{95}) can be in the range of minutes (Gabelle et al., 2012). Generally, poor mixing is responsible for oxygen starvation in the medium (Junker, 2004). Furthermore, often, glucose, pH, and temperature gradients are all likely to emerge in production-scale fermentation causing heterogeneities (Larsson et al., 1996a; Oosterhuis and Kossen, 1984). These heterogeneities are one of the primary concerns when dealing with process scaling in the microenvironment and these do affect the biomass growth.

Thus, it is crucial to understand the interactions between the liquid motion, the agitator speed and the power input into the bioreactor. Any small change of scale can affect these interactions. Nienow (1998) has illustrated many physical aspects of the agitation/agitator that required high considerations in any bioprocess (Nienow, 1998). Generally, the agitation affects the viscosity of the medium. For example, with low viscosities, the flow in the fermenter is turbulent when moving from small bench bioreactor to the largest scale. The Reynolds number is another way of clarifying the differences in flow regime as it also changes between scales. Its formulation is given by this equation

Reynolds number, Re = ρLND2 /μ = 104 where ρL is the growth medium density (kg/m3), μ is its viscosity (Pa s), D is the impeller diameter (m) and N, its speed (rev/s).

This number is used to analyse the fluid mechanics in the bioreactors across the scales. The differences in mixing properties will lead to differences in mass and heat transfer in the processes (Pollard et al., 2007). For instance, poor mixing working on a larger scale is aggravated by increasing vessel sizes. Accordingly, in larger reactors, many problems can occur which consist of bad disturbance of substrate and oxygen, leading to the formation of depletion zones with enhanced stress conditions. Besides, in the large volume of culture media, mixing times are longer and higher hydraulic pressure gradients influence the oxygen transfer rate measurement (Larsson et al., 1996b; Bylund et al., 2000; Bylund et al., 1998; Bylund et al., 1999; Enfors et al., 2001).Another important observation to be noted is that cells at the fermenter top are more exposed to excess glucose concentrations and oxygen limitations, whereas on the other hand, cells at the bottom are exposed to glucose starvation. For example, during the culture of *E. coli* in the presence of excess glucose concentrations acetate overproduction (overflow metabolism) resulted. However, the oxygen limitation induces the formation of ethanol, hydrogen, and other organic acids such as formate, lactate and succinate (Bylund et al., 2000; Xu et al., 1999; Castan and Enfors, 2001). These acids can be assimilated in oxygen rich zones; however, they lead to a temporary acidification of the microenvironment. Combined with a decreased transportation and elimination of carbon dioxide, detrimental metabolites and surplus heat generated by agitation and metabolic processes resulted in zonal overheating. Usually, in large scales, the lower mixing rates lead to the formation of zones with enhanced stress conditions. The subsequent activation of stress genes (a survey on bacterial stress proteins is given by Kwint et al. (2003), only partially protects the cell against detrimental stress effects) (Kwint et al., 2003). For instance, despite activation of heat stress genes like *E. coli* dnaK and clpB, reducing misfolding and aggregation of heat-sensitive proteins (Mogk et al., 1999), metabolic changes and damages such as translocation of proteins and membranes are observed upon heat exposure (Umakoshi et al., 1998). The unavoidable detrimental effects of the repeated and cyclic passing of different stress zones and the subsequent continuous activation and shut down of the corresponding stress genes are believed to lead to a completely altered physiology (Schweder et al., 1999, Enfors et al., 2001), with constant metabolic shifts, which ultimately reduce growth and productivity and increase by-product formation.

12.2.2 Mass Transfer of Oxygen (Mass Transfer, Morphology, and Rheology)

The transfer of oxygen into a fermentation broth has been studied since 1940. The volumetric mass transfer coefficient of oxygen (k_La) is used as a critical parameter for scaling purposes (Reuss, 1993) and maintaining k_La constant across scales is primordial, however it is not possible to conserve all parameters across scales, and a suitable compromise must be found.

Other important scaling parameters include the oxygen transfer rate (OTR), the volumetric power input (*P*/*V*) and the superficial gas velocity (*vg*). A maximum oxygen transfer rate must be achievable, and this rate is ultimately dependent on the mass transfer coefficient, $\mathbf{K_La}$ (1/s), and the driving force for mass transfer, **ΔC**, since **OUR=** $\mathbf{K_La \times \Delta C}$. For oxygen transfer, the driving force **ΔC** is the difference between the oxygen concentration in the air bubbles and that in the broth, which must always be held above the critical dO_2 value throughout the fermenter for the duration of the process.

The overall oxygen demand of the cells throughout the batch or fed-batch fermentation must be met by the oxygen transfer rate and the demand increases as long as the number of cells is increasing. Accordingly, for every mole of O_2 utilized, 1 mole of CO_2 is produced, and the respiratory quotient, RQ = 1 (Nienow, 2006).

Usually, the power requirement for agitation in large vessels is lower than that in small vessels, which is principally due to the effect of the higher pressure and the increased pressure from the liquid column. This will facilitate oxygen mass transfer. In order to achieve a high volumetric power input in small bioreactors, the agitation speed has to be very high, which resulted in higher shear rates in the smaller vessels. Accordingly, fermentation broths exhibited various flow behaviours (non-Newtonian, Newtonian, Pseudoplastic, a shear ticking, a shear thinning). For example, the viscosity of non-Newtonian broths is dependent on the shear rate and the difference in shear rate across scales will lead to differences in apparent viscosity (Meerman et al., 2004). This in turn affects k_La and explains why the mass transfer of filamentous broths is limited in large scale. Thus, mass transfer is strongly dependent on the viscosity of the fermentation broth (Magdouli et al., 2017). The rheology of the broth is thereby at the same time dependent on the biomass concentration and the morphology of the cells used as expression system (Cascaval et al., 2003). It is the complex relation between all these variables that makes scaling up/scaling down difficult.

In order to have proper oxygen supply, stirring and aeration are a prerequisite for almost all types of cells and even for cell cultures (Ma et al., 2006). However, agitation may also lead to changes in morphology which in turn will affect product formation (Smith et al., 1990, Jüsten et al., 1996). Therefore, with an aerated STR setup the question of shear sensitivity of the employed organisms arises, and there has been an ongoing discussion around the topic.

12.2.3 "Shear Damage"

Bacteria and yeast on the other hand appear to be organisms that are more resistant to mechanical stresses, and it has been suggested that their small size is the reason for their resistance to shear stress: they are smaller than the Kolmogorov micro-scale of turbulence (Nienow et al., 2010).

Therefore, it is believed that the morphology is not affected by mechanical mixing. Nienow (Nienow, 2009) reported no cell damage by agitation based on flow cytometry. Nevertheless, they reported a reduction in biomass concentration, but increase in viability when scaling up a process; this was attributed to the poor homogeneities in industrial scale reactors (Nienow, 2009).

"Shear damages" are frequently made to explain poor process performance when mechanical agitation and aeration are introduced into a bioreactor as compared to the non-agitated and non-sparged conditions in a shake flask or microtitre plate (Thomas, 1990). This latter has suggested that cells might be considered to be unaffected by fluid dynamic stresses if they were of a size smaller than the Kolmogoroff microscale of turbulence, λ_K. The microscale of turbulence is related to the local specific energy dissipation rate ε_T.

Generally, fermentation process optimization studies are focused on filamentous fungi since bacteria and yeast cultivation exhibit a Newtonian behavior due to the spherical shape of their morphology, which facilitates to work with. However, the most challenging expression systems are the filamentous microorganisms in terms of mass transfer and rheology because of their morphology that impacts directly the mass transfer. Generally, the morphology for a given bioprocess varies, cannot be generalized, and relies on the desired product (Gibbs et al., 2000). Moreover, the phenomena of pellet formation can occur and varied depending on the fermentation and the employed strain. Accordingly, this latter resulted from the aggregation of spores after inoculating the media (Grimm et al., 2004). Latter, the requirements for analysis methods to study the morphological changes is crucial, thus, particle size analysis like laser diffraction and focused beam reflectance (FBR) can be investigated in this aim to study the pelleted biomass (Grimm et al., 2004, Rønnest et al., 2012). The study of pellet-type morphology is very important, because it contributes to the simplification of downstream process. For instance, the Newtonian fluid behavior due to the biomass pellet formation required low aeration and agitation power input, meanwhile the pelleted morphology lead to nutrient concentration gradients within the pellet. Other to macroscopic level, the morphology influenced the production kinetics at the microscopic level, for example in the case of higher enzyme secretion by *A. oryzae* (Spohr et al., 1998). Consequently, the morphology of filamentous fungi had an impact on the productivity and it is affected by the outer appearance of the fungus, this finding was approved by Ratul et al. (2014), who found that the morphology of *Rhizopus oryzae* affected the fumaric acid production (Das and Brar, 2014). These observations were in agreement with Magdouli et al. (2017) who found that the morphology of Yarrowialiyolytica is ultimately correlated with lipid production (Magdouli et al., 2017). Usually, the shift in morphology and the pellet formation is due to the variation of disturbance of the system/equilibrium as well as operating factors (i.e. pH changes and salt composition) and in some case N-starvation led to the transition from pellet to free mycelial growth.

For non-Newtonian fluids behaviours (e.g. fungal fermentation) the study of morphological behaviour and at which shear rate the viscosity should be evaluated, might cause problems during scale up. Moreover, there is no reliable shear rate that can be estimated in the fermentation tank itself. In the case of stirred tank reactors, the shear rate varied along the fermenter, thus, it increased at the agitator tip and decreased when approaching the vessels walls. Hence, the simple method of shear rate calculation is to take the maximum or the average shear rate. Meanwhile, it is not clear which shear rate is governing the mass transfer processes. The calculation of shear rate is usually done empirically by the method of Metzner and Otto (1957) based principally on the Reynolds numbers in the laminar and transitional regime (Metzner and Otto, 1957). This method has been employed in both laboratory and full-scale fermenters (Stocks, 2013), although, however, it remains limited. Finally, the determination of shear rate and viscosity across scales is very challenging.

12.2.4 Measurements for Control

In industrial bioreactors, the sensors play an important role in the entire process and are limited to pH, dissolved oxygen (DO) and temperature sensors at a single location in vessels with large volumes, often with concentration gradients (Larsson et al., 1996a). Besides, the spatial heterogeneity in the vessel should be considered during the process, however, its determination is limited. Generally, during fermentation, the variables should be controlled and kept close to the set point and optimal value. Among standard methods employed PID controller is very representative. However, this method is still limited because of the time – variation of set points and the difficulty of their control. Another point to be taken into consideration is the *supervision* phenomena due to the possible changes which occurred during industrial fermentation; for example, substrate composition, moisture content, and temperature, where the set points are modified accordingly.

The supervision phenomena are critically difficult due to the problem of controlling important variables such as substrate, product, or by-product concentration, which are less controlled and sporadically monitored compared to other controlled variables such DO and pH. Besides, the models used to synthesize controllers that operate are close to the set point in industrial operations are rare. Consequently, the inquiry of complementary approaches such as hybrid models that focused on the efficient control design with minimal modeling are highly demanding.

12.2.5 Other Aspects

To obtain high product yields on a large scale, larger culture broth volumes and longer inoculation time were needed. However, in the case of recombinant hosts, the maintenance of the stability of the propagation of plasmids to daughter cells, is more difficult to control and less easy to be maintained on a larger scale. Accordingly, plasmid stability is governed by the plasmid properties (i.e. size and nucleotide sequence) and, by the genetic background of the host (Friehs, 2004) as well as by process parameters such as growth rates, substrate concentrations and temperature. Neubauer et al. (2003) have shown that high glucose concentrations diminish plasmid stability (Neubauer et al., 2003), meanwhile, Lin and Neubauer (2000) have mentioned that the rapid glucose oscillations favour plasmid stability and recombinant protein production rate (Lin and Neubauer, 2000). For this attempt, the application of runaway plasmids was investigated (Ansorge and Kula, 2000). This approach aimed to separate the replication and the expression of plasmids from the growth phase. More than 1000 copies of plasmids can be obtained. Other issues can be investigated such as the reduction of the generation time and generation numbers throughout the process, the deceleration of the metabolic speed, e.g. through a reduction in temperature, and finally the development of more stress-resistant microbial strains. Thus, the critical issue to be considered during the scale up is to develop a proper process design that improves the physiological conditions and greater enhances the metabolic accuracy by minimizing microbial stress exposure.

12.3 Microbial Tolerance

During the optimisation of fermentation process, it is crucial to increase tolerance of the strain to the desired product, this task can be performed at any stage of strain development (Ling et al., 2014). This will allow the direction of metabolic flux to the formation of a desired product. Generally, the improvement of microbial product tolerance is recommended to generate the product at a requisite level. Usually, the development of a product-tolerant strain can be done by a serial of cell subculture with the increase of the concentrations of product with or without mutagen treatment, followed by the identification of cells that grow faster (Utrilla et al., 2012).

By repeating the subculture process, the tolerance level can be increased. Improvement of product tolerance has also been achieved by using rational engineering approaches, such as the use of an efflux pump (Dunlop et al., 2011) or the manipulation of ionic membrane gradients of *S. cerevisiae* (Lam et al., 2014). The overexpression of genes encoding the L-valine exporter led to an increase in L-valine production titer by more than 40% in *E. coli* (Lam et al., 2014, Park et al., 2007). The engineering of efflux pumps to improve product tolerance and/or titer are a strategy that deserves investigation. However, the producer cells selected from serial sub-culturing or rational engineering do not necessarily overproduce the product. This is mainly due to the fact that enhanced product tolerance at high concentrations does not necessarily correlate with an ability to synthesize product at increased specific productivity and yield.

12.4 Phage Invasion

Many biotechnologically-important by-products are produced microbial cells which cover a wide range of applications in various sectors of environmental biotechnology. However, serious problems occurred in laboratories and factories when optimising the process on a large scale. Among potential problems that face the effectiveness of the process and cause hurdles for the by-product production, the prevalence of natural parasites such as bacteriophages during the fermentation process is very common (Callanan and Klaenhammer, 2002). Bacteriophages like all viruses are obligate intracellular parasites that need a host cell to multiply. They require a susceptible host cell that is actively growing to multiply (Lu et al., 2003). The bacteriophage infection resulted in detrimental problems (e.g. loss of bacterial culture and bioproduct formation, the spread of bacteriophage contamination throughout the laboratory or factory (Ogata, 1980).

This problem is critical, and many efforts should be made to avoid such operational problem that contributes to cell lysis of cultures even after extensive cleaning process. Accordingly, decontamination and disinfection process remain difficult and even impossible in large factories. Consequently, phase invasion is a serious problem and if phage is present during the process fermentation, it can spread rapidly throughout the plant and resist to decontamination process. This causes the persistence of phage for long periods and the reappearance of phage infections in the factory even after treatment (Primrose, 1990). Not only contamination but also phages affect the genetic material of the host, for instance upon infection, phages can insert their genetic material into the host chromosome, and lead to the formation of prophages, instead of lysing the cell (Los et al., 2004). Consequently, the growth rate is lowered and the efficiency of synthesis of bioproducts is decreased (Jones et al., 2000). To avoid cited problems, different issues need to be taken into consideration (e.g. the location of a bioprocessing factory, plant design) which have a crucial impact on the occurrence of infections. In the case of bioprocesses involving genetically modified E. coli, the factory should be far from sewage disposal plants to minimize the risk of coliphages propagation and dispersal, known to be present at high number in sewage plants. Besides this, the sterilisation and the control of sampling ports should be done carefully since these places may increase the propagation of unwanted phages (Los et al., 2004). In this regard, the decontamination process should be performed carefully with high precautions to prevent phage infection during fermentation process. A better knowledge of the fundamental properties, genomic taxa, and the mechanism of infection should be provided. Generally, infection occurs when the phage is adsorbed or attached to the bacterial surface, having a complementary site. Then, viral nucleic acid enters the cells where transcription and transduction of viral genes occurred. Then, viral lysozymes are synthesized and attack the peptidoglycan layer of bacterial cell walls to induce cell lysis and new phage particles are released into the environment (Holder and Bull, 2001; Budzik, 2003).

Recent debates focused on the development of effective methods of decontamination since a large number of phages was found to be resistant to the common processes of sterilisation such as drying, freezing and thawing and phages can survive for long periods (Atmar et al., 1993). Furthermore, if infection occurred during the bioprocess, the bioreactor should be shut down immediately since continued aeration of the culture medium will dispense large numbers of phages into the environment of the factory which make the clean-up process very difficult. Moreover, culture having excess foam is more sensitive to phage infection, so the control of foam addition during the fermentation is very important. Besides, if the inlet and the outlet of air are very close, their risks of contamination are high and different aeration system, e.g. filters, or compressors will be later contaminated (Primrose, 1990; Wunsche, 1989). The sterilization of the bioreactor with high temperatures is suggested and the bioreactor must be held at a high temperature for as long as possible. Besides, the phage assay should be carried out after each step of sterilization to ensure the absence of viable phages. Moreover, different air filters and compressors, air inlets and outlets must be extensively sterilized, because bacteriophages can pass through most of the filters, if not all. Besides, all exterior surfaces of the pilot line must be disinfected. All of these issues depend on the size and design of the plant. The washing of whole plant with hot water is suggested,although however, this seems to be not practicable. Furthermore, samples of the washing liquid should be subjected to a phage assay to determine the extent of infection and the efficiency of the washing procedure in phage reduction. To sum up, urgent measures must be taken during the bioprocess to prevent the phage invasion. Most of the proposed methods include:

i) Good laboratory hygiene;
ii) the employment of phage resistant mutants during the bioprocess; and
iii) the use to potential chemicals designed specifically to halt the multiplication and dispersal of phages.

As discussed earlier:

i) Good laboratory hygiene is a crucial issue in preventing phage invasion. All the equipment and apparatus should be washed and sterilized. The medium of the bioprocess (i.e. raw material) should be sterilized and the assay of phage infection should be done regularly. Another key factor in preventing phage evasion is the verification of the purity of the seed culture in the glycerol stocks and the get rid of the spoiled cultures by autoclaving (Los et al., 2004; Ogata, 1980);
ii) the employment of phage-resistant mutants involve the isolation of a potential culture with high resistance to phages destruction, this latter will be used after the process in the fermentation process. The choice of potential mutants should be carried out carefully and the selected strain should exhibit the same production capabilities as does the original strain. However, the examination of phage infection should be done herein regularly since some resistant mutants alone cannot cope with phage infections. Although mutants contribute to the stabilization of the fermentation process, it is still important to ensure that all phages are eliminated (Los et al., 2004; Jones et al., 2000);
iii) the Employment of chemicals that alter phage multiplication and they don't show a side effect on the by-product formation. For instance, the employment of divalent cations, chelating agents or DNA injection is very common. These chemicals will inhibit the phage multiplication without interfering with byproduct formation. Another option can be investigated in this aim which consists on the employment of specific antibiotics (Ogata, 1980) as well as some vitamins such as ascorbic acid that are reported to cause virus inactivation (Wang and van Ness, 1989). There is overwhelming literature that suggested that certain vitamins were capable of hindering bacteriophage infection (German and Dillard, 2006).

Higher temperatures are required since phages are sensitive to heat; however, some phages can resist to temperatures (Callanan and Klaenhammer, 2002). Generally, the sensitivity of bacteriophages to environmental factors correlates with the corresponding properties of the host. The employment of chemicals detergents and disinfectants is also common, for example sodium dodecyl sulfonate showed a potential phagocidal effect (Wunsche, 1989). UV treatment can be very helpful in keeping production rooms clean and disinfected (Los et al., 2004; Wunsche, 1989). UV rays are known to disrupt DNA phage and inactivate the phage. However, some phages have the ability to repair their own DNA damage (Kokjohn et al., 2005) which make this process not practicable.

12.5 Process Failures

The identification and genetic engineering of the potential organism, followed by prudent piloting certainly is very important to success with scale up fermentation. Therefore, the success of any biotechnological process lies in its capacity to be scaled up to the industrial level. So, the effective technology transfer from the lab scale to the large scale, required the development of more efforts and attention to several issues. Thus, a scale-up strategy that combines integrated teamwork with solid engineering efforts can go a long way despite its complexity and difficulty. This led to minimize the cost of the whole process on time and on budget. However, to achieve this goal, many bottlenecks and limitations in asepsis discussed in this chapter should be resolved.

To success the scale-up and the commercialization of any process technology; various points must be considered and managed in such as business, market, and technology risks. The risk identification, risk assessment and prioritization, and development and execution of risk mitigation plans are of importance to achieve a successful track in scale-up.

Among strategies, piloting is used for reducing the risk of failure at larger scale owing to many factors:

i) Operational issues (fouling, plugging, corrosion, etc.);
ii) recycle problems (build-up of impurities); and
iii) process capability issues (low yield and capacity, poor product quality, etc.).

At an early stage of scale up, the process failure is due to the inability to produce large quantities of a viable inoculum, overheating of the bio piles. Besides, the lack of understanding of the potential markets and process scale up can cause an early failure that generates latter negative impacts on the development of the technology. Other technical problems such as technical complexity, high installation and operating costs are again a major cause of failure. In this regard, the optimization of the fermentation process is required to avoid process failure. Once the feasibility of the production has been demonstrated, the fermentation can be succeeded. Other points have to be taken carefully during the large-scale fermentation such as the stability of the strain. Accordingly, most of expressions systems based on plasmids are sometimes unstable and depended with the host characteristics (Posno et al., 1991), culture conditions (Koizumi et al., 1985; Reinikainen and Virkajärvi, 1989). Thus, a better documentation of the strain and the cloning procedure of the strain is required and approved by a competent quality assurance service. These considerations impact strongly on the scale up as well as the downstream process. Moreover, the quality of the communication between developers of the fermentation and the downstream process is crucial and contributes to the success of the scaling up the process.

One of the most important factors of success is to define the variables that could affect the productivity or the quality of the product. Statistical data analysis enables the visualization of the interactions among experimental variables, leading to predictions of the experimentation (Murphy, 1977). The factors that affected the scale up process are the number of generations, the mutation probability, and medium sterilization, the quality of temperature and pH regulations, agitation, aeration and pressure. Consequently, it is pivotal to the preparation for the scaling-up of a process to first scale-down to the pilot scale of the conditions of culture (Kwanmin, 1989). The process should be characterized and validated by some controls to evaluate the robustness of the procedure as well as its reliability and reproducibility. All the equipment must be validated for good manufacturing practices like the qualification of the installation, the validation of cleaning in place and sterilizing in place, maintenance and calibration plan (Naglak et al., 1994).

Lack of technological knowledge, insufficient management capacity, high construction costs and the absence of maintenance services contributed to failure process. The lack of success can be explained by inadequate technology and know-how, and a strong focus on quantitative implementation targets. Furthermore, the scale up of fermentation depends on power input or airflow rate per unit volume, as well as the type of impeller. The difficulty in scaling up a fermentation process is that it is unlikely to simulate simultaneously in geometry, fluid kinetics and fluid dynamics.

It is imperative to develop a feasible approach that considers various parameters, which are useful to the scale up of a process. Added to these aspects, a well-planned pilot plant design and operation is essential to minimize the risk of failure. A laboratory development program that combines both the core technology and the necessary data for a pilot plant to handle separation steps, waste streams, and material handling operations is required to meet different pilot goals. These latter must be clearly stated to prevent commercial failure.

12.6 Potent Inhibitors (e.g. Substrate Inhibition)

In any fermentation system, diverse interactions between the environmental conditions and the growth rate of the microorganisms occurred. Recently, trends have been increased towards the control of fermentation systems to optimize the biomass or product productions. Thus, the control of substrate was shown to improve the final biomass concentration, productivity rate or other kinetic parameters of the fermentation. Practically, many attempts at substrate concentration control are ineffective.

The ultimate limit to cell population growth could be simplistically considered to be the availability of major substrates. Substrate availability is the primary constraint on growth. Accordingly, during fed batch process, high cell density is required to increase the volumetric productivity of many by-products as nucleic acids (Elsworth et al., 1968), amino acids (Forberg and Haggstrom, 1987) and heterologous recombinant proteins (Riesenberg and Schulz, 1991). So, the key feature in such process is the continuous feeding of the concentrated growth limiting substrate (i.e. the carbon source). Consequently, specific growth rate is increased and many problems that occur during unlimited batch processes are resolved which are mainly related to catabolic regulation, oxygen limitation and heat generation (Minihane and Brown, 1986). Besides, the "overflow" caused by toxic concentrations of metabolic biproducts can also be avoided. The overflow has been reported for many strains such as *Saccharomyces cerevisiae* and *E. coli* respectively (George et al., 1993, Andersson et al., 1996). In *S. cerevisiae*, the overflow metabolism, named also as "Crabtree Effect", the biproduct (i.e. ethanol) exhibited an inhibitory effect on the production of acetate. Many commercial fermentation processes are operated in fed-batch mode, where a limiting carbon source substrate is dosed to the fermenter during the fed batch phase (Lee et al., 1999). Most of the cited examples in literature are of lactic acid production which can be improved by selectively removing the product from the fermentation broth, thereby reducing the effects of product inhibition on the production strain at high titers (Schmidt et al., 2000).

In batch fermentation, overflow metabolism can be omitted by the use of a slowly metabolizable carbon source such as glycerol (Elsworth et al., 1968), however, this method is not practical and doesn't show higher efficiency, but the preferred method is the use of a fed-batch process where growth can easily be controlled by substrate feed rate (Magdouli et al., 2014).

The timing of the addition and its position should be controlled. The rapid dispersal of the feed is envisaged, and the optimal position of feed was found to be near the impeller, so that the high concentrations of the feed can be reduced to the desired value (Nienow, 1998; Nienow, 2006). Most industrial processes use surface additions (reduction of contamination and pipe blockage), as well as to ensure mechanical stability (Nienow, 1998). Consequently, the design and the bioreactor configurations should be designed according to issues discussed below.

Another issue to be considered is the control of oxygen during the feed batch. Oosterhuis et al. (1985) have mentioned that at a large scale, major differences in dO_2 are found in batch fermentations. The concept of "regime analysis" developed by Oosterhuis et al. (1985), who proved such differences (Oosterhuis et al., 1985). For a long time, radial flow Rushton turbines were used to satisfy oxygen mass transfer. Today, Rushton turbines were replaced by hydrofoil impellers to enhance the mass transfer, bulk mixing, high homogeneity and consequently fermentation performance (Buckland et al., 1988).

The hypertonic environment caused by excessive levels of substrate (glucose) is responsible of the weakness of cell viability and the fermentation ability of yeast. For instance, during the very high gravity fermentation, when the concentration of the substrate is higher than a certain limit, the production for example in this case (ethanol) and yeast concentration will not increase even with additional substrate (Thomas and Ingledew, 1992). Usually, when the substrate concentration is higher, the microorganism accelerates its metabolism to increase the consumption rate of substrate in order to reply to the higher substrate concentration. Therefore, the growth of microorganism can be affected and lowered due to the damage induced by hypertonic solution on membrane fluidity and enzyme activity. This phenomenon represents the suppression of yeast growth and reproduction induced by hypertonic solution that contains higher substrate concentration. Under these conditions, growth parameters will be affected, for instance, Yx/s decreased as the substrate increased and the additional carbon source will be consumed by strain not for growth but to for survival in the overly hypertonic microenvironment. Moreover, the long exposure to a hypertonic solution can lead to a severe decrease in cell membrane fluidity, making it more difficult for the substrate to enter and for the product to exit the cell (Thomas and Ingledew, 1992). Consequently, the additional carbon sources were consumed by cell to maintain the activity of the transport system of essential materials instead of being fermented to the final product (e.g. ethanol). Thus, the study of the substrate concentration is of great importance in the large-scale study and the most appropriate substrate dosage should be optimised to maximise the product formation that will not affect the yeast fermentation capacity.

In addition to hypertonic solutions caused by higher substrate concentration, it may create a decrease in membrane fluidity which may lead to the accumulation of ethanol and other toxic metabolites in the cells, which may cause biological damage to the yeast inducing damage to its transport and metabolism system added to the excessive loss of water from the cell might result in cell atrophy and organelle dehydration (Thomas and Ingledew, 1992).

Besides, the higher concentration may alter the activity of the intracellular enzyme activities, reducing substrate utilization and ethanol fermentation as well as influencing yeast growth and reproduction. For instance, hexokinase (HK) and pyruvate kinase (PK) are the key enzymes of the Embden-Meyerhof-Parnas (EMP) pathway. HK converts glucose into glucose-6- monophosphate in the first step of the EMP pathway. This step is incremental in determining the speed of the subsequent reactions. PK catalyzes the conversion of phosphoenolpyruvate to pyruvic acid. Therefore, the energy supply of the cell is mainly determined by the activities of PK and HK. In this study, the highly concentrated substrate greatly restricted the normal expression of HK and PK and thus directly reduced the flux intensity of the EMP pathway. The metabolic disorder of glucose is also expected to inhibit downstream ethanol fermentation and yeast growth (Rossi et al., 2003; Sridhar et al., 2000). Moreover, because fermentation is a comprehensive system that is affected by various factors, the over accumulation of the product during the later period of VHGF also synergistically inhibits the yeast's growth and growth activity. This may happen in response to excess product interfering with key enzymes of the glucose transport system and damaging the cell structure, particularly the membrane components and permeability (Fan, 2008). Further exploration of the inhibitory effect of the product alone is required.

Rather than alcoholic fermentation, Moulin et al. (1980) have observed this phenomenon of substrate inhibition during the fermentation of glucose by *Candida pseudotropicalis* strain when the concentration of sugar is above 100 g/L (Moulin et al., 1980). However, under the same conditions, lactose was less inhibitory than glucose. In both cases, there was a synergistic action between ethanol and the carbon substrate. However, the inhibitory action of the substrates appeared only at high concentrations and remained weak compared with the action of ethanol.

12.7 Case Studies: Biofuels (Biodiesel, Ethanol) Enzymes (Novozymes), Antibiotics, Platform Chemicals

12.7.1 Biofuels (Biodiesel, Ethanol)

The cost analysis is based on several factors like biomass transportation distance, type of biomass used, type of processing technology used, efficiencies of different processing steps, type of biofuel, and coproducts production.

The price of biodiesel can vary considerably, depending on the process used for its production and of the type of the raw material used in this process (Ong et al., 2012). Biodiesel costs about 1.5 times more than diesel depending on the source of raw material oils (Prokop, 2002). Krawczyk reported that 70–95% of the total biodiesel production cost arises from the cost of raw materials (Krawczyk, 1996).

In large scale biodiesel production, base catalytic process is most often applied in order to have fast reaction rate (less than 1 h for base catalytic process and several hours for acid catalytic process) and low requirement of the catalyst (0.3–4% w/w of the oil for base versus 5–25% w/w of the oil for acid catalytic process) (Kywe and Oo, 2009; Prueksakorn and Gheewala, 2006). For example, in the case of microalgae cultivation aimed at biodiesel production, the choice and the configuration of the bioreactor is crucial in the large-scale production of biodiesel, where tubular photo bioreactor is considered suitable for large scale cultivation (Sanchez Miron et al., 1999).

For biodiesel production, the scale up depends on many key factors such as raw material, its pre-treatment, the technology used for bio production, the recovery of the product and separation of biofuels from fermentation broth, extraction, transesterification is the key step of biodiesel production.

Life cycle analysis (LCA) is often used to assess the net environmental impact of these processing steps (Martinez-Hernandez et al., 2013, Sanchez Miron et al., 1999, Uihlein and Schebek, 2009). The challenges are in carrying out accurate LCA analysis depending on the data that is collected from the group of aligned processes that will be used in the biorefinery. Many companies are taking the LCA very seriously to assess the environmental impact so that they could make decisions to adjust the process or areas that is needed to be focused on to reduce emissions. Depending on the emissions estimated by LCA this will influence the cost of establishing the biorefinery.

Several bottlenecks that are being faced by big corporations to commercialize the biofuel production technology (pre-treatment, hydrolysis, microbial fermentation, and biofuel separation) (Figure 12.2). The choice of pre-treatment and lignocellulosic biomass could be decided based on the availability of a sufficient quantity of catalyst and feedstock in that region. The aligned technologies are currently scaled up to establish pilot plants to demonstrate the feasibility. Simultaneously, biomass logistics and techno-economic evaluations are carried out to assess the technology readiness level (TRL). Then assessments are made regarding the environmental impact of using different technologies. Once appropriate feedstock, pre-treatment, and enzymes are combined to produce cheap sugars, the choice of biofuels and biochemicals depends on the market demand and more importantly the biofuel policy defined by the local and federal

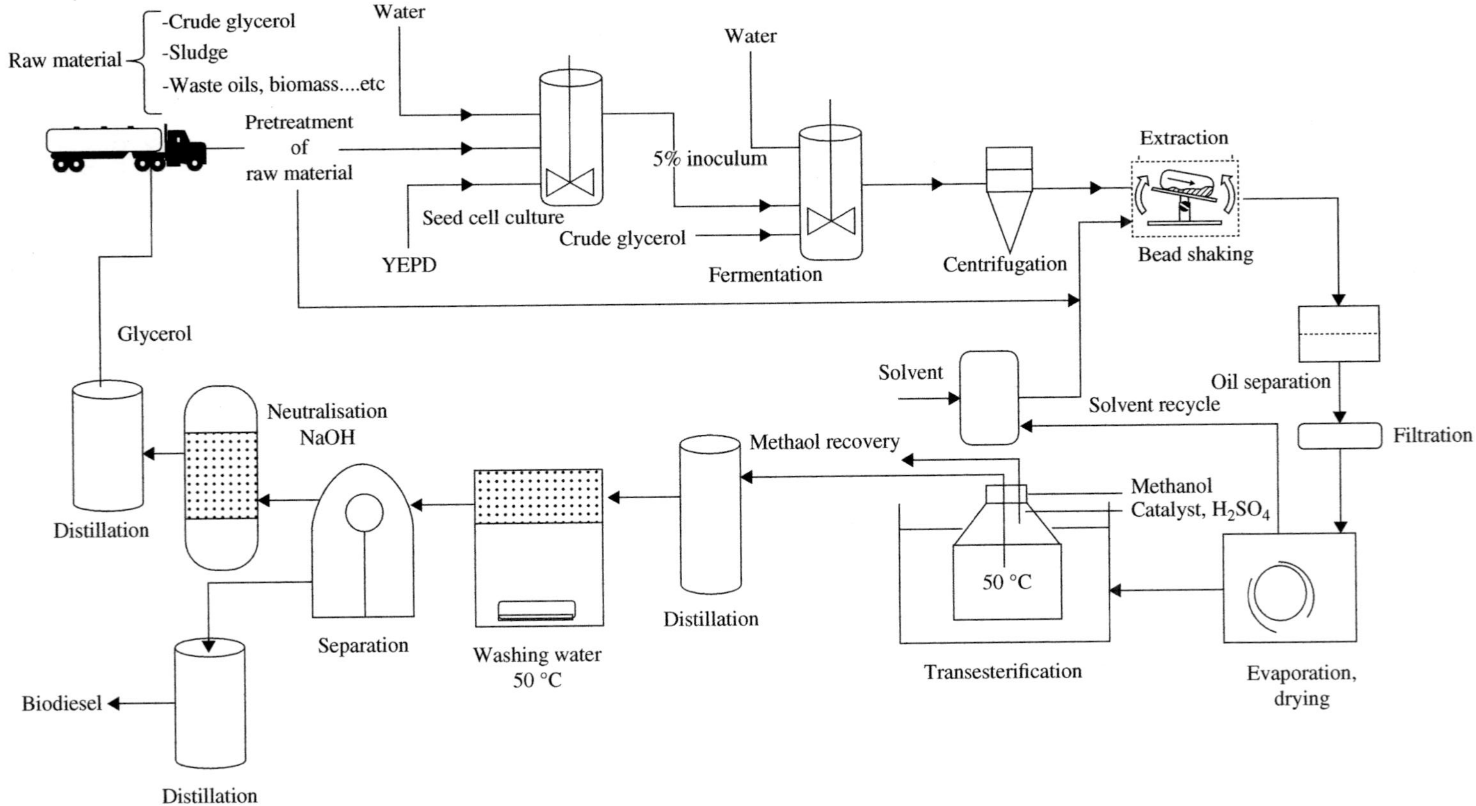

Figure 12.2 Flow diagram of Biodiesel production.

government. Furthermore, in order to compete with the cost of petroleum fuels, the cost of biofuel processing should be kept as low as possible using energy efficient technologies and using less water. Producing as many coproducts as possible in a biorefinery will help to reduce the cost of biofuel production.

It is important that a biorefinery should be established in an appropriate location that has good water resources, access to feedstocks, and energy that is needed to process the feedstock.

12.7.2 Enzymes (Novozymes)

Enzyme technology represents an interdisciplinary field, recognized by the Organization for Economic Cooperation and Development (OECD) and an important component of the sustainable industrial development (Bull et al., 1998). Thus, enzymes were known to catalyze chemical reactions with high specificities and high rate enhancements and covered many industrial commercial applications. Owing to metagenome screening and genome mining, many enzymes were discovered, and their field of application was extended to cover many sectors such as the manufacture of margarine, beer, yogurt, concrete, leather, textiles and ethanol, etc. Recently, technical enzymes had revenues of nearly US $1.2 billion in 2011, which is expected to reach US$1.5 billion in 2015 and US$1.7 billion in 2016. The highest sales are expected to be in the biofuels (bioethanol) market (World Enzymes, 2011). Besides, the use of enzymes for foods and beverages is expected to reach US$1.3 billion by 2015. For a long time, chemical synthesis of products and pharmaceuticals has been applied in the industry, however, this latter presented several disadvantages such as low catalytic efficiency, lack of enantiomeric specificity for chiral synthesis, need for high temperature, low pH and high pressure and required high concentrations of organic solvents which leads to the generation of organic waste and pollutants. In this regard, the use of enzymes for these applications was a promising alternative to limit these disadvantages. Besides, enzymes were found to work under mild reaction conditions (e.g. temperature, pH, atmospheric conditions) without the need for the protection of substrate functional groups, have a long half-life, a high stereo-selectivity yielding stereo- and regio-chemically-defined reaction products at an acceleration of 10^5 to 10^8 -fold, and, in addition, they work on unnatural substrates (Magdouli et al., 2014).

Enzymes are also used directly in products such as laundry detergents, where they help remove stains and enable low-temperature washing North America and Europe are the largest consumers of industrial enzymes although the Asia Pacific region will undergo a rapid increase in enzyme demand in China, Japan and India, reflecting the size and strength of these country's economies. Many industrial enzymes markets were developed and the largest market in the industry was Novozymes followed by DSM, and DuPont. Successful application of enzymatic processes in the chemical industry depends mainly on cost competitiveness with the existing and well-established chemical methods (Tufvesson et al., 2011).

Accordingly, naturally-occurring microorganisms are the most productive producers of enzymes. However, not all bacteria and fungi are suitable for the industrial production of enzymes. Only enzymes that are easy to handle, can be grown in huge tanks without light, don't produce unwanted side-effects, and have a very high growth rate, are selected. Most wild microorganisms found are not well suited to domestication in large fermentation tanks. However, like most things in life, the ideal microorganism is hard to come by. The use of enzymes as detergent additives represents a major application of industrial novoenzymes. Proteases, lipases, amylases, oxidases, peroxidases and cellulases are added to detergents to catalyze the breakdown of chemical bonds in the presence of water. To be suitable, they must be active under thermophilic (60 °C) and alkalophilic (pH 9–11) conditions, as well as in the presence of the various components of washing powders.

Proteases constitute over 60% of the global market for enzymes and they are widely used to produce detergents, pharmaceuticals, foods, leather, silk and agrochemical products. In laundry detergents, they account for approximately 25% of the total worldwide sales of enzymes. The first detergent containing a bacterial protease ("Biotex") was introduced by Novo Industry A/S (now Novozymes) in 1956. It contained an alcalase produced by *Bacillus licheniformis*. In 1994, Novo Nordisk introduced Lipolase™, the first commercial recombinant lipase for use in a detergent, by cloning the *Humicola lanuginose* lipase into the *A. oryzae* genome. In 1995, Genencor International introduced two bacterial lipases, one from *Pseudomonas mendocina* (Lumafast™), and another from *Pseudomonas alcaligenes* (Lipomax™). An enzyme added recently to detergents is Mannaway™, a *Bacillus* mannanase which removes food stains containing guar gum (Kirk et al., 2002).

Most of Novozymes' enzymes are produced by the *Bacillus subtilis* and the fungus *Aspergillus oryzae* or via recombinant enzymes in hosts. Both have a huge capacity for producing enzymes and are completely harmless for humans. Generally, Novozymes exploits their productivity through submerged fed-batch fermentation. Over the last decade, Novozymes' business has grown substantially and the key to success for the biotechnological process was to optimize the fermentation processes. Novozymes uses scale-up/scale-down from production to laboratory as a tool to explore fermentation process. For this purpose, several studies were carried out and have led to advances in large scale fermentation processes. For example, an upgraded agitation system was recently installed, enabling a process to be scaled down effectively and revealing an oxygen transfer limitation on a small scale. Comparison of production-prepared and laboratory-prepared main media led to the identification of a limiting adjunct in the production process. Poor outgrowth in one host strain was tied to an anomaly in the propagation procedure. Trials were made to replace chemicals and improve the efficiency of products enzymes showing high stability and making them the most environmentally friendly solution for industrial manufacturing.

When the biochemical reaction is over, the enzyme is ready to affect the same reaction on another molecule again and again. Given the right conditions, the enzyme can go on and on for as long as needed. In some production processes, this lowers costs.

When industrial enzymes leave a production plant with the waste water, the used enzymes create no hazardous waste. They do not last long in the surrounding environment as they are broken down by microorganisms found in nature.

Before scaling up the fermentation process, the screening of suitable microorganisms is required as a first step in the process, Novozymes has developed several methods for finding the right bacterium or fungus. The selection method is based on automated processes that can test the right enzyme in different substrates. After the screening step, the Novozymes' researches focused on finding the gene that instructs the microorganism to produce the enzyme. Thus, Novozymes has the world's biggest toolbox of biotech-based screening technologies, together with some of the best scientists in the business, for finding and recreating the exact genetic code of the desired enzyme. And once the genetic code has been isolated, the researchers can improve the enzyme even further in many different ways.

After the identification of the gene, this latter is inserted into the production microorganism using modern biotechnology in order to produce larger quantities of enzymes under much safer conditions than the original microorganism. After that, the optimal conditions for the production organism must be determined (i.e. nutrients, temperature, oxygen level and pH) to get most efficient production.

Here, the challenge will be to predict enzyme properties, such as substrate range or thermostability, from (modeled) structure. Characterization of enzymes with respect to substrate range, kinetic parameters, and optimal reaction conditions will still be necessary, and may benefit from broadly applicable high-throughput methods.

During the fermentation process, the desired enzyme is grown on a small-scale. Once everything is up and running, large-scale production is performed in three-storey fermentation tanks. Finally, after the fermentation process, novozymes have developed a separation method of enzymes in a large drum-filter. Following a series of other filtration processes, the enzymes were separated from the water using a simple evaporation process. The resultant waste of wax, nutrients and microorganisms after the recovery process, was used as fertilizer on local farms after heat and chalk pre-treatments process. Finally, the enzymes have been formulated as either liquid or granulate, they are packed in cans or bags and are then ready to leave the factory.

In the textile industry, enzymes are applied to develop cleaner processes and reduce the use of raw materials and production of waste. The application of cellulases for denim finishing and laccases for decolorization of textile effluents and textile bleaching are the most recent commercial advances (Araujo et al., 2008). An alternative enzymatic process based on a pectate lyase, in the manufacturing of cotton has been recently developed (Tzanov et al., 2001). This process is performed at much lower temperatures with less water, which is more advantageous compared to the classical method.

Moreover, enzymes are also used in a wide range of agro-biotechnological processes, such as enzyme-assisted silage fermentation, bioprocessing of crops and crop residues, fibre processing, and production of feed supplements to improve feed efficiency. Especially the latter application, which includes the use of phytases to improve the efficiency of nutrient utilization and to reduce waste, is a rapidly growing sector (Novozymes, 2002). The feed enzyme market now amounts to US$150 million. In fact, 65% of poultry and 10% of swine feed already contain enzymes such as carbohydrases or phytase. Several developments have started to tie in the agricultural sector with the chemical and pharmaceutical industries. Plants are being modified by genetic engineering for the production of polymers and pharmaceuticals such as antibodies or for improved nutritional value, for example, by increasing lysine or carotenoid content (Fraser et al., 2001). Only recently, ProdiGene announced the scale-up of trypsin production in recombinant plants (www.prodigene.com), while other enzymes such as laccase may soon be produced in plants as well (Hood, 2002). Here, the price of cellulase needed to convert cellulosic biomass to fermentable sugars is a major factor. Therefore, the US Department of Energy awarded US$32 million to Genencor and Novozymes to reduce the price of cellulase by a factor of ten, which could make bioethanol production and many other sugar-based fermentations economically viable (Russo, 2001).

12.7.3 Antibiotics

The discovery of penicillin fuelled the exploration and exploitation of microorganisms as a potential source for antibiotics (Rokem et al., 2007, Zhu et al., 2014). This led to a boost in the development of industrial and biotechnological techniques. Latter, these techniques were subsequently transferred to processes for a large variety of products (Richter et al., 2014). Accordingly, submerged fermentation was widely applied in large-scale penicillin production, which was latter investigated for others bio-based chemicals. The titer and the conversion yields of antibiotics production have been greatly improved by brute-force genetics and process optimization during the scaling up fermentation. Recently, other industrial strains rather than *Penicillium notatum* has shown to produce 100,000 times more penicillin than Fleming's original strain (Rokem et al., 2007). The world market of β-lactam antibiotics reached around US$15 billion that represented 65% of the whole antibiotic market. Besides, the scale of penicillin fermenters could hit 30 to 100 miles gallons, operated via a fed-batch process (Elander, 2003). Kardos and Demain (2011) have reported that of the US$42 billion market for antibacterial pharmaceuticals, penicillin alone accounted for 16% (Kardos and Demain, 2011). Generally, fed-batch process is highly recommended for the production of industrial products and to avoid practical bottleneck in the fermentation process (i.e. inhibition of substrate excess). The insufficient mixing and mass transfer limitations give rise to environmental gradients such as substrate, temperature, oxygen, and carbon dioxide gradients, which considerably impaired product amounts, qualities, and yields and latter affected anticipated economic benefits (Lara et al., 2006, Takors, 2012). Furthermore, the study of the complex interaction between the cellular machinery and the extracellular environment is needed especially when an improved strain is transferred from the laboratory scale to that of industrial practice (Lapin et al., 2010, Schmalzriedt et al., 2003).

Ye et al. (2005) proposed that the multi-scale study based on the correlations and analyses of multiple online parameters is very efficient approach to solve fundamental open questions during industrial scale-up (Ye et al., 2005). The multi-scale method is widely applied for the optimization of industrial bioprocesses for penicillin, erythromycin, chlortetracycline, inosine, and guanosine productions (Zhang et al., 2004). But, owing to the complexity of the fermentation system, the multi-scale method is unable to deduce unequivocal relationships between bioprocess variables, product formation, and microbial growth. Therefore, it is rather difficult to play a conclusive role in the fermentation optimization process (Zhang et al., 2006). Generally, for a successful antibiotic production, the study of the in vivo kinetic models and metabolic network describing the biochemical reactions is recommended. This data will a comprehensible image on the in vivo kinetics of enzymes and transporters involved in the product formation pathway (Kresnowati et al., 2008; Mashego et al., 2006; Zhang et al., 2006). In this regard, some substrates or metabolites from the penicillin biosynthetic pathway were found as a perturbation agent. For example, Douma et al. (2012) have found that the perturbation experiments with phenylacetic acid and penicillin-G during chemostat cultures of *P. chrysogenum* at two time scales, provided more insight in the mechanisms of transport of the precursor phenylacetic acid and the product penicillin-G (Douma et al., 2012). Thus, the inquiry of metabolic data is highly recommended. Previous attempts have been carried out to develop mathematical models to describe growth, cell differentiation, and penicillin production (Birol et al., 2002a; Birol et al., 2002b; Henriksen et al., 1996). Thus, the optimization and the control of industrial fermentation processes will be based on mathematic modeling which quantitatively characterizes the dynamic conditions of the fermentation broth over the fermentation process (Bannari et al., 2012). Meanwhile, these models did not allow the description of the biochemical reactions inside cells. Therefore, little has been gathered about the intracellular metabolism of perturbed cells and the whole cell system remained a black box. In addition, with respect to the relation between the specific growth rate and the specific penicillin production rate, some models were established, and parameters estimated from steady-state chemostat cultivations, which were not suited to the prediction of a fed-batch fermentation where the growth rate is changing all the time (Van Gulik et al., 2000). To address this issue, Douma et al. (2010) established a dynamic gene expression regulation model in which the expression level of isopenicillin N synthetase exerted near total control in the biosynthetic pathway towards penicillin. This model combined with process mass balances could accurately describe the biomass and penicillin production for all stages of chemostat culture and fed-batch cultivation (Douma et al., 2010b). To further quantitatively understand the intracellular metabolism of the perturbed cells, it is suggested that the details of cellular reactions should be simplified, and much attention should be paid to the variations of different dynamic pools in cells, such as assumed substrate pool, storage pool, precursor pool, enzyme pool, and energy pool, when the cells are exposed to different extracellular stimuli. In accordance, kinetic models are needed to describe the biochemical reactions as a set of differential equations instead of algebraic relations as in steady state. A set of differential equations can be further coupled to the CFD model to establish a mathematical model that helped to improve the strain and establish a reasonable design of the fermenter (Moilanen et al., 2006; Wu, 2012).

12.7.4 Platform Chemicals

Synthetic biology and metabolic engineering approaches offer great promise to improve the efficiency of fermentation and to expand the product spectrum beyond native products ethanol, acetate and butanediol to a range of higher-value fuels and commodity chemicals (Council, 2015, Latif et al., 2014). Generally, Acetyl-CoA is the key factor and the starter in all chemicals produced. Generally, microorganisms convert sugars to new products like farnesene, isobutanol, 1,3-propanediol (1,3-PDO), and 1,4-butanediol (1,4-BDO) (Burk and VanDien, 2016; Cho et al., 2014; George et al., 2015).

Although ethanol (and potentially other short-chain alcohols) produced by gas fermentation can serve as an additive or potential replacement to gasoline, it is not considered to be an energy dense enough molecule to serve as a jet fuel. Jet fuels are typically aliphatic and contain 8–16 carbon atoms (Atsonios et al., 2015). Molecules this large can be made from syngas substrates via coupled biological processes (see above), but it may be more efficient to convert the small alcohols to jet fuel chemically by employing a process called alcohol-to-jet (ATJ). To obtain jet fuel from these small alcohols, ethanol and butanol (separated from the fermentation broth) are first dehydrated to corresponding alkenes ethylene and butylene (Harvey and Meylemans, 2011; Zhang and Yu, 2013). The alkenes are then oligomerized to a desired range of sizes (i.e. number of carbon atoms) that fit the end application (Harvey and Meylemans, 2011; Heveling et al., 1998). Reaction conditions are tuned to avoid extensive polymerization to thermoplastics like poly(ethylene) when fuels are the desired product (Atsonios et al., 2015). There are currently multiple industrial ATJ projects for both ethanol and butanol conversions to jet fuel (Lane, 2015a; LanzaTech, 2015). The ATJ fuels have undergone and continue to undergo rigorous testing to establish their use as drop-in fuels for commercial aircraft (Luning Prak et al., 2015). Successfully meeting these standards will expand the volume of fuels that may be sourced from renewable resources, such as biomass, and aid in keeping fossil fuels typically needed for aviation fuel in the ground. Another promising approach to convert ethanol (and other short-chain alcohols) to jet (or biodiesel) fuels is via extractive catalytic upgrading directly from the fermentation broth (Baer et al., 2016; Bormann et al., 2014; Sreekumar et al., 2015). This technology requires strains that co-produce acetone (Becker et al., 2012; Köpke et al., 2015; Köpke et al., 2012; Schiel-Bengelsdorf and Dürre, 2012), which harbors a nucleophilic α-carbon amenable to C–C bond formation with the electrophilic alcohols. As mentioned, alkenes derived from gas fermentation alcohols are sources for ubiquitous thermoplastic and rubber products. Ethylene made from the dehydration of ethanol can be more fully polymerized beyond oligomers for ATJ fuel and converted to poly(ethylene). Poly(butadiene) is another polymer used in the mass production of products such as nylon or automobile tires. The monomer of this synthetic rubber, 1,3-butadiene, can be formed from gas fermentation products ethanol (Makshina et al., 2012) and 2,3-butanediol (Duan et al., 2015). These processes have been reviewed more in depth elsewhere (Makshina et al., 2014). 1,3-butadiene is also considered as a precursor for nylon production (Magdouli et al., 2017). LanzaTech is currently a part of multiple collaborations to develop gas fermentation production of 2,3- butanediol for chemical conversion to 1,3- butadiene (Köpke and Havill, 2014).

Various strategies were investigated to minimize the cost production of drop in chemicals. For instance, the use of multi-step bioprocesses by the combinations of pure cultures, co-cultures, and/or mixed cultures with natural and/or genetically modified strains is very promising. The potential advantages of co-cultures and mixed cultures (i.e. the increase of substrate utilization, and the improvement product yields etc.) were detailed in Magdouli et al. (2016). For example, undesirable inhibitory or toxic by-products of a metabolic pathway can be consumed by a second microbe, or two host cells can work together to ensure the optimal environment for all pathway enzymes by effectively dividing the pathway (Zhang, 2015; Zhou et al., 2015).

The contamination is the technical factor found during the scale up. The mixed cultures permit the reduction of contamination by using foreign bacteria and allows the resistance to bacteriophages as well as the oxygen removal and the production of growth factors (Angenent and Wrenn, 2008). Although the advantages cited before, the difficulty of understanding the interaction between microbes in the fermentation resulted in the reduction of reproducibility due to challenges in maintaining population dynamics, and inability to optimize parameters for each strain.

12.8 Conclusions

The development of the fermentation process and its optimization is seriously challenged by a lack of the knowledge about scale up and other issues such as the influence of morphology on broth rheology and mass transfer. Furthermore, CFD is now reaching a level of maturity such that CFD can start supporting the process scientist in addressing some of the engineering challenges, that is CFD can be increasingly used to test new ideas, both with respect to reactor operation and design, and to support scaling up and scaling down bioreactors.

Rather than the points discussed in this chapter, other points will not be discussed in the current chapter which are related to biological aspects such genetic instability (Sipiczki, 2011), population heterogeneity (Lencastre Fernandes et al., 2011), or the modeling of metabolism at single cells level (Lapin et al., 2010). Herein, we assume that the environmental conditions put a selective pressure on the phenotype of the cells, and therefore, by controlling the macroscopic variables of the fermentation, it is assumed possible to reduce and control the variability in the cellular population.

Acknowledgments

Thanks to all the contributing authors. The views and opinions expressed in this chapter are those of the authors.

References

Adkesson, D.M., Alsop, A.W., Ames, T.T. et al. (2004) Purification of biological-produced 1,3-propanediol. *International Patent WO2004101479.*

Andersson, L., Yang, S., Neubauer, P. and Enfors, S.-O. (1996) Impact of plasmid presence and induction on cellular responses in fed-batch cultures of Escherichia coli. *J. Biotechnol.*, 46, 255–263.

Angenent, L.T. and Wrenn, B.A. (2008) Optimising mixed-culture bioprocessing to convert wastes into bioenergy. *in Bioenergy*, A.Demain, C.S.Harwood, and J.D.Wall *(eds)* Washington,DC: American Society of Microbiology, 179–194.

Ansorge, M.B. and Kula, M. (2000) Production of recombinant L-leucine dehydrogenase from Bacillus cereus in pilot scale using the runaway replication system E. coli[pIET98]. *Biotechnol Bioeng*, 68, 557–562.

Araujo, R., Casal, M. and Cavaco-Paulo, A. (2008) Application of enzymes for textiles fibers processing. *Biocatal. Biotechnol.*, 26, 332–349.

Atmar, R.L., Metcalf, T.G., Neill, F.H. and Estes, M.K. (1993) Detection of enteric viruses in Oysters by using the Polymerase Chain reaction. *Appl. Environ. Microbiol.*, 59, 631–635.

Atsonios, K., Kougioumtzis, M.-A.D., Panopoulos, K. and Kakaras, E. (2015) Alternative thermochemical routes for aviation biofuels via alcohols synthesis: Process modeling,techno-economic assessment and comparison. *Appl. Energy*, 138, 346–366.

Baer, Z.C., Bormann, S., Sreekumar, S. et al. (2016) Co-production of acetone and ethanol with molar ratio control enables production of improved gasoline or jet fuel blends. *Biotechnol. Bioeng.*

Bannari, R., Bannari, A., Vermette, P. and Proulx, P. (2012) A model for cellulase production from Trichoderma reesei in an airlift reactor. *Biotechnol. Bioeng.*, 109, 2025–2038.

Becker, F.U., Grund, G., Orschel, M. et al. (2012) Cells and Method for Producing Acetone. *US2A1. Washington, DC: U.S. Patent and Trademark Office.*

Birol, G., Ündey, C. and Cinar, A. (2002a) A modular simulation package for fed-batch fermentation: penicillin production. *Comput. Chem. Eng.*, 26, 1553–1565.

Birol, G., Ündey, C., Parulekar, S.J. and Çınar A. (2002b) A morphologically structured model for penicillin production. *Biotechnol. Bioeng.*, 77, 538–552.

Bormann, S., Baer, Z.C., Sreekumar, S. et al. (2014) Engineering Clostridium acetobutylicum for production of kerosene and diesel blend stock precursors. *Metab. Eng.*, 25, 124–130.

Buckland, B.C., Gbewonyo, K., Dimasi, D. et al. (1988) Improved performance in viscous mycelial fermentations by agitator retrofitting. *Biotechnol. Bioeng.*, 31, 737–742.

Budzik, J.M. (2003) Phage isolation and investigation. *Dartmouth Undergraduate J. Sc.*, 3, 37–43.

Bull, A., Marrs, B. and Kurane, R. (1998) Biotechnology for clean industrial products and processes. Towards industrial sustainability. *OECD Publications*, 7–139.

Burk, M.J. and Vandien, S. 2016. Biotechnology for chemical production: challenges and opportunities. *Trends Biotechnol.*, 34, 187–190.

Bylund, F., Castan, A., Mikkola, R. et al. (2000) Influence of scale-up on the quality of recombinant human growth hormone. *Biotechnol. Bioengin.*, 69, 119–128.

Bylund, F., Collet, E., Enfors, E.O. and Larsson G. (1998) Substrate gradient formation in the large scale lowers cell yield and increases byproduct formation. *Bioproc. Eng.*, 18, 171–180.

Bylund, F., Guillard, F., Enfors S.O. et al. (1999) Scale down of recombinant production: a comparative study of scaling performance. *Bioproc. Engin.*, 20, 377–289.

Callanan, M. and Klaenhammer, T.R. (2002) Bacteriophages in industry. *Encyc. of Life Sc.*, 1–8.

Cascaval, D., Oniscu, C. and Galaction, A. (2003) Rheology of fermentation broths 2. Influence of the rheological behavior on biotechnological processes. *Rev. Roum. Chim.*, 48, 339–356.

Castan, A. and Enfors, S.O. (2001) Formate accumulation due to DNA release in aerobic cultivations of Escherichia coli. *Biotechnol. Bioengin.*, 77, 324–328.

Cervera, A.E., Petersen, N., Eliasson Lantz, A. (2009) Application of near-infrared spectroscopy for monitoring and control of cell culture and fermentation. *Biotechnol. Prog.*, 25, 1561–1581.

Cho, C., Choi, S.Y., Luo, Z. W., and Lee, S.Y. (2014) Recent advances in microbial production of fuels and chemicals using tools and strategies of systems metabolic engineering. *Biotechnol. Adv.*, 33, 1455–1466.

Council, N.R. (2015) Industrialization of Biology: A Roadmap to Accelerate the Advanced Manufacturing of Chemicals. Washington,DC: National Academies Press.

Das, R.K. and Brar, S.K. (2014) Enhanced Fumaric Acid Production from Brewery Wastewater and Insight into the Morphology of Rhizopus oryzae 1526. *Appl. biochem. biotechnol.*, 172.

Douma, R.D., Deshmukh, A.T., De Jonge, L.P. et al. (2012) Novel insights in transport mechanisms and kinetics of phenylacetic acid and penicillin-G in Penicillium chrysogenum. *Biotechnol. Prog.*, 28, 337–348.

Douma, R.D., Verheijen, P.J., De Laat, W.T. (2010b) Dynamic gene expression regulation model for growth and penicillin production in Penicillium chrysogenum. *Biotechnol. Bioeng.*, 106, 608–618.

Duan, H., Yamada, Y. and Sato, S. (2015) Efficient production of 1,3-butadiene in the catalytic dehydration of 2,3-butanediol. *Appl. Catal. A Gen.*, 491, 163–169.

Dunlop, M.J., Dossani, Z.Y., Szmidt, H.L. et al. (2011) Engineering microbial biofuel tolerance and export using efflux pumps. *Mol. Syst. Biol.*, 7, 487.

Elander, R.P. (2003) Industrial production of β-lactam antibiotics. *Appl. Microbiol. Biotechnol.*, 61, 385–392.

Elsworth, R., Miller, G., Whitaker, A. et al. (1968) Production of E. coli as a source of nucleic acids. *J. Appl. Chem.*, 17, 157–166.

Enfors, S.O., Jahic, M., Rozkov, A. et al. (2001) Physiological responses to mixing in large scale bioreactors. *J. Biotechnol.*, 85, 175–185.

Fan, Y.F. (2008) Breeding of yeast high producing ethanol and fermentation condition with dynamics. *Fujian Normal University: Fujian*, 128–135.

Forberg, C. and Haggstrom, L. (1987) Effects of cultural conditions on the production of phenylalanine from a plasmid-harboring Escherichia coli strain. *Appl. Microbiol. Biotechnol.*, 26, 136–140.

Fraser, P.D., Roemer, S., Kiano, J.W. et al. (2001) Elevation of carotenoids in tomato by genetic manipulation. *J. Sci. Food. Agric.*, 81, 822–827.

Friehs, K. (2004) Plasmid copy number and plasmid stability. *Adv. Biochem. Eng. Biotechnol.*, 86, 47–82.

Gabelle, J.C., Augier, F., Carvalho, A. (2012) Effect of tank size on kLa and mixing time in aerated stirred reactors with non-Newtonian fluids. *Can. J. Chem. Eng.*, 89, 1139–1152.

George, K.W., Alonso-Gutierrez, J., Keasling, J.D., and Lee, T.S. (2015) Isoprenoid drugs, biofuels, and Chemicals-Artemisinin,farnesene, and beyond. *in Biotechnology of Isoprenoids., eds* J. Schrader *and* J. Bohlmann (Cham: *Springer International Publishing*), 355–389.

George, S., Larsson, G., and Enfors, S.-O. (1993) A scale-down two-compartment reactor with controlled substrate oscillations: metabolic response of Saccharomyces cerevisiae. *Bioproc. Eng.*, 9, 249–257.

German, J. and Dillard, C. (2006) Composition, structure and absorption of milk lipids: A source of energy, fat soluble nutrient and bioactive molecules. *Food Sc. Nutr.*, 46, 57–92.

Gernaey, K.V., Bolic, A., and Svanholm, B. (2012) PAT tools for fermentation processes. *Chem. Today*, 30, 38–43.

Gibbs, P.A., Seviour, R.J., and Schmid, F. (2000) Growth of filamentous fungi in submerged culture: Problems and possible solutions. *Crit. Rev. Biotechnol.*, 20, 17–48.

Grimm, L.H., Kelly, S., Hengstler, J. et al. (2004) Kinetic studies on the aggregation of Aspergillus niger conidia. *Biotechnol. Bioeng.*, 87, 213–218.

Harvey, B.G. and Meylemans, H.A. (2011) The role of butanol in the development of sustainable fuel technologies. *J. Chem. Technol. Biotechnol.*, 86, 2–9.

Henriksen, C.M., Christensen, L.H., Nielsen, J., and Villadsen J. 1996. Growth energetics and metabolic fluxes in continuous cultures of Penicillium chrysogenum. *J. Biotechnol* 45, 149–164.

Heveling, J., Nicolaides, C.P., and Scurrell, M.S. (1998) Catalysts and conditions for the highly efficient,selective and stable heterogeneous oligomerisation of ethylene. *Appl. Catal. A Gen.*, 173, 1–9.

Holder, K.K. and Bull, J.J. (2001) Profiles of adaptation in two similar viruses. *Genetics.*, 159, 1393–1404.

Hood, E.E. (2002) From green plants to industrial enzymes. *Enzyme. Microb. Technol.*, 30, 279–283.

Johnson, I.S. (1983) Human insulin from recombinant DNA technology. *Science*, 219, 632–637.

Jones, D.T., Shirley, M., Wu, X., and Keis, S. (2000) Bacteriophage infections in the industrial acetone butanol (AB) fermentation process. *J. Mol. Microbiol. and Biotechnol.*, 2, 21–26.

Junker, B.H. (2004) Scale-up methodologies for Escherichia coli and yeast fermentation processes. *J. Biosci. Bioeng.*, 97, 347–364.

Jüsten, P., Paul, G.C., Nienow, A.W., and Thomas, C.R. (1996) Dependence of mycelial morphology on impeller type and agitation intensity. *Biotechnol. Bioeng.*, 52, 672–684.

Kardos, N. and Demain, A.L. (2011) Penicillin: the medicine with the greatest impact on therapeutic outcomes. *Appl. Microbiol. Biotechnol.*, 92, 677–687.

Kirk, O., Borchert, T.V., and Fuglsang, C.C. (2002) Industrial enzyme applications. *Curr. Opin. Biotechnol.*, 13, 345–351.

Koizumi, J.-I., Monden, Y., and Aiba, S. (1985) Effects of the temperature and the dilution rate on the copy number of recombinant plasmid in continuous culture of Bacillus staerothermophilus (pLP11). *Biotechnol. Bioeng.*, 27, 721–728.

Kokjohn, T.A., Schrader, J.J., and Schrader, H.S. (2005) http://www.isb.vt.edu/brarg/brasym94/kokjohn.html.

Köpke, M., Al-Sinawi, B., Jensen, R.O. et al. (2015) Microorganisms and Methods for the Production of Ketones. *US20150152445. Washington,DC:U.S.Patent and Trade mark Office.*

Köpke, M. and Havill, A. (2014) Lanza Tech's route to bio-butadiene. *Catal. Rev.*, 27, 7–12.

Köpke, M., Simpson, S., Liew, F.M., and Hen, W. (2012) Fermentation Process for Producing Isopropanol Using a Recombinant Microorganism. *US20120252083A1. Washington, DC: U.S. Patent and Trademark Office.*

Krawczyk, T. (1996) Biodiesel-alternative fuel makes inroads but hurdles remain. *Inform*, 7, 801–829.

Kresnowati, M.T.A.P., Van Winden, W.A., Van Gulik, W.M., and Heijnen, J. J. 2008. Energetic and metabolic transient response of Saccharomyces cerevisiae to benzoic acid. *FEBS J.*, 275, 5527–5541.

Kwanmin, J.J. (1989) Scale-down techniques for fermentation. *Biopharm.*, 2, 30–39.

Kwint, K., Nachin, L., Diez, A., and Nyström, T. (2003) The bacterial universal stress protein: Function and regulation. *Curr. Opin. Microbiol.*, 6, 140–145.

Kywe, T. and Oo, M. (2009) Production of biodiesel from Jatropha oil (Jatropha curcas) in pilot plant. *WASET*, 14, 78–86.

Lam, F.H., Ghaderi, A., Fink, G.R., and Stephanopoulos, G. (2014) Biofuels. Engineering alcohol tolerance in yeast. *Science*, 346, 71–75.

Lane, J. (2015a) Alcohol to renewable jet fuel: the Digest's 2015 8-Slide Guide to Gevo's ATJ,"in Biofuels Dig. *Available online at: http://www.biofuelsdigest.com/bdigest/2015/12/16/alcohol-to-renewable-jet-fuel-the-digests-2015-8-slide-guide-to-gevos-atj/.*

Lanzatech (2015) Available on lineat:http://www.lanzatech.com/ innovation/markets/fuels/[Accessed Jully, 2017]. *Fuels.*

Lapin, A., Klann, M., and Reuss, M. (2010) Multi-scale spatio-temporal modeling: Lifelines of microorganisms in bioreactors and tracking molecules in cells. *Adv. Biochem. Eng.*, 121, 23–43.

Lara, A.R., Galindo, E., Ramírez, O.T., and Palomares, L.A. (2006) Living with heterogeneities in bioreactors. *Mol. Biotechnol.*, 34, 355–381.

Larsson, G., Törnkvist, M., Wernersson, E.S. et al. (1996a) Substrate gradients in bioreactors: Origin and consequences. *Bioprocess. Eng.*, 14, 281–289.

Larsson, G., Törnkvist, M., Wernersson, E.S. et al. (1996b) Substrate gradients in bioreactors: origin and consequences. *Bioproc. Engin*, 14, 281–289.

Latif, H., Zeidan, A.A., Nielsen, A.T., and Zengler, K. (2014) Trash to treasure: production of biofuels and commodity chemicals via syngas fermenting microorganisms. *Curr. Opin. Biotechnol.*, 27, 79–87.

Lee, J., Lee, S.Y., Park, S., and Middelberg, A.P.J. (1999) Control of fed-batch fermentations. *Biotechnol. Adv.*, 17, 29–48.

Lencastre Fernandes, R., Nierychlo, M., Lundin, L. et al. (2011) Experimental methods and modeling techniques for description of cell population heterogeneity. *Biotechnol. Adv.*, 29, 575–599.

Lin, H.Y. and Neubauer, P.I. (2000) Influence of controlled glucose oscillations on a fed-batch process of recombinant Escherichia coli. *J. Biotechnol.*, 79, 27–37.

Ling, H., Teo, W., Chen, B. et al. (2014) Microbial tolerance engineering toward biochemical production: from lignocellulose to products. *Curr. Opin. Biotechnol.*, 29, 99–106.

Los, M., Czyz, A., Sell, E. et al. (2004) Bacteriophage contamination: is there a simple method to reduce its deleterious effect in the laboratory cultures and biotechnological factories? *J. Appl. Genet.*, 45, 111–120.

Lu, Z., Breidt, F., Plengvidhya, V., and Fleming, H.P. (2003) Bacteriophage ecology in commercial sauerkraut fermentations. *Appl. Environ. Microbiol.*, 9, 3192–3202.

Luningprak, D.J., Jones, M.H., Trulove, P. et al. (2015) Physical and chemical analysis of alcohol-to-jet(ATJ) fuel and development of surrogate fuel mixtures. *Energy and Fuels*, 29, 3760–3769.

Lütke-Eversloh, T. and Bahl, H. (2011) Metabolic engineering of Clostridium acetobutylicum: Recent advances to improve butanol production. *Curr. Opin. Biotechnol.*, 22, 634–647.

Ma, N., Mollet, M., and Chalmers, J.J. (2006) Aeration, mixing and hydrodynamics in bioreactors. *in: Ozturk, S., Hu, W. (Eds.), Cell Culture Technology for Pharmaceutical and Cell-Based Therapies, Taylor & Francis.*

Magdouli, S., Brar, S.K. and Blais, J.F. (2016) Co-culture for lipid production: Advances and challenges. *Biomass Bioenergy*, 92, 20–30.

Magdouli, S., Brar, S.K., and Blais, J.F. (2017) Morphology and rheological behaviour of Yarrowia lipolytica: Impact of dissolved oxygen level on cell growth and lipid composition. *Process. Biochem.*, https://doi.org/10.1016/j.procbio.2017.10.021.

Magdouli, S., Yan, S., Tyagi, R.D., and Surampalli, R.Y. (2014) Heterotrophic Microorganisms: A Promising Source for Biodiesel Production. *Crit. Rev. Environ. Sci. Technol.*, 44, 416–453.

Makshina, E.V., Dusselier, M., Janssens, W. et al. (2014) Review of old chemistry and new catalytic advances in the on-purpose synthesis of butadiene. *Chem. Soc. Rev.*, 43, 7917–7953.

Makshina, E.V., Janssens, W., Sels, B.F., and Jacobs, P.A. (2012) Catalytic study of the conversion of ethanol into 1,3-butadiene. *Catal. Today* 198, 338–344.

Martinez-Hernandez, E., Campbell, G., and Sadhukhan, J. (2013) Economic value and environmental impact (EVEI) analysis of biorefinery systems. *Chem. Eng. Res. Des.*, 91, 1418–1426.

Mashego, M.R., Van Gulik, W.M., Vinke, J.L. et al. (2006) In vivo kinetics with rapid perturbation experiments in Saccharomyces cerevisiae using a second-generation Bio Scope. *Metab. Eng.*, 8, 370–383.

Meerman, H.J., Kelley, A.S., and Ward, M. (2004) Advances in protein expression in filamentous fungi. in: Baneyx, F. (Ed.), Protein Expression Technologies. *Current Status and Future Trends, 1st Edn., Horizon Bioscience*, pp. 345–385.

Metzner, A.B. and Otto, R.E. (1957) Agitation of non-Newtonian fluids. *AIChE J.*, 3, 3–10.

Minihane, B.J. and Brown, D.E. (1986) Fed-batch culture technology. *Biotechnol. Adv.*, 4, 207–218.

Mogk, A., Tomoyasu, T., Goloubinoff, P. et al. (1999) Identification of thermolabile Escherichia coli proteins: Prevention and reversion of aggregation by DnaK and ClpB. *EMBO J.*, 18, 6934–6949.

Moilanen, P., Laakkonen, M., and Aittamaa, J. (2006) Modeling aerated fermenters with computational fluid dynamics. *Ind. Eng. Chem. Res.*, 45, 8656–8663.

Moulin, G., Broze, H., and Galzy, P. (1980) Inhibition of alcoholic fermentation by substrate and ethanol. *Biotechnology and bioengineering*, 22, 2375–2381.

Moyer, A.J. (1948) Method for production of penicillin. *US patent 2442141*.

Murphy, T.D. (1977) Design and analysis of industrials experiments. *Chem. Eng.*, 6, 168–182.

Naglak, T.J., Keith, M.G., and Omstead, D.R. (1994) Validation of fermentation processes. *Biopharm.*, 7, 28–36.

Neubauer, P., Lin, H.Y., and Mathiszik, B. (2003) Metabolic load of recombinant protein production: inhibition of cellular capacities for glucose uptake and respiration after induction of a heterologous gene in Escherichia coli. *Biotechnol. Bioeng.*, 83, 53–64.

Nienow, A.W. (1998) Hydrodynamics of stirred bioreactors. In: Fluid mechanics problems in biotechnology. R. Pohorecki (ed.). *App. Mech. Rev*, 51, 3–32.

Nienow, A.W. (2006) Reactor engineering in large-scale animal cell culture. *Cytotechnology*, 50, 9–33.

Nienow, A.W. (2009) Scale-up considerations based on studies at the bench scale in stirred bioreactors. *J. Chem. Eng. Jap.*, 42, 789–796.

Nienow, A.W., McLeod, G., and Hewitt, C.J. (2010) Studies supporting the use of mechanical mixing in large scale beer fermentations. *Biotechnol. Lett.*, 32, 623–633.

Novozymes (2002) AS: Annual Report 2001. Bagsvaerd, Denmark.

Ogata, S. 1980. Bacteriophage contamination in industrial process. *Biotechnol. Bioeng.*, 22, 177–193.

Ong, H.C., Mahlia, T.M.I., Masjuki, H.H., and Honnery, D. (2012) Life cycle cost and sensitivity analysis of palm biodiesel production. *Fuel*, 98, 131–139.

Oosterhuis, N.M.G. and Kossen, N.W.F. (1984) Dissolved oxygen concentration profiles in a production scale bioreactor. *Biotechnol. Bioeng.*, 26, 546–550.

Oosterhuis, N.M.G., Kossen, N.W.F., Olivier, A.P.C., and Schenk, E.S. (1985) Scaledown and optimization studies of the gluconic acid fermentation by Gluconobacter oxydans. *Biotechnol. Bioeng.*, 27, 711 - 720.

Ostergaard, S., Olsson, L., and Nielsen, J. (2000) Metabolic engineering of Saccharomyces cerevisiae. *Microbiol. Mol. Biol. Rev.*, 64, 34–50.

Park, J.H., Lee, K.H., Kim, T.Y., and Lee, S.Y. (2007) Metabolic engineering of Escherichia coli for the production of L-valine based on transcriptome analysis and in silico gene knockout simulation. *Proc. Natl. Acad. Sci. USA*, 104, 7797–7802.

Pollard, D.J., Kirschner, T.F., Hunt, G.R. et al. (2007) Scale up of a viscous fungal fermentation: Application of scale-up criteria with regime analysis and operating boundary conditions. *Biotechnol. Bioeng.*, 96, 307–317.

Posch, A.E., Herwig, C., and Spadiut, O. (2013) Science-based bioprocess design for filamentous fungi. *Trends. Biotechnol.*, 31, 37–44.

Posno, M., Leer, R.J., Van Luijk, N. et al. (1991) Incompatibility of Lactobacillus vectors with replicons derived from small cryptic Lactobacillus plasmids and segregational instability of the introduced vectors. *Appl. Environ. Microbiol.*, 57, 1822–1828.

Primrose, S.B. (1990) Controlling bacteriophage infection in industrial processes. *Adv. Biochem. Eng. Biotechnol.*, 43, 1–10.

Prokop, T. (2002) Imperial western products. *Communication. P.*, 14970. Chandler: St., Coachella.

Prueksakorn, K., and Gheewala, S.H., (2006) Energy and greenhouse gas implications of biodiesel production from Jatropha curcas L. *In: The 2nd Joint International Conference on "Sustainable Energy and Environment (SEE 2006)", Bangkok, Thailand; 21–23 November 2006.*

Reinikainen, P. and Virkajärvi, I. (1989) Escherichia coli growth and plasmid copy numbers in continuous cultivations. *Biotechnol.Lett.*, 11, 225–23.

Reuss, M. (1993) Oxygen transfer and mixing: Scale-up implications. In: H.J. Rehm and G. Reed (eds), *Biotechnology, Vol.* 3, *2nd Edn., VCH* pp. 185–213.

Richter, L., Wanka, F., Boecker, S. et al. (2014) Engineering of Aspergillus niger for the production of secondary metabolites. *Fungal. Biol. Biotechnol*, 1, 1–13.

Riesenberg, D. and Schulz, V. (1991) High cell density cultivation of E. coli at controlled specific growth rates. *J. Biotechnol.*, 20, 17–28.

Rokem, J.S., Lantz, A.E., and Nielsen, J. (2007) Systems biology of antibiotic production by microorganisms. *Nat. Prod. Rep.*, 24, 1262–1287.

Rønnest, N.P., Stocks, S.M., Lantz, A.E., and Gernaey, K.V. (2012) Comparison of laser diffraction and image analysis for measurement of Streptomyces coelicolor cell clumps and pellets. *Biotechnol. Lett.*, 34, 1465–1473.

Rossi, F.G., Ribeiro, M.Z., Converti, A. et al. (2003) *Enzyme Microb. Technol.*, 32, 107–113.

Russo, E. (2001) Turning trash into treasure. Can organic waste become the nation's next big power source? *The Scientist*, 15, 1–4.

Sanchez Miron, A., Contreras Gomez, A., Garcia Camacho, F. et al. (1999) Comparative evaluation of compact photobioreactors for large-scale monoculture of microalgae. *J. Biotechnol.*, 70, 249–70.

Scarff, M., Arnold, S.A., Harvey, L.M., and McNeil, B. (2006) Near Infrared spectroscopy for bioprocess monitoring and control: Current status and future trends. *Crit. Rev. Biotechnol.*, 26, 17–39.

Schiel-Bengelsdorf, B. and Dürre, P. (2012) Pathway engineering and synthetic biology using acetogens. *FEBS. Lett.*, 586, 2191–2198.

Schmalzriedt, S., Jenne, M., Mauch, K., and Reuss, M. (2003) Integration of physiology and fluid dynamics. Process integration in biochemical engineering. *Springer, Berlin,*, pp 19–68.

Schmidt, A.S., Garde, A., Klinke, H.B. et al. (2000) Lactic acid production from wheat straw: Effect of pretreatment conditions. *In: Proceedings of the First World Conference and Exhibition on Biomass for Energy and Industry. Sevilla, Spain.*

Schmidt, F.R. (2005) Optimization and scale up of industrial fermentation processes. *Appl. Microbiol. Biotechnol.*, 68, 425–435.

Schweder, T., Krüger, E., Xu, B. et al. (1999) Monitoring of genes that respond to process related stress in large-scale bioprocesses. *Biotechnol. Bioeng.*, 65, 151–159.

Shu, P. and Johnson, M.J. (1948) Citric acid. *J. Ind. Eng. Chem.*, 40, 1202–1205.

Sipiczki, M. (2011) Diversity, variability and fast adaptive evolution of the wine yeast (Saccharomyces cerevisiae) genome - a review. *Ann. Microbiol.*, 61, 85–93.

Smith, J.J., Lilly, M.D., and Fox, R.I. (1990) The effect of agitation on the morphology and penicillin production of Penicillium chrysogenum. *Biotechnol. Bioeng.*, 35, 1011–1023.

Spohr, A., Dam-Mikkelsen, C., Carlsen, M. et al. (1998) On-line study of fungal morphology during submerged growth in a small flow-through cell. *Biotechnol. Bioeng.*, 58, 541–553.

Sreekumar, S., Baer, Z.C., Pazhamalai, A. et al. (2015) Production of an acetone-butanol-ethano lmixture from Clostridium acetobutylicum and its conversion to high-value biofuels. *Nat. Protoc.*, 10, 528–537.

Sridhar, J., Eiteman, M.A., and Wiegel, J.W. (2000) *Appl. Environ. Microbiol.*, 66, 246–251.

Stocks, S.M. (2013) Industrial enzyme production: Process scale up/scale down. *in:* McNeil, B., Archer, D., Giavasis, I., Harvey, L. *(Eds.), Microbial Production of Food Ingredients, Enzymes and Nutraceuticals, Woodhead Publishing 2013, pp.* 144–172.

Takors, R. (2012) Scale-up of microbial processes: impacts, tools and open questions. *J. Biotechnol.*, 160, 3–9.

Thomas, C.R. (1990) Problems of shear in biotechnology. In *Chemical Engineering Problems in Biotechnology*, M.A. Winkler, (ed.), pp 23–94. Elsevier Applied Science, United Kingdom.

Thomas, K. and Ingledew, W.J. (1992) *Ind. Microbiol. Biotechnol.*, 10, 61–68.

Tufvesson, P., Lima-Ramos, J., Nordblad, M., and Woodley, J.M. (2011) Guidelines and cost analysis for catalyst production in biocatalytic processes. *Org. Process Res. Dev.*, 15, 266–274.

Tzanov, T., Calafell, M., Guebitz, G.M., and Cavaco-Paulo, A. (2001) Bio-preparation of cotton fabrics. *Enzyme Microb. Technol.*, 29, 357–362.

Uihlein, A. and Schebek, L. (2009) Environmental impacts of a lignocellulose feedstock biorefinery system: an assessment. *Biomass. Bioenergy*, 33, 793–802.

Umakoshi, H., Kuboi, R., Komasawa, I. et al. (1998) Heat Induced translocation of cytoplasmic -galactosidase across inner membrane of E. coli. *Biotechnol. Prog.*, 14, 210–217.

Utrilla, J., Licona-Cassani, C., Marcellin, E. et al. (2012) Engineering and adaptive evolution of Escherichia coli for D-lactate fermentation reveals GatC as a xylose transporter. *Metab. Eng.*, 14, 469–476.

Van Gulik, W.M., De Laat, W., Vinke, J.L., and Heijnen, J.J. (2000) Application of metabolic flux analysis for the identification of metabolic bottlenecks in the biosynthesis of penicillin-G. *Biotechnol. Bioeng.*, 68, 602–618.

Wang, Y. and Van Ness, B. (1989) Site specific cleavage of supercoiled DNA by ascorbate /Cu(II). *Nucl. acid research*, 17, 6915–6926.

Woolston, B.M., Edgar, S., and Stephanopoulos, G. (2013) Metabolic engineering: Past, and future. *Annu. Rev. Chem. Biomol. Eng.*, 4, 259–288.

World Enzymes (2011) Cleveland, Ohio, United States of America. 12–26.

Wu, B. (2012) Integration of mixing, heat transfer, and biochemical reaction kinetics in anaerobic methane fermentation. *Biotechnol. Bioeng.*, 109, 2864–2874.

Wunsche, L. (1989) Importance of bacteriophages in fermentation processes. *Acta Biotechnol.*, 9, 395–419.

Xu, B., Jahic, M., Bomsten, G., and Enfors, S. O. (1999) Glucose overflow metabolism and mixed acid fermentation in aerobic large-scale fed-batch processes with Echerichia coli. *Appl. Microbiol. Biotechnol.*, 51, 564–571.

Ye, X.H., Chu, J., Zhuang, Y.P., and Zhang, S.L. (2005) Multi-scale methodology: a key to deciphering systems biology. *Front. Biosci.*, 10, 961–965.

Zhang, M. and Yu, Y. (2013) Dehydration of ethanol to ethylene. *Ind. Eng. Chem. Res.*, 52, 9505–9514.

Zhang, S.L., Chu, J., and Zhuang, Y.P. (2004) *A multi-scale study of industrial fermentation processes and their optimization. Biomanufacturing. Springer*, Berlin, pp 97–150.

Zhang, S.L., Ye, B.C., Chu, J. et al. (2006) From multi-scale methodology to systems biology: to integrate strain improvement and fermentation optimization. *J. Chem. Technol. Biotechnol.*, 81, 734–745.

Zhang, T. (2015) More efficient together. *Science*, 350, 738–739.

Zhou, K., Qiao, K., Edgar, S., and Stephanopoulos, G. (2015) Distributing a metabolic pathway among a microbial consortium enhances production of natural products. *Nat. Biotechnol.*, 33, 377–383.

Zhu, H., Swierstra, J., Wu, C. et al. (2014) Eliciting antibiotics active against the ESKAPE pathogens in a collection of actinomycetes isolated from mountain soils. *Microbiology*, 160, 1714–1725.

13

Future Market and Policy Initiatives of New High Value Products

Ha Thi Thanh Pham[1,2], *Maria Puig-Gamero*[3], *Luz Sanchez-Silva*[3], *Paula Sánchez*[3], *José Luis Valverde*[3], *Michele Heitz*[1], *and Antonio Avalos Ramirez*[2]

[1] *Université de Sherbrooke, Canada*
[2] *Centre National en Électrochimie et en Technologies Environnementales, Shawinigan, Québec, Canada*
[3] *Universidad de Castilla La Mancha, Spain*

13.1 Introduction

The biotechnology to produce bio-based products is developing very quickly over the entire world. Some products dominate the scene and it is expected that between 2030 and 2050 bio-based products will be competitive in terms of volume and price in comparison with traditional products obtained mainly by chemical synthesis or derived from fossil carbon sources. This rapid development leads to the need for the governments of several countries to establish policies in order to promote and regulate the production of bio-based products. Civil, government and private sectors participate in the development of policy initiatives for bio-products, in order to protect the health and the environment, and especially to regulate internal and global markets. Among all the kind of bio-products, the production of biofuels is the best example of the establishment of policy initiatives.

13.2 Market Analysis, Market Trends and Statistics

13.2.1 Biofuels

Biofuels are among the bio-products that present the highest increase in installed production because they will contribute to the transition of non-fossil energy sources, helping to mitigate the climate change. The biofuels market has increased mainly in response to transport sector energy demand. Currently, 2% of the total transport fuels are satisfied by biofuels; it is expected that this percentage will increase with the development of the specific technologies (optimization of second and third generation biofuels) (BP, 2014).

Microbial Sensing in Fermentation, First Edition. Edited by Satinder Kaur Brar, Ratul Kumar Das, and Saurabh Jyoti Sarma.

Nowadays, the three most developed biofuels are bioethanol, biodiesel and biogas. Bioethanol is typically blended at a maximum allowed percentage of 10% (E10) into fossil fuels. Technical modifications in engines are required when adding more than 10% bioethanol. This fact can be a limiting factor for bioethanol market growth which could cause a halt in demand until engines can use blends with higher ethanol content (Trumbo and Tonn, 2016; Santamaría and Azqueta, 2015; Qiu et al., 2014).

Currently, the US is the largest producer of bioethanol in the world, followed by Brazil, and, a long way behind, by the EU and China (Su et al., 2015). Worldwide, the bioethanol of first generation is the most produced.For example, corn ethanol is the most recognized form of biofuels in the United States. In China, ethanol was essentially produced from grains but nowadays it is produced from cassava, sweet potato or sorghum. In the EU, the principal sources to produce bioethanol are sugar or starch, and in Brazil is the sugar cane. Figure 13.1 shows the capacity of bioethanol production for main producer countries. There has been a continuous increase in bioethanol production since 2000. From 2006 to 2012, the US bioethanol production increased from 300 to 900 thousand barrels per day. However, Brazil produced the same amount as the US in 2006, but its production has kept more or less stable over time. Until now, one of the main barriers for biofuels market is its high cost.

On the other hand, the commercial biodiesel production has been mainly established in the US, Europe and Asia (Su et al., 2015). Figure 13.2 shows the biodiesel production for the main producer countries. Before 2005, the EU was the largest manufacturer of biodiesel in the world (90% of the worldwide production). From this year on, the production of biodiesel in the US and Asia progressively increased. Thus in 2011, the EU produced only 39% of the worldwide biodiesel. Anyway, a progressive increase in biodiesel production has been observed in recent years.

Currently, the conventional biofuels dominate the market. Governments are funding many research projects to promote the development of second and third

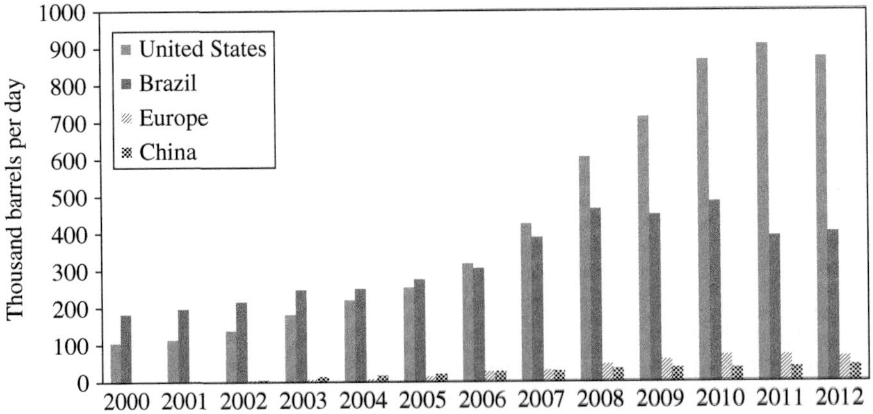

Figure 13.1 Main bioethanol producers in the world. *Source:* EIA, 2016. (*See insert for color representation of this figure.*)

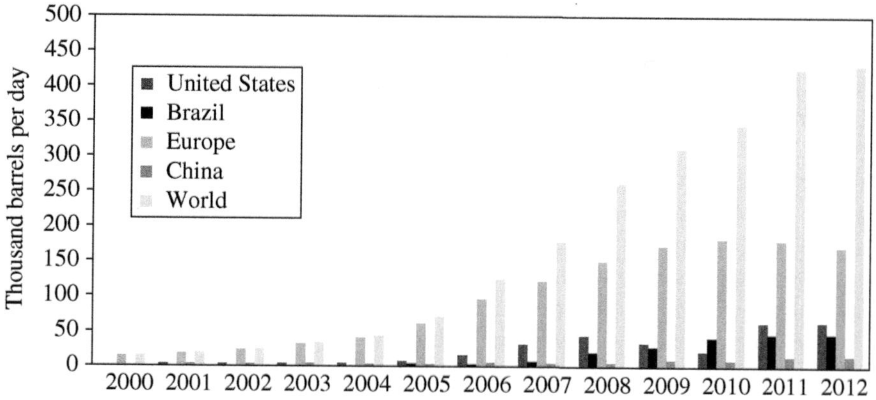

Figure 13.2 Main biodiesel producers and total world production. *Source:* EIA, 2016. (*See insert for color representation of this figure.*)

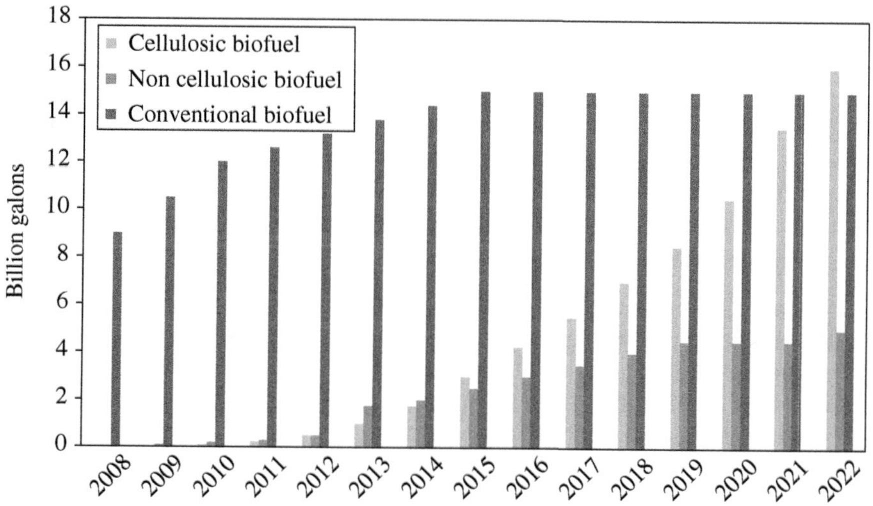

Figure 13.3 Evolution of biofuels production. *Source:* (EIA, 2016). (*See insert for color representation of this figure.*)

generation biofuels to avoid the use of food crops as energy sources. Figure 13.3 shows the evolution and perspectives of different kind of biofuels. It is expected that cellulosic biofuels production will rapidly increase up to 2022 while conventional biofuels will remain stable. Biofuels production is predicted to increase by 1.4% per year (EIA, 2016). This largely depends on local biomasses, which must be available and exclusive to biofuel production. For example, in Europe, the biomass production dedicated to energy is expected to increase from 72 10^6 tonnes in 2004 to 220 10^6 tonnes in 2020 (Dumitru, 2015).

Biogas (bio-methane) is becoming increasingly important. It can be used as a combustible to obtain heat and electricity or as a vehicle fuel (Soccol et al., 2016). In North America the biogas production is concentrated in US, where there are more than 2000 plants. In Canada there are around 100 biogas plants representing 3% of the total natural gas consumption. Germany is the largest producer of biogas in Europe, with around 5800 plants in 2010. The network of biogas production in this country is based in small and medium plants, with an energetic potential of less than 150 MWe for around 50% of plants and only 7% with capacity higher than 500 MWe (Pascual et al., 2011). In Asia, China is the main biogas producer. The biogas production in this country had increased from 499.5 10^6 m^3 in 2005 to 1077 10^6 m^3in 2012 (Soccol et al., 2016).

Since rural regions are not well deserved by energy networks and agri-wastes are a good source for biogas production, national governments are trying to encourage young farmers to invest in the biogas sector. The objective of this initiative is to help them to be self-sufficient for their energy requirements. The biogas produced could be used to produce heat, power and as vehicle fuel in farms (IEA, 2016).

In the case of urban residual biomasses, some countries are phasing out the discharge of organic matter. For example, the province of Quebec in Canada will ban the access of organic wastes to landfills and incineration in 2020 in the aime of recycling materials. These initiatives in waste management will promote the construction of new biogas plants worldwide to valorize organic waste into biogas through anaerobic digestion technology (from industrial, institutional and municipal sources). In addition, the development of natural gas networks to deserve industrial and transport sectors will encourage bio-methane production.

13.2.2 Bio-surfactants

Bio-surfactants are bio-based products currently acquiring a sustainable growth of market and becoming potential alternatives to chemical surfactants for a wide range of application in agricultural, pharmaceutical, food, cosmetics and detergent industries (Marchant and Banat, 2012; De Almeida et al., 2016). The demand of bio-surfactant in different application fields increases because of environmental concern about utilization of chemical surfactants, as well as for the unique advantageous properties of bio-surfactants and the advances in biotechnology for high scale production. As illustrated in the Table 13.1, the global market of bio-surfactants has increased from 1,735 million US$ in 2011 to 1,870 million US$ in 2013, and it is expected that this market will reach up to 2,477 million US$ in 2023 at an annual average growth rate of 4.3% (CAGR) during 2014–2020 (Research and Market, 2016).

Table 13.1 Bio-surfactant market: total production volume and income

Year	2011	2013	2020	2023
Production volume, kiloton (kT)		344	462	524
Income, US$	1,735	1,870	2,308	2,477

Source: Adapted from Research and Market, 2016.

The bio-surfactant market can be geographically segmented into North America (United States, Canada and Mexico), Europe (Germany, United Kingdom, France, Italy and Rest of Europe), Asia-Pacific and the Rest of the World (ROW) (Grand View Research, 2015). Europe is the largest regional market with the consumption of 178.9 kT in 2013 accounting for over 50% of global consumption and is expecting to dominate the market in the following years with 53.3% of global bio-surfactants market revenue share in 2018. North America is the second consumer of bio-surfactants with around 25% of the overall market. The leading proportion of bio-surfactant demand particularly on green cleaner products in Europe and North America is due to the consumer preference to bio-based products over chemical ones and the stringent regulations of environment. Asia Pacific (mainly China and India) is considered as the fastest growing region for the production and consumption of bio-surfactants due to growing application industry in textile, agriculture and personal care products of the two most populated countries in the world.

Taking into account the kind of bio-surfactants, the molecules which predominate in the current market are Rhamnolipids, Sophorolipids, Methyl Ester Sulfonates (MES), Alkyl Polyglucosides (APG), Sorbitan Esters, Sucrose Esters and others. As illustrated in Figure 13.4, rhamnolipids was the largest consumed bio-surfactant of 34 %, followed by Methyl Ester Sulfonate (MES) accounting for 33% of the 344 kT of the global market in 2013. Whereas combined demand of APG, Sorbitan Esters and Sucrose Esters was estimated at 115 kT in 2013. Previous analysis indicates that rhamnolipids presented the highest consumption increase gaining with over 8.5% CAGR and corresponding to an increase of 4 million US $ in revenue by 2015 (Grand View Research, 2015).

The domination of these key biosurfactants in the market is driven by their applications on various sectors of household detergents, personal care, food processing, pharmaceuticals, lubricant, fuel additives, oilfield chemicals and agriculture. Household detergent is the largest application of biosurfactants consuming 153.5 kT in 2013 corresponding to 44.6% of market size, followed by personal care with more than 10% and food processing application with over 11%

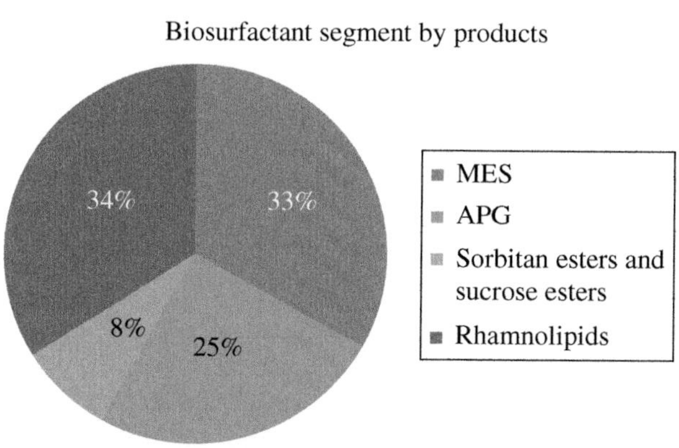

Figure 13.4 Most used bio-surfactants by kind of molecule. *Source:* Adapted from Grand View Research, 2015. (*See insert for color representation of this figure.*)

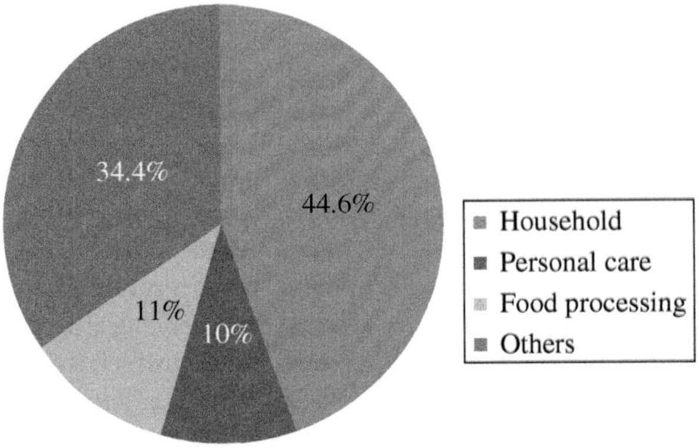

Figure 13.5 Consumption of bio-surfactants according to segment of applications. *Source:* Adapted from Radiant Insights, 2014. (*See insert for color representation of this figure.*)

(Figure 13.5). Growing demand for sustainable products in cleaning applications contributes to the continuous increase in biosurfactants for household detergents. Personal care application is expected to gain a significant share owing to the increasing demand in Asia Pacific and the growing consumer awareness of using bio-based personal care products. The demand for biosurfactants in personal care is expected to reach 50.7 kT by 2020, growing at a CAGR of 4.5% from 2014 to 2020 (Radiant Insights, 2014).

In perspective of competitiveness, the biosurfactant manufacture is principally concentrated by certain companies such as BASF Cognis, Henkel AG & Co. KGaA, Soliance, MG Intobio, Ecover, Kemin Industries, Inc., Jeneil Biosurfactant Company, Saraya and Urumqui Unite. Their products account for over 80% of market share in 2013. BASF Cognis is the largest manufacturer of biosurfactant, contributing with a market share over 20% in 2013 (Grand View Research, 2015).

Biosurfactant market size is still small compared to synthetically chemical surfactants. It is estimated that the production of synthetic surfactants will reach a volume of 24,037 10^6 tonnes and represent 42.12 billion US $ by 2020 (Brumano et al., 2016). Lack of cost competitiveness of biosurfactants against chemical surfactants is one of the most important barriers of biosurfactant production for commercialization. High cost of production is attributed to low availability and high cost of feedstock in addition to low yield of production. Development of hyper producing strains, optimization of cultivation conditions and utilization of cost-effective alternative raw materials are the efforts that the biosurfactant manufacturers are making to ensure biosurfactants are competitive with petroleum-based counterparts.

There are numerous factors that drive the market size of biosurfactants. The most important reason is the consumer preference for biobased products over petroleum-based products which can cause negative impacts on environment and human health. In Europe and North America, the second main reason is the strict regulations of environment on production and consumption of products. In Asia Pacific, the rapidly growing application industry of textile, agriculture and personal care is a driver to boost the market.

13.2.3 Enzymes

Being used as biocatalysts to accelerate reactions and to produce desired products, enzymes have a diversity of applications in multiple industries such as food, beverage, detergent, pharmaceutics, biotechnology, textile, agriculture, and pulp and paper. The development of protein engineering and the optimization of fermentation processes have increased the diversity of enzymes commercially available and their performance. This contributes with the widely utilization of enzymes as biocatalysts in manufacturing processes in order to replace chemical catalysts for green technology development, as well as additives for clean products, foods, environmental remediation products, etc.

As shown in Table 13.2, the global market of enzymes is increasing substantially due to the growth of enzyme demand by industries. The volume of global market for enzymes was 5,727 kT in 2013 and is expected that it would reach up to 7,716 kT in 2020, growing at a CAGR of 4.5% from 2014 to 2020. This corresponds to an increase of around 43% of global market from 2013 to 2020; from 5,726 to 8,194 million US $ (Grand View Research, 2014).

North America is the biggest enzyme consumer, followed by Europe (Table 13.3). North America was the leading enzymes market with a demand of 1,931.5 kT(2,053.3 million US $) in 2013. It is expected that enzymes market will reach to 2,548 kT in 2020 for this region, presenting a CAGR of 5.1% during a six-year period. However, the enzymes market in North America and Europe is estimated to increase slower than Asia Pacific, which will be the most dynamic region from 2013 to 2020 for the growth of enzymes demand related to a substantial increase in industrial activities in this region (Grand View Research, 2014).

Table 13.2 Enzymes market based on total production volume

	2013	2020
Production volume, kT	5,727	7,716
Revenue, Million US $	5,726	8,194

Table 13.3 Regional market of enzymes

	2013		2020	
Region	Production volume, kilo tonnes	Revenue, USD $	Production volume, kilo tonnes	Revenue, USD $
North America	1,931.5	2,053.3	2,548.0	2,916.4
Europe	1,856.9	2,073.3	2,420.2	2,905.4
Asia Pacific	1,421.3	1,346.5	2,069.7	2,132.1
ROW	771.0	737.2	1,145.5	1,187.7

The types of enzymes produced can be broadly categorized according to their main function, such as carbohydrases, proteases, and lipases. The proteases comprise the largest selling product segment with about 60% of market share, followed by carbohydrases, phytases and lipases presented fast growing demand (Kumar et al., 2014).

Based on the application, the market is segmented in various sectors including food processing and beverages, household care, bioenergy, pharmaceutical and biotechnology, animal feed and others. The industry which consumes the highest quantity of enzymes is the food processing sector (amylase, cellulase, xylanase, pectinase, protease and lipase) in order to reach quality food targets. According to a new market report published by Transparency Market Research (2013), the food additives market revenue was 28.2 billion US $ in 2011 and is expected to reach up to 36 billion US $ in 2018, growing at a CAGR of 3.6% from 2012 to 2018 (Transparency Market Research, 2013). The enzyme market for technical applications will also expand to meet the needs of other industries. The demand of cleaning enzymes for detergent industry will increase gradually as the innovated product lines are continuously developed requiring more-effective new enzymes. Currently, the innovative technology for production of second generation of biofuels and bio chemicals is encouraged, thus the market of biofuel enzymes for biomass conversion will be growing continuously (BCC Research report, 2015). The specialty enzyme market for application in pharmaceutics, research and biotechnology, diagnostics and biocatalysts is estimated to increase, particularly in the research and development sector, and in pharmaceutical manufacture (Marketsandmarkets, 2014). The enzymes for animal feed industry dominates a sustainable market share with other enzymes since the use of enzymes for animal feeding is becoming important to enhance the animal production efficiency.

The microbial fermentation-based enzymes are leading the trend to dominate the source of enzyme production. High enzyme production yield from microorganism culture and specific activities and stability of enzymes are the main advantages of microbial enzymes over plant- and animal-based enzymes. In addition, the advance in the development of fermentation technology as well as the engineering of robust enzyme producing strains allows the production of various types of high level, purified and well-characterized enzymes at a high scale for commercialization.

In terms of competitiveness, the enzyme market is currently dominated by some huge producers of industrial enzymes including Novozymes, DSM, DuPont, Dyadic International Inc., Codexis Inc., AB Enzymes GmbH, Roche Holdings AG, BASF, Life technologies, Advance Enzyme Technologies Ltd, and Affymetrix Inc. Novozymes is the market leader with a share of 48% of the global market, followed by DSM and DuPont (Research and Market, 2015).

13.3 Public Mobilization Initiatives and Government Policies

13.3.1 Public Mobilization Initiatives

Since the consumer awareness about sustainable production and consumption is rising, more and more consumers are adopting the use of bio-based products instead of chemically synthesized products. For this change, they are willing to pay even higher prices for bio-based products. This is also observed in the increase of renewable energy uses by a highly aware population, for example the boost for electric car use in the entire world. These consumers' behavior is the initiative driving force for the enlargement of bio-products market size.

13.3.2 Government Policies

The governments of countries from regions of North America, Europe and Asia Pacific are making efforts to establish policies to develop and regulate the production of bio-based products. There is a lot of investment into research, technology development and innovation to support the production of bio-based products on a large scale. The policies relating to climate change mitigation and sustainable production and consumption, industrial policy and employment growth are also amended to approve the bio-based products market. This policy development is involved in different areas such as agriculture, environment, transport, energy, and health.

13.3.3 Regional Policy Development for Growing Bio-based Production

North America and Europe are the pioneers to develop the policies to promote bio-based product markets in order to accelerate the consolidation of a sustainable bio-based economy in coming years. In Europe, the policy responsible sectors have been created conditions to facilitate the commercialization of bio-based products as follows: launch an initiative as a new policy approach to support the development of bio-based product markets with high economic and social value; creation of lead markets and other measures to enhance market demand for ecofriendly bio-based products, in order to exploit the positive environmental impact of bio-based products, and the promotion of circular economy inside the EU.

In North America, particularly the US, the policy development of bio-based product market is focused on the support for bio-based production and innovation technologies as follows: to provide product parity and early stage support in bio-renewable tax policy, and increase funding through grants and other programs to support bio-based production. Among bio-based products, the best example of public policies around the world is the biofuel sector, which is presented below.

13.4 Regulations and Conformity – Case of Biofuels

Currently, the market of biofuels is limited by socioeconomic and political challenges. In this sense, national energy policy plays an important role in the development of the biofuels industry because biofuels are less competitive than traditional fuels and they present high production costs due to immature technology (Trumbo and Tonn, 2016).

In recent years, countries have authorized laws and regulations promoting renewable energies. They include financial supports for the use of biofuels and the construction of production plants.

An essential factor for the biofuel production is the consolidation of raw biomass markets. The main driving forces for the development of biomass dedicated to biofuel production in most countries are:

1) To reduce dependence on oil imports and guarantee oil security;
2) to create more employment opportunities and develop agriculture;
3) to develop markets and competitiveness in bio-economy sectors; and
4) to develop new technologies (Su et al., 2015).

The US has established key legislations and regulations in order to facilitate the biofuels development. The Energy Policy Act of 2005 (Malmedal et al., 2007) set a renewable fuel standard (RFS). The first RFS required about 28 10^9 liters of biofuel to be blended into gasoline in 2012. It was expanded by the Energy Independence and Security Act of 2007 (Ruester et al., 2013) which prohibits bioethanol derived from corn starch. In addition, there are different regulations that establish tax credits for biodiesel and ethanol blending. Finally, the production of cellulosic biofuels is being encouraged in this country. On the other hand, there are important projects and programs to develop biofuels such as biomass Multi-Year Program Plan, the bio-refinery assistant program and many others.

In the same way, the EU has set renewable energy goals, taxes subsidies and R&D projects. The Renewable Energy Directive (European Commission Directive, 2009) requires all vehicle fuel to be powered by at least 10% of biofuels in 2020. In October 2012, the EC published a recommendation on Indirect Land Use Change whose goal was the transition from conventional biofuels to biofuels from non-food feedstock. After 2020, first generation biofuels will not be supported anymore and will be limited. A further amendment: The Renewable Energy Directive in 2015, limits the use of conventional biofuels derived from crop plant up to 7% and forces members states to implement biofuels from non-food feedstock of at least 0.5% in transport energy in 2020 (Su et al., 2015). Finally, "European Advanced Biofuels Flight path" is an initiative to achieve an annual production of two 10^6 tonnes of biofuels for aviation in 2020.

In Asia area, China has not changed its bioethanol policy in the last years. The government have placed a cap on bioethanol from grain. Nowadays, China has five bio-ethanol plants licensed for fuel production that use feedstock as corn, wheat and cassava. China government has total control over fuel distribution; in other words, if the government does not approve biofuels distribution, biofuels cannot be imported (Su et al., 2015). One of the most important projects in China consists of the design, the construction, and the commissioning of a plant which could produce 12 106 tonnes of aviation biofuel in 2020. Chinese government has invested in biogas production because it is thought that it can give an economical revenue benefit. Nowadays, there are more than 91,000 biogas projects in China. Finally, Indian government approved the National Policy on biofuels in 2009. This policy tries to promote the use of renewable energy resources as an alternative to traditional fuel. Its main purpose is to implement 20% of biofuels in 2017. The aim of the government at the end of 2020 is to construct 8000 new large scale biogas plants (Soccol et al., 2016).

13.5 Global Marketing and Competitiveness in Biofuel Sector

Biofuels production must not only be environmentally sustainable but also have social and economic impacts and depend on factors such as technology complexity, production capacity and feedstock costs. At present, feedstock is the main component of production costs of conventional biofuels, accounting for 45 to 70% of the total costs. In advanced biofuels the main component of the cost is the capital one (35–50%) followed by the feedstock one (25-40%). Nowadays, net price in the market of biofuels is not competitive against fossil fuels. However, indirect costs caused by fossil fuel consumption should be considered, such as the increase of health systems expenses, the increase of cost reconstruction after destructive meteorological events and the economic and social costs of governments and population caused by climatic migration. All of these are consequences of climatic change accelerated by carbon dioxide accumulation in the atmosphere, mainly issued by the combustion of fossil fuels. The environmental benefits derived from the use of biofuels should promote governments to take more ambitious actions and invest more in either research or building plants for their production.

Finally, in order to protect local businesses and markets, many countries impose duties on products which are being dumped on their national market. For example, the EU will impose anti-dumping duties on imports of biodiesel from Argentina and Indonesia in order to positively affect the economical and operational performances of European biofuels producers.

References

BCC (2015) Global Markets and Technologies for Biofuel Enzymes.

BP (2014) BP statistical review of world energy 2014. In: Energy, BSRoW, editor, *BP statistical review of world energy*. London, UK; 2014.

Brumano, L.P., Soler, M.F., and da Silva, S.S. (2016) Recent advances in sustainable production and application of biosurfactants in Brazil and Latin America. *Industrial Biotechnology*, 12 (1), 31–39.

De Almeida D.G., Da Silva, R. de C.F.S., Luna, J.M. et al. (2016) *Frontier in Microbiology*, 7, 1–14.

Dumitru, M. (2015) Development trends of Biogas. *Development*, 15.

EIA. web page https://www.eia.gov/ consulted on October 2016.

European Commission Directive (2009) Promotion of the use of energy from renewable sources and amending and subsequently repealing Directives 2001/77/EC and 2003/30 Off J Eur Union, 5 (2009), p. 2009.

Grand View Research (2015) Biosurfactants Market Analysis By Product (Rhamnolipids, Sophorolipids, MES, APG, Sorbitan Esters, Sucrose Esters) And Segment Forecast To 2020.

Grand View Research (2014) Fermentation chemical market analysis and segment forecast to 2020.

IEA. web page http://www.iea.org/ consulted on October 2016.

Kumar V., Singh D., Sangwan P., and Gill P.K. (2014) Global scenario of industrial enzyme market. Book chapter, *Industrial enzymes, trends scope and relevance.*

Malmedal, K., Kroposki, B., and Sen, P.K. (2007) Energy Policy Act of 2005. *IEEE Industry Applications Magazine*, 13, 14–20.

Marchant R. and Banat I.M., (2012) Biosurfactants: a sustainable replacement for chemical surfactants? *Biotechnol. Lett.*, 34(9):1597–605.

Markets and markets (2014) Specialty Enzymes Market by Source, Type, Application & Geography - Global Trends & Forecasts to 2018.

Pascual, A., Ruiz, B., Gómez, P. et al. (2011) Situación y potencial de generación de biogás. Estudio Técnico PER 2011–2020. Instituto para la Diversificacion y Ahorro de la Energia, Spain, p. 104.

Qiu, C., Colson, G., and Wetzstein, M. (2014) An ethanol blend wall shift is prone to increase petroleum gasoline demand. *Energy Economics*, 44, 160–165.

Radiant Insights, Inc. (2014) Biosurfactants Market Size, Share, Growth, Trend & Research Report 2020. Hexa Research, US.

Research and Market (2016) Global Biosurfactants Market - Segmented by Product Type and Geography - Trends and Forecasts (2015–2020).

Research and Market (2015) World Enzymes Market - Opportunities and Forecasts, 2014–2020.

Ruester, S., Schwenen, S., Finger, M., and Glachant, J.-M. (2013) A post-2020 EU energy technology policy: revisiting the Strategic Energy Technology Plan. *Energy Policy*, 66, 209–217.

Santamaría, M. and Azqueta, D. (2015) Promoting biofuels use in Spain: A cost-benefit analysis. *Renewable and Sustainable Energy Reviews*, 50, 1415–1424.

Soccol, C.R., Brar, S.K., Faulds, C., and Ramos, L.P. (2016) *Green Fuels Technology: Biofuels*, Springer International Publishing.

Su, Y., Zhang, P., and Su, Y. (2015) An overview of biofuels policies and industrialization in the major biofuel producing countries. *Renewable and Sustainable Energy Reviews*, 50, 991–1003.

Transparency Market Research (2013) Enzymes Market - Global Industry Analysis, Size, Share, Growth, Trends, and Forecast, 2012–2018.

Trumbo, J. L. and Tonn, B.E. (2016) *Biofuels: A sustainable choice for the United States' energy future? Technological Forecasting and Social Change*, 104, 147–161.

Index

Microbial Sensing in Fermentation, First Edition. Edited by Satinder Kaur Brar, Ratul Kumar Das, and Saurabh Jyoti Sarma.

C

d

h

t

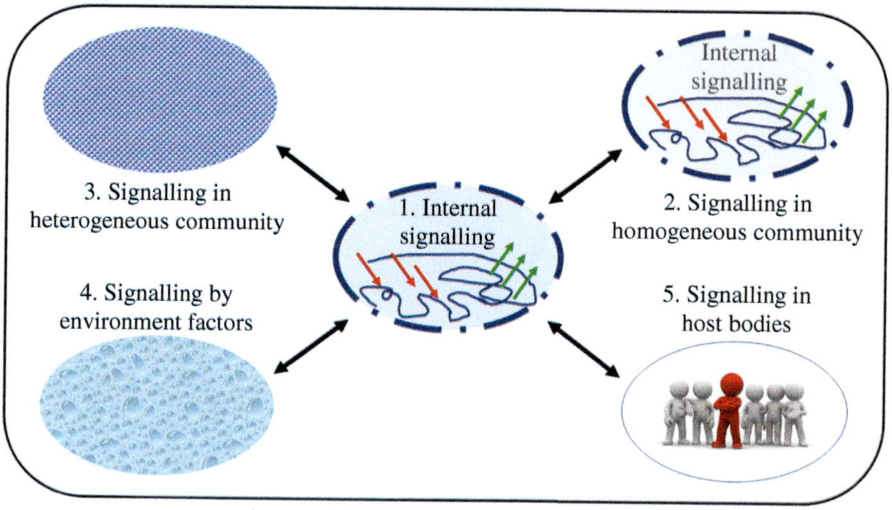

Figure 1.2 Microbial sensing classification according to signal origin.

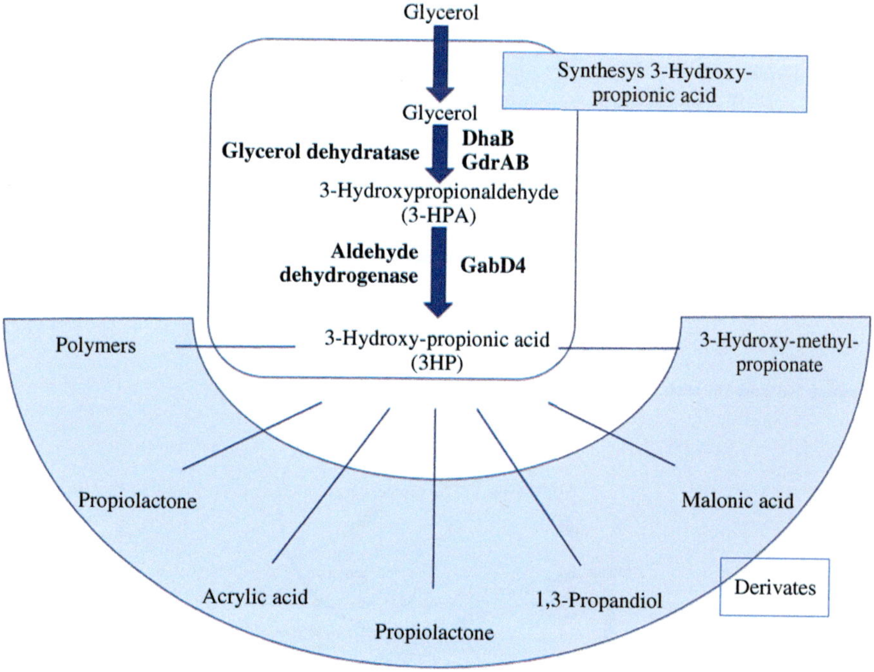

Figure 5.1 Genetically engineered metabolic pathway for the production of 3-hydroxy-propionic acid in *E. coli*.

Microbial Sensing in Fermentation, First Edition. Edited by Satinder Kaur Brar, Ratul Kumar Das, and Saurabh Jyoti Sarma.

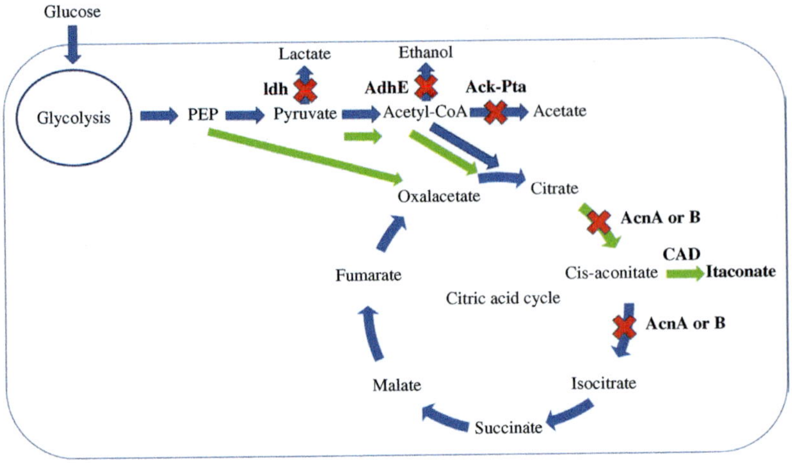

Figure 5.3 Genetically engineered metabolic pathway for the production of itaconic acid in *E. coli. Source:* modified from Harder et al., 2016.

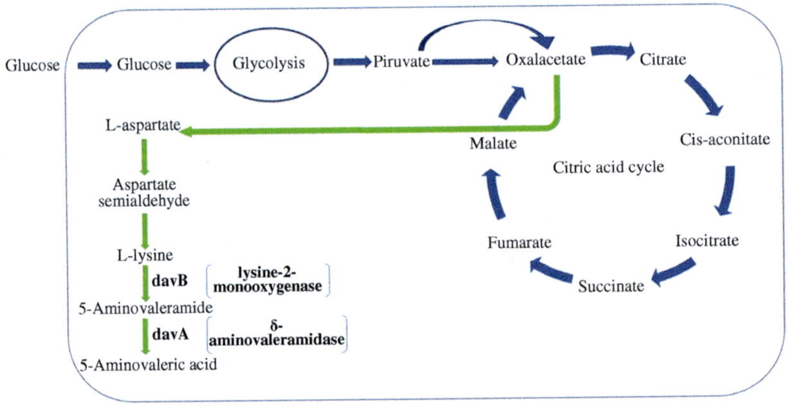

Figure 5.8 Genetically engineered metabolic pathway for the production of 5-aminovalerate in *E. coli. Source:* modified by Cho et al., 2015.

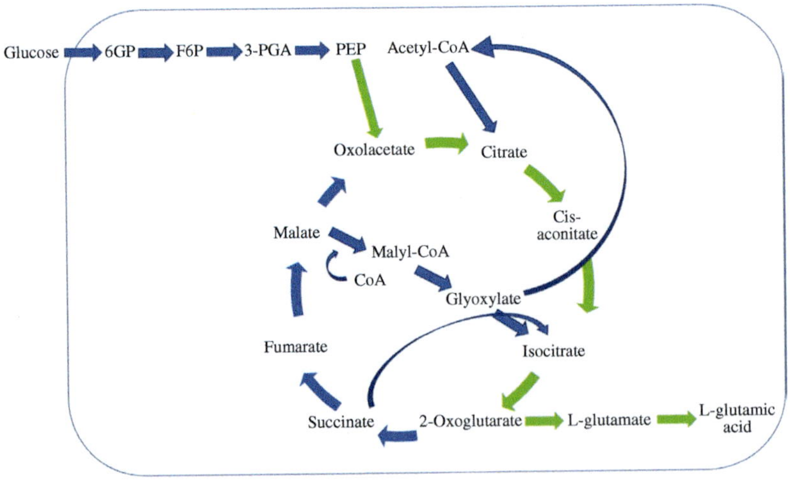

Figure 5.9 Genetically engineered metabolic pathway for the production of glutamic acid in *E. coli. Source:* modified from Vuoristo et al., 2016.

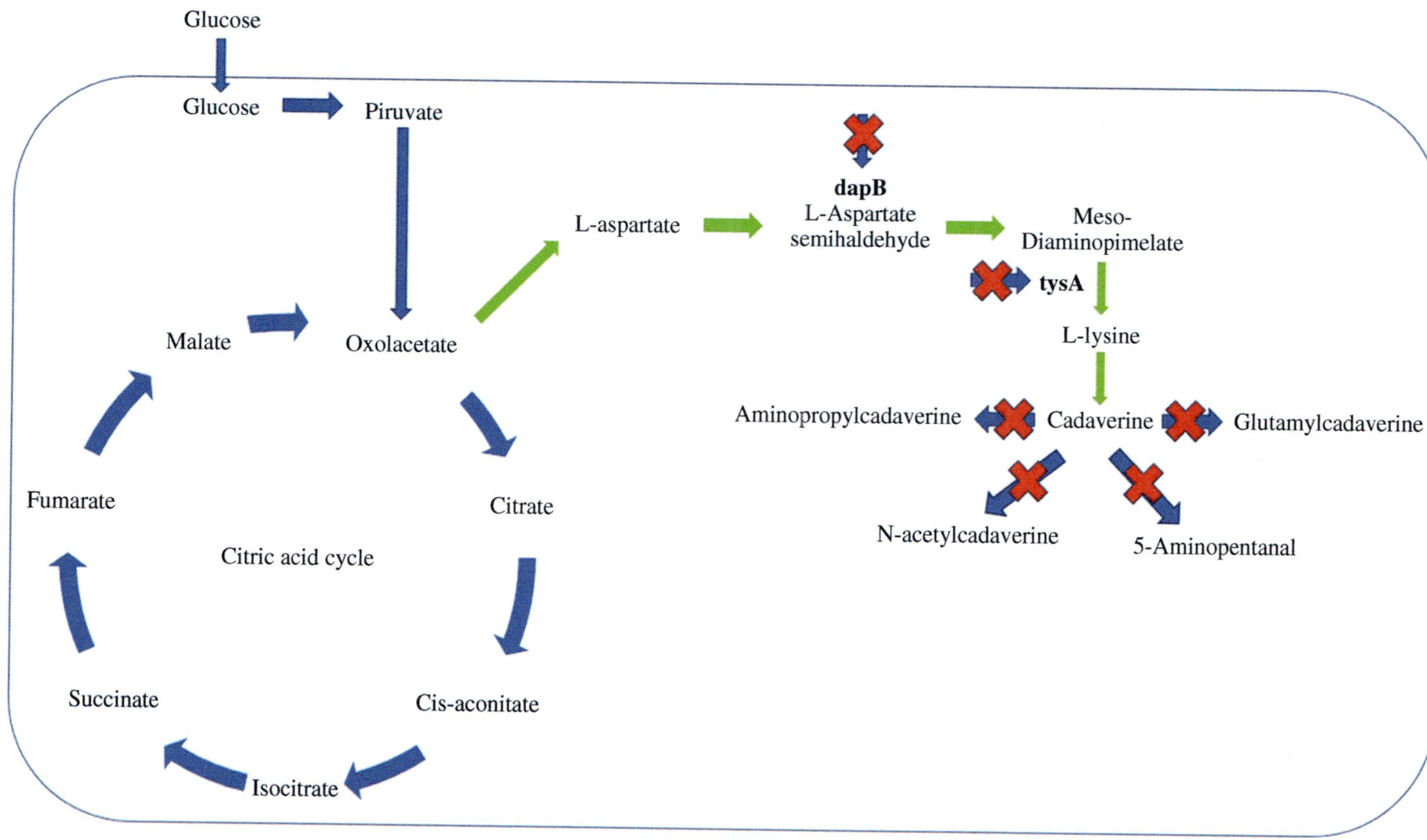

Figure 5.10 Genetically engineered metabolic pathway for the production of cadaverine in *E. coli*. *Source:* modified from Qian et al., 2011.

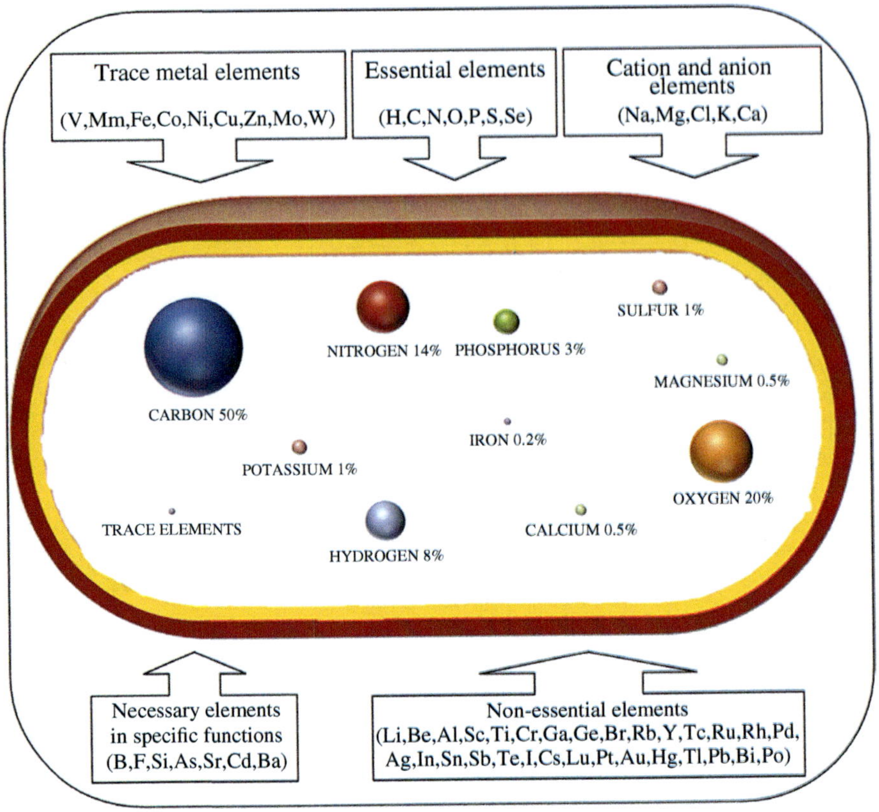

Figure 7.1 Chemical elements and their relative abundance in microbial cells.

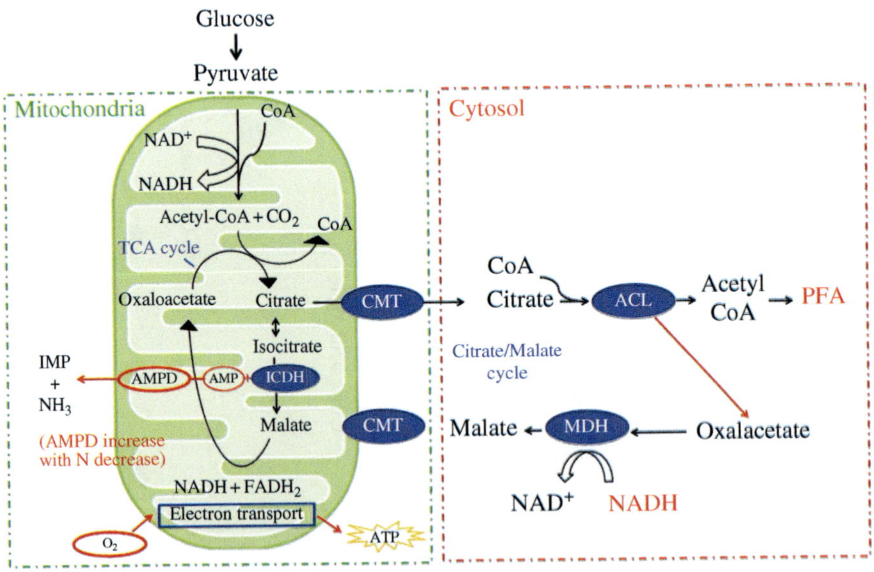

Figure 7.2 General mechanism of lipid accumulation by oil-producing microorganisms. AMPD = AMP deaminase; ICDH = isocitrate dehydrogenase; CMT = citrate–malate translocase; ACL = ATP citrate lyase; PFA = Polyunsaturated fatty acids; MDH = malate dehydrogenase. *Source:* Adapted from Lin (2015).

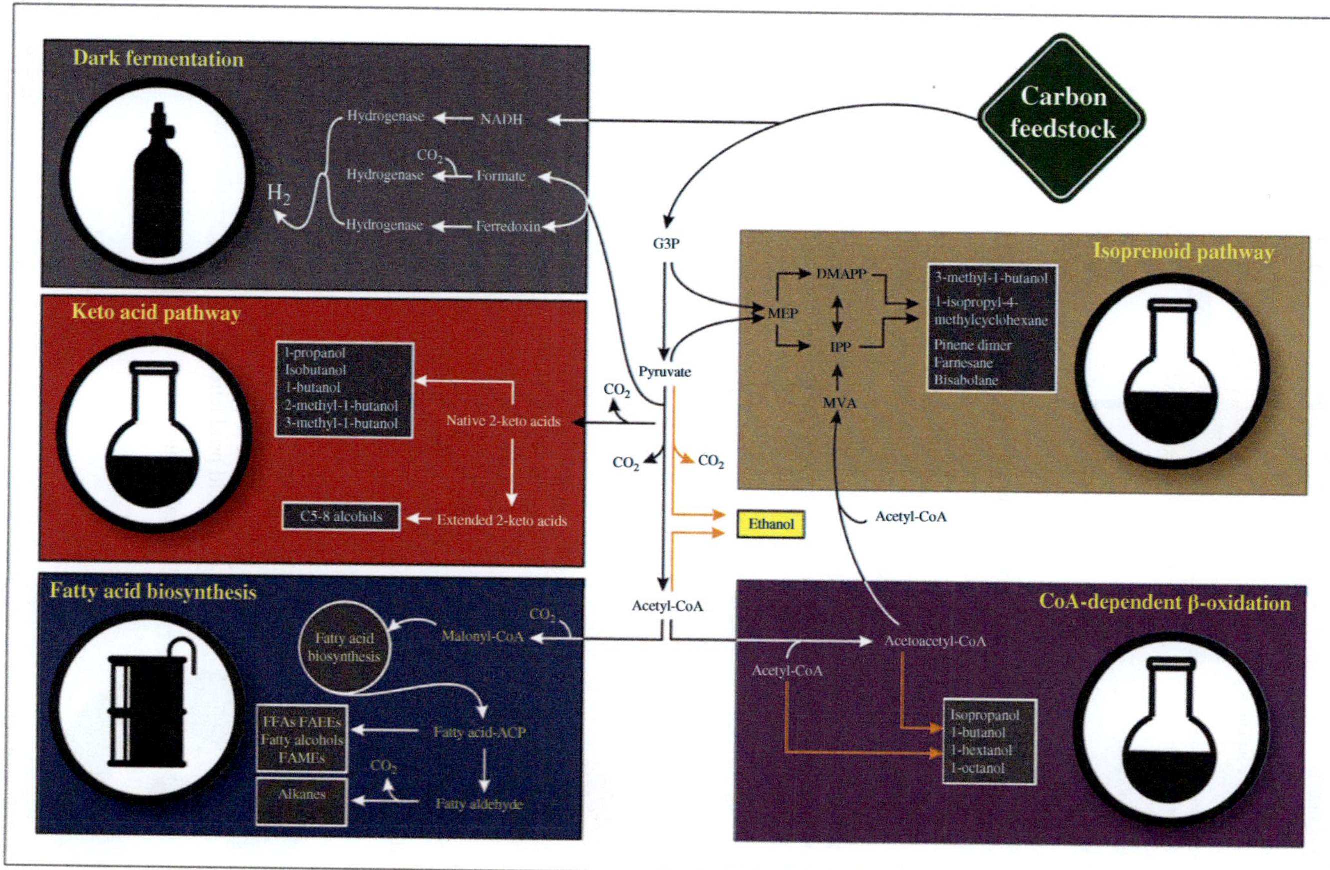

Figure 7.3 Biosynthetic pathways of biofuels (G3P: glyceraldehyde 3 phosphate; DMAPP: dimethylallyl pyrophosphate; MEP: methylerythritol 4 phosphate; IPP: Isopentenyl pyrophosphate; MVA: mevalonate; FFAs: free fatty acids; FAEEs: fatty acid ethyl esters; FAMEs: fatty acid methyl esters).

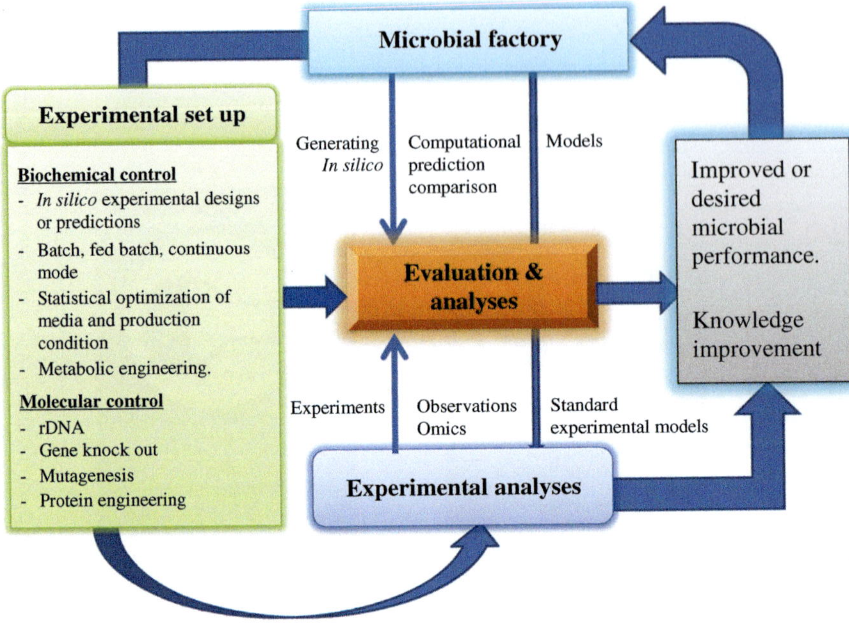

Figure 9.1 Schematic representation of biochemical and molecular events which can be explored for controlling or manipulating microbial performances. *Source:* Adapted from Lee et al., (2005).

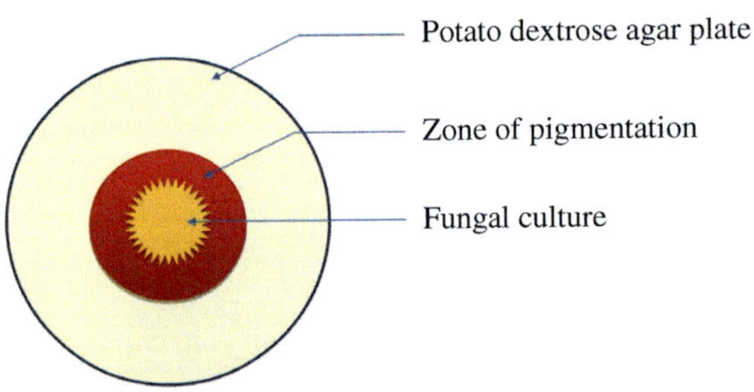

Figure 10.1 Screening of pigment producers.

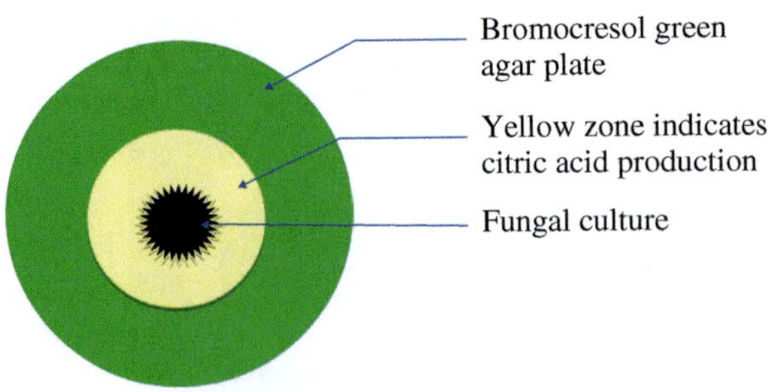

Figure 10.2 Screening of citric acid producers.

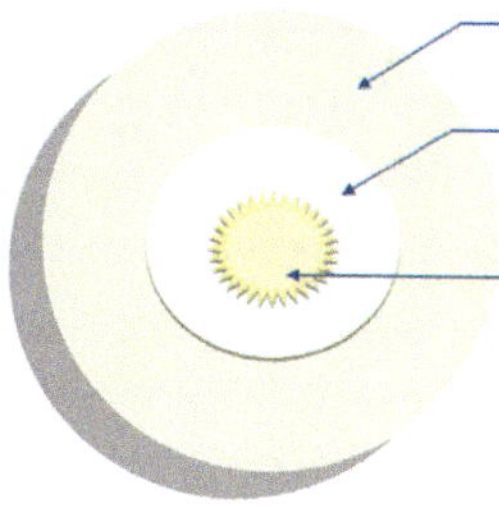

Figure 10.3 Screening of organic acid producers.

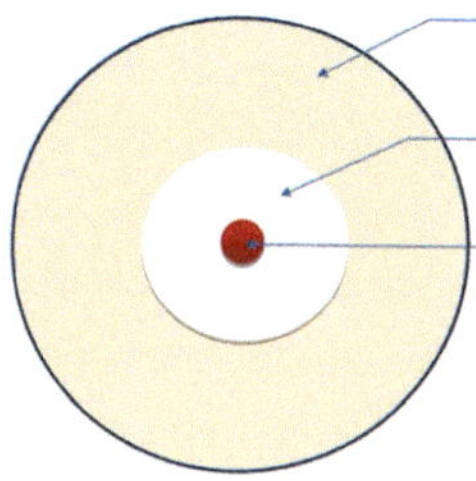

Figure 10.4 Screening of antimicrobial agent producers – Agar Disc diffusion method.

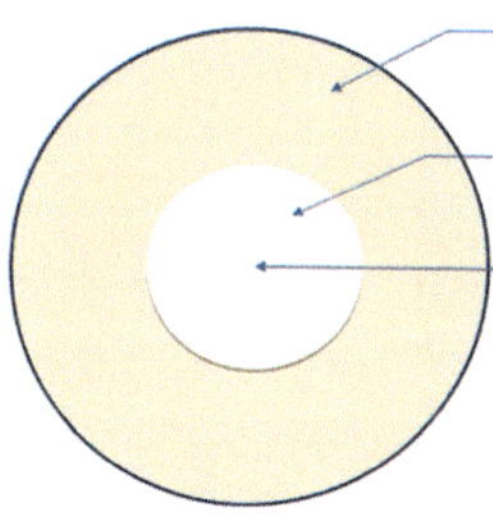

Figure 10.5 Screening of antimicrobial agent producers – Agar well diffusion method.

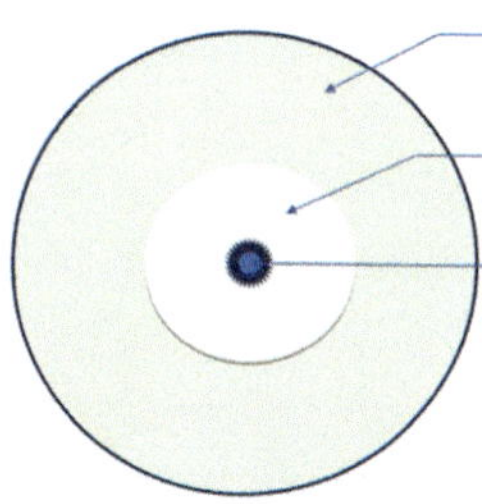

Figure 10.6 Screening of antimicrobial agent producers – Agar plug diffusion method.

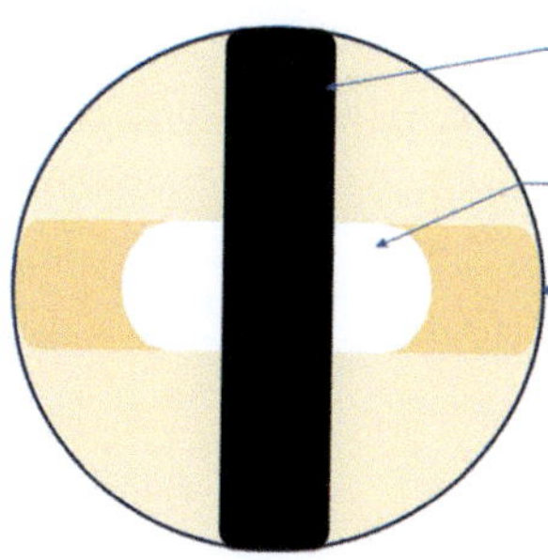

Figure 10.7 Screening of antagonistic agent producers – Cross streak method.

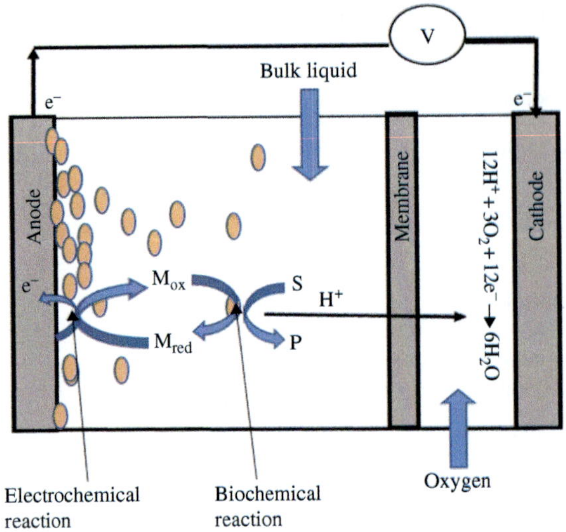

Figure 11.1 Principle of the Oxygen Sensor based on the respiration activity.

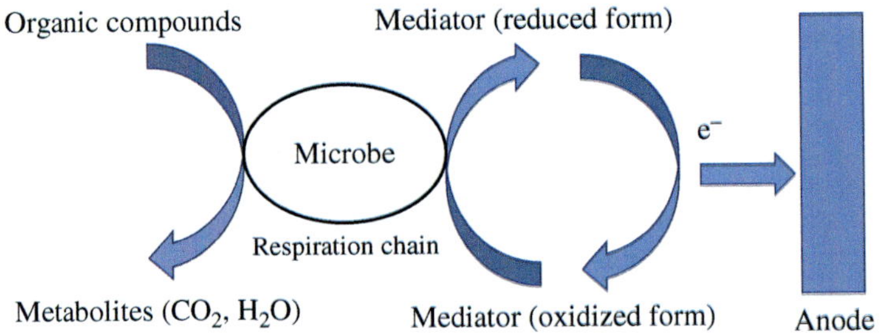

Figure 11.2 Principle of the Oxygen Sensor based on the Electron Transfer Measuring Systems.

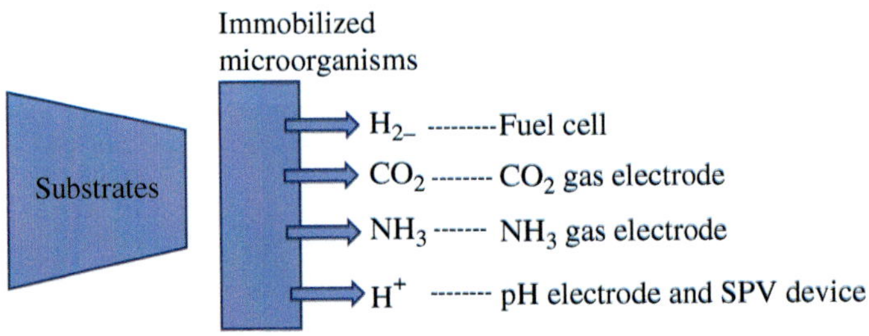

Figure 11.3 The metabolites produced by the microbes can also use as sensors.

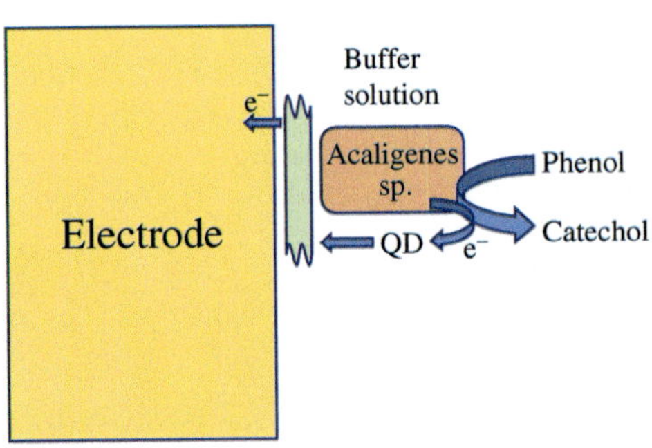

Figure 11.4 Application of microbe sensor in the quality control of red wine. *Source:* Adapted from Kim et al., 2011.

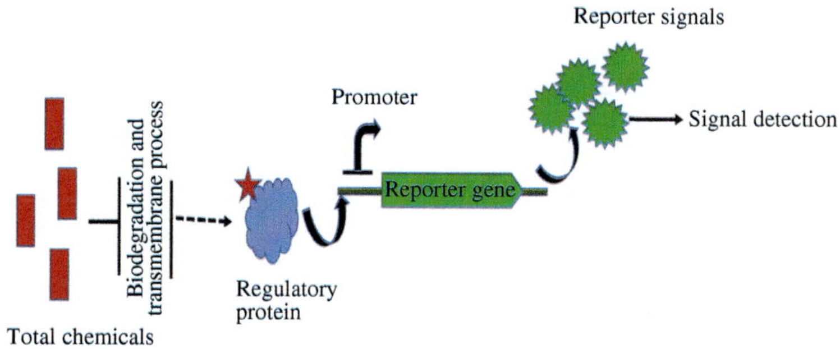

Figure 11.5 The regulation, construction and organization of the Genetically engineered microbial whole-cell biosensors. *Source:* Adapted from He et al. (2016).

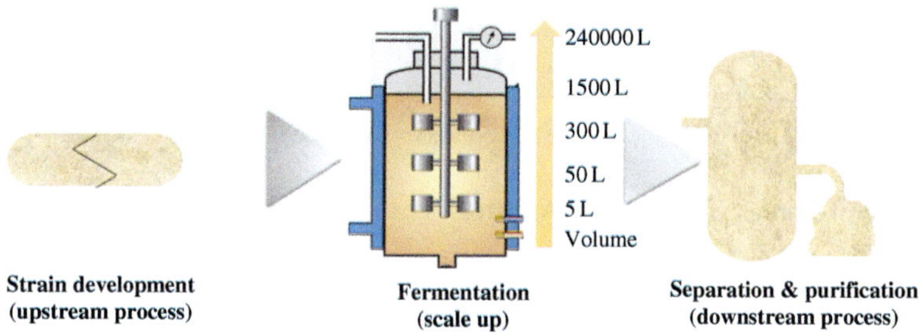

Figure 12.1 General schemes of bioprocess stages.

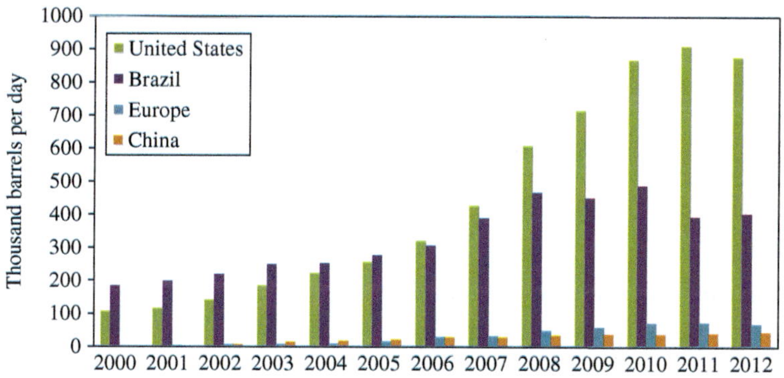

Figure 13.1 Main bioethanol producers in the world. *Source:* EIA, 2016.

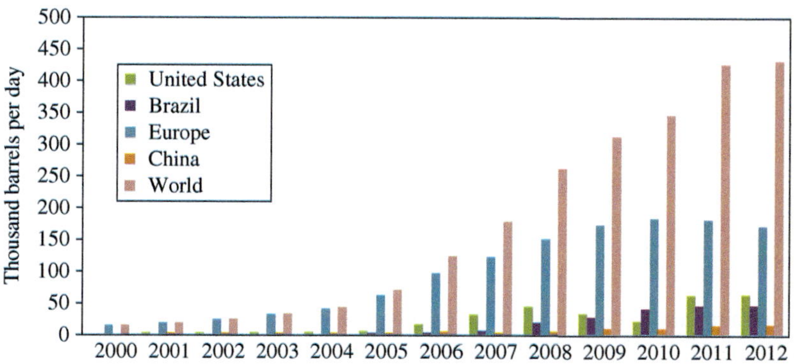

Figure 13.2 Main biodiesel producers and total world production. *Source:* EIA, 2016.

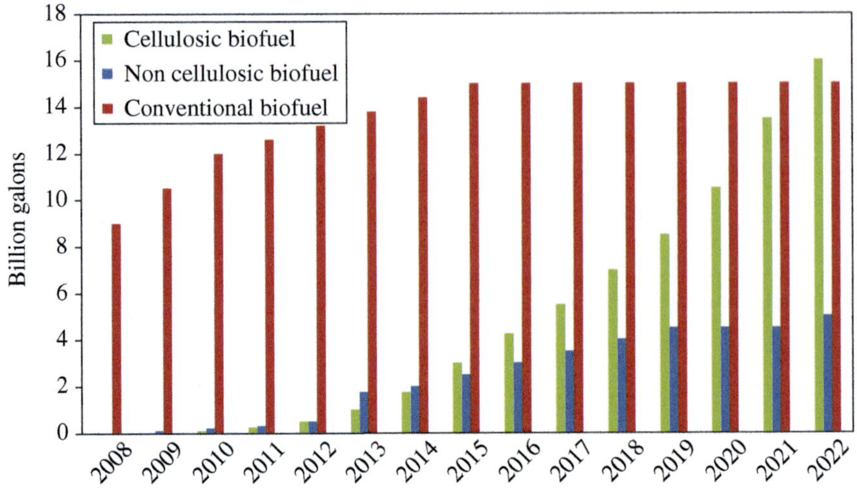

Figure 13.3 Evolution of biofuels production (EIA, 2016).

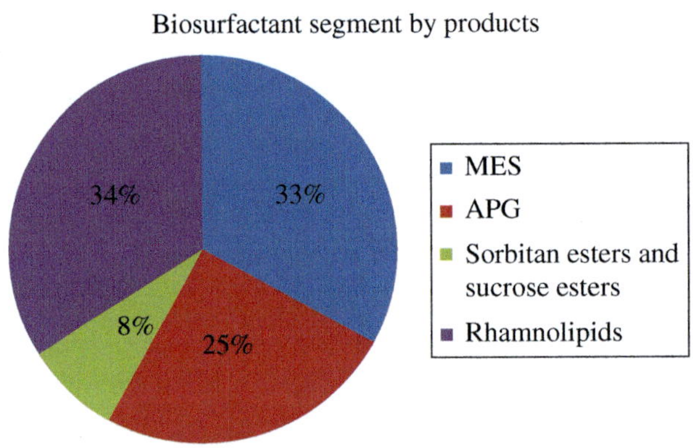

Figure 13.4 Most used bio-surfactants by kind of molecule. *Source:* Adapted from Grand View Research, 2015.

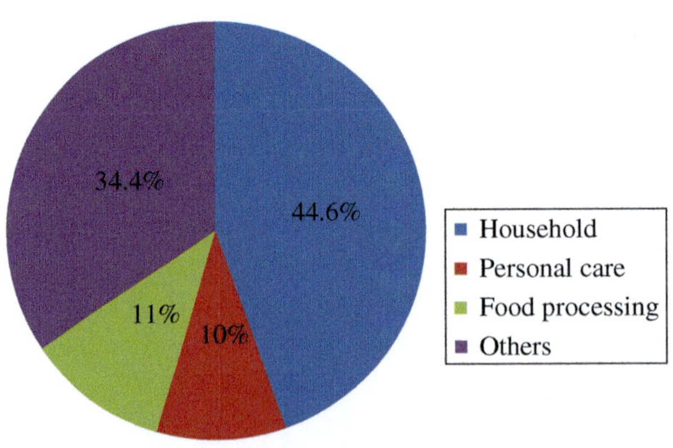

Figure 13.5 Consumption of bio-surfactants according to segment of applications. *Source:* Adapted from Radiant Insights, 2014.

Printed and bound by CPI Group (UK) Ltd, Croydon, CR0 4YY
17/09/2026
14964303-0002